Supply Chain Management with SAP APO™

Jörg Thomas Dickersbach

Supply Chain Management with SAP APO™

Structures, Modelling Approaches
and Implementation of SAP SCM™ 2008

3rd edition

 Springer

Dr. Jörg Thomas Dickersbach
E-mail: dickersbach@gmx.de

ISBN 978-3-540-92941-3   e-ISBN 978-3-540-92942-0
DOI 10.1007/978-3-540-92942-0
Springer Dordrecht Heidelberg London New York

Library of Congress Control Number: 2009926885

Cover design: WMXDesign GmbH, Heidelberg

Printed on acid-free paper

Springer is part of Springer Science+Business Media (www.springer.com)

Preface

This book rather addresses the question 'how to implement SAP APO™' than 'why to implement SAP APO™' and is written for people who are involved in SAP APO™ implementations. It is based on the SAP APO™ release SAP SCM™ 2008. The aim of this book is to provide the reader with the necessary background to start with first own steps in the system in the right direction by explaining the architecture and some basic structures of SAP APO™ and introducing common modelling approaches.

Although there are already several books published about SAP APO™ and there is a detailed documentation of the functions in the system, we have experienced a distinct need for explanations regarding the structure and the interaction of systems, modules and entities. The understanding of the possibilities and necessities on entity level is the basis for the modelling and the implementation of the business processes. This book mentions additionally many issues which have a great relevance in implementations, but are not mentioned in the literature.

In our experience with SAP APO™ projects we noticed an ever greater need (which remains more often than not unaware for much too long) to clarify the implications of the SCM approach for the implementation projects. Since SCM projects with SAP APO™ differ significantly from SAP ERP™ projects, there are some typical traps in which even experienced SAP ERP™ project managers are apt to fall which cause severe problems up to project failure. Especially in the first chapter common mistakes in SCM projects are pointed out.

The book does not claim to describe all SAP APO™ functionalities and modelling possibilities – since the modelling approaches are nearly unlimited and the product is still evolving, this would be impossible. Instead the focus is set on explaining common approaches especially for the high tech, the consumer goods and the chemical industries. Not included into the scope of this book are the scenarios and functionalities especially for automotive industry, repetitive manufacturing and aerospace and defence, and some other functionalities as VMI to third party customers, container resources and campaign planning.

Since the focus of the book lies on the practical use of SAP APO™, SCM theory in general as well as in connection with SAP APO™ is not within the scope of this book. Therefore instead of the SCM literature the SAP notes of the online service system (OSS) are quoted. Working with the OSS is anyhow inevitable for any implementation project and an important source for information.

Compared to the first edition this book contains additional topics (as transportation planning, interchangeability, bucket-oriented CTP and scheduling of complex job chains) and many updates in the functionality – representing two years' development.

Finally I would like to thank Jens Drewer and Claus Bosch, who helped me a lot during the whole project (the chapter about transportation and shipment scheduling was contributed by Jens Drewer), Bernd Dittrich for his help and comments on transportation planning, and Dr. Stephan Kreipl, Anita Leitz, Bernhard Lokowandt, Armin Neff, Stefan Siebert, Uli Mast and Christoph Jerger for their corrections and comments.

Germany,   Jörg Thomas Dickersbach
March 2009

Contents

PART I – OVERVIEW

1 SCM Projects with SAP APO™.. 3
  1.1 The Supply Chain Management Approach....................................... 3
  1.2 Supply Chain Management Projects with SAP APO™ 6
  1.3 SAP APO™ Project Management and Peculiarities 7

2 SCM Processes and SAP APO™ Modules 9

3 SAP APO™ Architecture 15
  3.1 Technical Building Blocks of SAP APO™................................... 15
  3.2 Master Data Overview... 17
  3.3 Model and Version 23
  3.4 Planners 24
  3.5 Order Categories... 25
  3.6 Pegging 25
  3.7 Data Locking 28

PART II – DEMAND PLANNING

4 Demand Planning.. 33
  4.1 Demand Planning Overview.. 33
    4.1.1 Demand Planning Process 33
    4.1.2 Planning Levels and Consistent Planning 34

4.2 Data Structure for Demand Planning 36
   4.2.1 Characteristics, Key Figures and Structure Overview.......... 36
   4.2.2 Planning Object Structure and Planning Area...................... 37
   4.2.3 Configuration of the Planning Object Structure 38
   4.2.4 Configuration of the Planning Area.................................... 39
   4.2.5 Disaggregation.. 42
   4.2.6 Organisation of Characteristic Value Combinations 43
   4.2.7 Time Series.. 45
4.3 Planning Book, Macros and Interactive Planning 46
   4.3.1 Planning Book 46
   4.3.2 Macros 52
   4.3.3 Fixing of Values 56
4.4 Statistical Forecasting.. 57
   4.4.1 Basics of Forecasting... 57
   4.4.2 Data History.. 58
   4.4.3 Univariate Forecast Models...................................... 58
   4.4.4 Multiple Linear Regression (MLR).............................. 60
   4.4.5 Forecast Execution 60
   4.4.6 Life Cycle Planning.. 64
4.5 Promotion Planning 67
4.6 Dependent Demand in Demand Planning.................................. 72
4.7 Collaborative Forecasting... 75
4.8 Background Planning... 77
4.9 Release of the Demand Plan 79
   4.9.1 Topics for the Demand Plan Release............................. 79
   4.9.2 Forecast Release 79
   4.9.3 Forecast After Constraints.. 81
   4.9.4 Transfer to SAP ERP™.. 83

5 Forecast Consumption and Planning Strategies............................. 85
5.1 Make-to-Stock 85
5.2 Make-to-Order.. 86
5.3 Planning with Final Assembly.. 87
5.4 Planning Without Final Assembly... 89
5.5 Planning for Assembly Groups 91
5.6 Technical Settings for the Requirements Strategies 92

PART III – ORDER FULFILMENT

6 Order Fulfilment Overview 97

7 Sales   99
  7.1 Sales Order Entry... 99
  7.2 Availability Check Overview 100
    7.2.1 Functionality Overview for the Availability Check 100
    7.2.2 ATP Functionality for Document Types 101
  7.3 Master Data and Configuration 104
    7.3.1 Master Data for ATP 104
    7.3.2 Basic ATP Configuration 105
    7.3.3 Time Buckets and Time Zones.. 107
  7.4 Product Availability Check... 109
    7.4.1 Product Availability Check Logic 109
    7.4.2 Product Availability Check Configuration 111
  7.5 Allocations... 114
    7.5.1 Business Background and Implications............................ 114
    7.5.2 Configuration of the Allocation Check 116
    7.5.3 Allocation Maintenance and Connection to DP 120
    7.5.4 Collective Product Allocations.. 121
  7.6 Forecast Check 121
  7.7 Rules-Based ATP.. 122
  7.8 Transportation and Shipment Scheduling.................................. 132
  7.9 Backorder Processing 134

8 Transportation Planning.. 145
  8.1 Transportation Planning Overview.. 145
  8.2 Master Data and Configuration 147
    8.2.1 Master Data for TP/VS... 147
    8.2.2 Geo-Coding 152
    8.2.3 Configuration of the CIF 154
  8.3 TP/VS Planning Board 154
  8.4 TP/VS Optimisation... 156
  8.5 Scheduling 159
  8.6 Carrier Selection 160
  8.7 Collaboration 161
  8.8 Release and Transfer to SAP ERP™... 162
  8.9 Dynamic Route Determination 162

PART IV – DISTRIBUTION

9 Distribution and Supply Chain Planning Overview........................... 165
  9.1 Distribution and Supply Chain Planning Scenarios.................... 165
  9.2 Applications for Distribution and Supply Chain Planning 167
  9.3 Order Cycle for Stock Transfers...................................... 171
  9.4 Integration of Stock Transfers to SAP ERP™........................... 173
  9.5 SNP Planning Book 178

10 Integrated Distribution and Production Planning 183
  10.1 Cases for Integrated Planning...................................... 183
  10.2 SNP Optimiser.. 184
    10.2.1 Basics of the Supply Network Optimiser 184
    10.2.2 Optimiser Set-up and Scope 185
    10.2.3 Costs and Constraints 187
    10.2.4 Discretisation.. 189
    10.2.5 Technical Settings 191
  10.3 Capable-to-Match 193
    10.3.1 CTM Planning Approach 193
    10.3.2 Prioritisation, Categorisation and Search Strategy 195
    10.3.3 CTM Planning 198
    10.3.4 CTM Planning Strategies 203
    10.3.5 Supply Distribution 206

11 Distribution Planning... 207
  11.1 Master Data for Distribution Planning.............................. 207
  11.2 SNP Heuristic 210
  11.3 Planned Stock Transfers 210
  11.4 Stock in Transit.. 212
  11.5 Storage and Handling Restrictions 213
  11.6 Sourcing.. 215

12 Replenishment 217
  12.1 Deployment 217
    12.1.1 Deployment Overview... 217
    12.1.2 Deployment Heuristic.. 218
    12.1.3 Deployment Optimisation 225
  12.2 Transport Load Builder.. 227

PART V – PRODUCTION

13 Production Overview 235
    13.1 Production Process Overview 235
    13.2 Applications for Production Planning 239
        13.2.1 Scenario and Property Overview 239
        13.2.2 Lot Size 245
        13.2.3 Scrap 247
    13.3 Feasible Plans 250
    13.4 Master Data for Production 252
        13.4.1 Production Master Data Overview 252
        13.4.2 Resource for SNP 256
        13.4.3 PDS and PPM for SNP 257
        13.4.4 Resources for PP/DS 262
        13.4.5 PDS and PPM for PP/DS 267
        13.4.6 Integration to PP-PI 273
    13.5 Dependencies to the SAP ERP™ Configuration 275

14 Rough-Cut Production Planning 277
    14.1 Basics of Rough-Cut Production Planning 277
    14.2 SNP Heuristic 280
    14.3 Capacity Levelling 282
    14.4 SNP Optimisation for Production Planning 283
    14.5 Capable-to-Match (with SNP Master Data) 285
    14.6 Scheduling in SNP 286
    14.7 Integration to PP/DS and SAP ERP™ 289
        14.7.1 Process Implications of Rough-Cut
                and Detailed Planning 289
        14.7.2 Integration to PP/DS 290
        14.7.3 Integration to SAP ERP™ 294

15 Detailed Production Planning 295
    15.1 Basics of PP/DS 295
        15.1.1 Order Life Cycle and Order Status 295
        15.1.2 Scheduling and Strategy Profile 297
        15.1.3 Planning Procedure 299
        15.1.4 Real Quantity 300

15.2 Heuristics for Production Planning.. 300
    15.2.1 Concept of Production Planning and MRP Heuristic....... 300
    15.2.2 Production Planning Heuristics.................................... 301
    15.2.3 MRP Heuristic.. 304
    15.2.4 Net Change Planning and Planning File Entries.............. 305
    15.2.5 Mass Processing 305
15.3 Consumption Based Planning... 308
15.4 Material Flow and Service Heuristics.................................... 309
15.5 Tools for Visualisation and Interactive Planning 311
    15.5.1 Product View... 311
    15.5.2 Product Overview... 314
    15.5.3 Product Planning Table 315
15.6 Reporting 316
15.7 Special Processes for Production Planning............................ 319
    15.7.1 MRP Areas 319
    15.7.2 Production in a Different Location................................ 321
15.8 Capable-to-Match (with PP/DS Master Data) 322

16 Sales in a Make-to-Order Environment 325
16.1 Process Peculiarities and Overview.................................... 325
16.2 Capable-to-Promise 327
    16.2.1 Steps Within the CTP Process.................................... 327
    16.2.2 Configuration of the CTP Process 328
    16.2.3 Problems with Time-Continuous CTP 330
    16.2.4 Bucket-Oriented CTP 332
    16.2.5 Interactive CTP.. 335
    16.2.6 Limitations for CTP.. 335
16.3 Multi-level ATP.. 336
    16.3.1 Steps Within the Multi-level ATP Process 336
    16.3.2 Configuration of the Multi-level ATP............................ 338
    16.3.3 Limitations for Multi-level ATP 339

17 Detailed Scheduling.. 341
17.1 Planning Board 341
17.2 Basics of Detailed Scheduling... 346
    17.2.1 Scheduling Strategies 346
    17.2.2 Error-Tolerant Scheduling.. 348
    17.2.3 Finiteness Level.. 349
17.3 Scheduling Heuristics 349
17.4 Sequence Dependent Set-up 353

17.5 Sequence Optimisation 356
17.5.1 Optimisation as Part of the Planning Process 356
17.5.2 Optimisation Model and Scope 357
17.5.3 Optimisation Controls Within the Optimisation Profile ... 359
17.5.4 Handling and Tools for Optimisation 365

18 Production Execution 367
18.1 Planned Order Conversion 367
18.2 ATP Check and Batch Selection 369
18.3 Production Order Handling 370

19 Modelling of Special Production Conditions 373
19.1 Alternative Resources 373
19.2 Modelling of Labour 377
19.3 Overlapping Production 378
19.4 Fixed Pegging and Order Network 379
19.5 Push Production 383

PART VI – EXTERNAL PROCUREMENT

20 Purchasing 387
20.1 Purchasing Overview 387
20.1.1 Process Overview 387
20.1.2 Order Life Cycle and Integration to SAP APO™ 387
20.2 Suppliers and Procurement Relationships 391
20.3 Supplier Selection 392
20.4 Scheduling Agreements 394
20.5 Supplier Capacity 397

21 Subcontracting 401
21.1 Subcontracting Process Overview 401
21.2 Modelling of the Production at the Receiving Plant 402
21.3 Modelling of the Production at the Supplier 405
21.4 Subcontracting Process Variants 408
21.5 Subcontracting in SNP 411

PART VII – CROSS PROCESS TOPICS

22 Stock and Safety Stock 415
    22.1 Stock Types in APO™ 415
    22.2 Safety Stock 416

23 Interchangeability 421
    23.1 Interchangeability Overview 421
    23.2 Interchangeability in DP 422
    23.3 Interchangeability in SNP 423
    23.4 Interchangeability in PP/DS 424
    23.5 Interchangeability in ATP 426

24 Exception Reporting 427
    24.1 Basics of Alert Monitoring 427
    24.2 Alert Types 429
    24.3 Alert Handling 433
    24.4 Alert Calculation in the Background 435
    24.5 Supply Chain Cockpit 435

PART VIII – SYSTEM INTEGRATION

25 Core Interface 439
    25.1 Overview of the Core Interface 439
    25.2 Configuration of the Core Interface 439
    25.3 Integration Models and Data Transfer 442
    25.4 Master Data Integration 446
    25.5 Transactional Data Integration 450
    25.6 Operational Concept 452
        25.6.1 Organisation of the Integration Models 452
        25.6.2 Organisation of the Data Transfer 453
        25.6.3 Data Consistency 454
        25.6.4 Queue Monitoring 454

26 Integration to DP 461
    26.1 Data Storage in Info Cubes 461
    26.2 Data Loading Structures 463
    26.3 Data Upload 466

APPENDIX

References.. 471

Abbreviations... 479

Transactions and Reports.. 481

Index... 495

Part I – Overview

1 SCM Projects with SAP APO™

1.1 The Supply Chain Management Approach

For a long time the focus in logistics projects has been on the optimisation of certain logistic functions – e.g. the optimisation of the transportation and distribution structure – usually with small concern to the adjacent processes and to the complete product portfolio. The supply chain management approach differs from this by grouping products with similar properties (from a logistics point of view) to a supply chain and taking all the processes – in SCOR terminology: plan, source, make, deliver – per supply chain into account. Figure 1.1 visualises the different approaches to structure the logistics processes within a company.

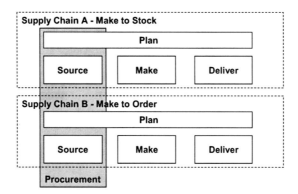

Fig. 1.1. The supply chain approach

The main differentiator for supply chains is the production strategy, that is if a product is created according to a specific customer demand (make to order) or anonymously (make to stock). Other criteria for separate supply chains might be different customer groups or product properties as the shelf life or the value.

J. T. Dickersbach, *Supply Chain Management with APO*,
DOI: 10.1007/978-3-540-92942-0_1, © Springer-Verlag Berlin Heidelberg 2009

The advantage of the supply chain approach is that the processes are examined from the point of view how they contribute to the targets of the supply chain management (e.g. low operating costs, flexibility and responsiveness or delivery performance). Therefore the integration between different logistical functions, for instance sales planning and production planning, is stronger within the focus of the supply chain management approach. In many cases the transparency between different logistical functions and between planning and execution offers already a significant potential for optimisation. The next step is to extend the supply chain approach beyond the limits of a single company and regard the entire value chain from the raw material to the finished product for the consumer. In this area the collaborative processes gain increasing importance.

• *Beer Game*
The beer game illustrates the importance of the transparency within the supply chain in a playful way. The supply chain for the beer game consists of a retailer, a wholesaler, a distributor and a factory. Each round a customer order is placed at the retailer, the retailer places his order at the wholesaler and so on unto the factory. The factory finally creates production orders. The goods flow is modelled by deliveries from the factory to the distributor, from the distributor to the wholesaler and so on to the customer. Each delivery requires goods movements across two fields and takes therefore two rounds. The production – the time between the creation of the production order and the goods receipt at the factory – requires three rounds. Figure 1.2 shows the structure for the order flow and for the goods flow.

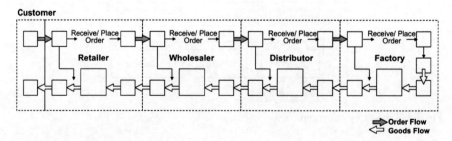

Fig. 1.2. The beer game

The customer orders are given and represent a steady demand with one increase of the level, as shown in figure 1.3. The game starts at a steady state with initial stock, orders at all levels and deliveries. The time lag between placing the order and receiving the supply, the insecurity about the future orders of the partner on the demand side and the insecurity regarding the stock outs at the partner on the supply side usually cause overreactions for the own orders, which destabilise the supply chain. This behaviour is known as the 'bullwhip effect'. Figure 1.3 shows the result of a game which was played with experienced sales and logistics managers. The orders of each team – retailer, wholesaler, distributor and factory – are displayed, and the amplitude of the changes in the order quantity increases with the distance to the customer.

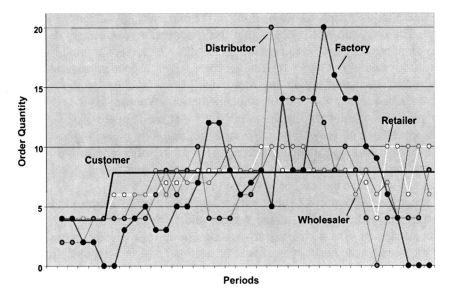

Fig. 1.3. The bullwhip effect

The impression at the factory site is that the customer demand is completely arbitrary. It is evident that a visibility of the customer demand across the supply chain helps to prevent this kind of destabilisation of the supply chain. To improve the transparency both a change of the processes and a system which enables the data transparency is necessary.

1.2 Supply Chain Management Projects with SAP APO™

A successful supply chain management project requires more than the implementation of a planning tool. The belief that the implementation of SAP APO™ solves all problems is in fact one of the major causes for the failure of SCM projects.

SAP APO™ is able to support SCM processes by visualising and processing data with a set of algorithms, but the adaptation to the particular business requirements has to be done in the particular implementation project. The prerequisite for this is that the requirements are clearly defined. No planning tool is able to provide the results you always wanted to have but never really thought of how to get them. Even if detailed requirements regarding the use of a planning tool exist, exaggerated expectations are a major risk for any SAP APO™ implementation. We strongly recommend keeping the solution as simple as possible – at least for the first step. To our experience all projects which aimed too high – by modelling too many constraints, avoiding manual planning steps at any price and including too many business areas, countries or plants – were significantly prolonged and had to reduce their scope in the end nevertheless.

Ideally a SCM project starts with a business case to define the targets and quantify the benefits of the project and is followed by a high level process design. The high level process design defines which processes are in the scope, whether they are local or a global and is used to define the according roles and responsibilities. Depending on the impact of the organisational changes, change management gains increasing importance to support the acceptance of the new processes and thus indirectly of the new planning system.

Another case is the implementation of SAP APO™ as a replacement of the existing planning systems due to support problems and/or strategic IT decisions. To our experience these cases are less apt to compromise regarding their expectations.

Since the supply chain management projects can significantly affect and change the company, a strong commitment by the sponsor in a sufficiently high position is necessary.

1.3 SAP APO™ Project Management and Peculiarities

SAP APO™ projects differ significantly from SAP ERP™ implementation projects, because the planning processes are usually more complex and less standardised and the integration aspects have an increased importance. The possibilities to model processes across modules and systems are quite numerous and the technical aspects of the system and the data integration do play an important part. The challenges for the project management in SAP APO™ implementations are mainly to define an appropriate project scope, avoid dead ends in the modelling approaches, ensure the integration of the processes and plan the necessary tests with sufficient buffers for changes (e.g. after the stress test).

Since SAP APO™ offers many functions and possibilities, it is very tempting to stretch the scope by including too many functions, constraints and business areas, countries and plants, so that the project becomes too complex to be successful. Therefore both in the definition of the scope as well as in the functional requirements a clear prioritisation is necessary. Generally we recommend a roll-out approach instead of a 'big bang' scenario, that is to divide the scope of a big implementation into several smaller implementations. The roll-out approach has the advantage of increased acceptance by a faster success and decreases the risk of running into a dead end (because of organisational incompatibilities, inappropriate modelling, insufficient master data quality ...).

We strongly recommend starting any SAP APO™ implementation project with a more or less extensive feasibility study. The benefits of the feasibility study is an increased security regarding the modelling approach and a basis for the definition of the scopes for the pilot and the following roll-out projects as well as for the project planning. The result of the feasibility study has to be a prioritisation of the functions and the business areas and an agreement about the scope and the modelling approach. To avoid misunderstandings due to working on a high level of abstraction and to ensure the feasibility of tricky modelling approaches, we recommend creating an integrated prototype in the systems already at this stage.

In general the benefits of advanced planning systems like SAP APO™ are the data transparency to support SCM decisions and the possibility to apply complex planning algorithms and optimisation techniques to improve the plan. Though optimisation techniques are often placed in the foreground, in most cases the main benefit is derived from the data transparency, and the application of processes which require a consistent global data basis – e.g. a coordinated sales and operations planning, demand visibility and forecast collaboration, global inventory management – is usually already a big challenge for a company. Often however a master data harmonisation is required before these benefits might apply – the quality of the master data is a severe threat for any SCM project and has therefore to be carefully examined during the feasibility study. Another issue regarding the use of optimisation techniques which should not be underestimated is whether the result is understandable and hence accepted by the planner. Though there are processes where optimisation tools provide a clear advantage, awareness for the implications is necessary – especially in the first implementation steps.

One main advantage of SAP APO™ to its competitors is its property regarding the integration to SAP ERP™. Nevertheless both the importance and the difficulties of the integration to the execution system(s) are usually underestimated. The integration is not just a simple exchange of data but an alignment of planning and execution processes, the more the scope of the planning moves towards execution.

SCM processes are often modelled across modules and across systems. To avoid the risk of unfeasible interfaces (both from a process as from a data point of view), the project organisation has to be according to the processes and not according to the systems and modules (especially if the implementation project contains both SAP ERP™ and SAP APO™).

Typically an implementation project contains the phases project preparation, blueprint, realisation, test, go live preparation and support after go live. For the blueprint phase of a SAP APO™ implementation it is absolutely necessary to perform a prototyping in parallel, because the processes are too complex to design without system feedback. During the entire project the basis support has an increased importance due to the more complex system landscape and additional, new technologies as the live cache and SAP BI™, which require administration.

A challenge for the project management is to plan sufficient buffers for adjustments and corrections after the integration test and the stress resp. the performance test. Especially the performance test has to be as early as possible, since the result might cause the procurement of additional hardware or even a redesign of some processes. Another important issue which is often neglected is the system management concept, which defines the requirements and procedures for the system administration, e.g. for backup and recovery, for downtimes and for upgrades.

SAP offers a set of services to check and review the implementation projects at different stages both from a technical and a process modelling point of view. Using these services might help recognising problems earlier.

2 SCM Processes and SAP APO™ Modules

The focus of this book is the SCM processes within a company. Though the possibilities of collaboration with customers and suppliers and the according processes are mentioned as well, the focus is on the SCM within a company, because to our experience in this area there is still the biggest potential for most companies. From a company's point of view a supply chain usually consists of

- customers,
- distribution centres (DCs),
- plants and
- suppliers.

There might be several levels for distribution (e.g. regional and local DCs) or several levels of production, if one plant produces the input material for another plant. Another characteristic of a supply chain is whether sourcing alternatives exist. Multiple sourcing is common for suppliers, and in many cases alternatives exist for production and distribution as well. Common variants in the distribution are direct shipments from the plant to the customer (instead from the local DC) depending on the order size. The most common supply chain processes cover the areas

- demand planning,
- order fulfilment (sales, transportation planning),
- distribution (distribution planning, replenishment, VMI),
- production (production planning, detailed scheduling, production execution) and
- external procurement (purchasing, subcontracting).

In cases of multiple sources for internal procurement an integrated approach for distribution and production planning may be favourable as described in chapter 10. The difference between distribution planning and replenishment is that the purpose of distribution planning is to propagate the demand in the network to the producing (or procuring) locations. Therefore distribution takes place from short- to medium-term or even long-term and requires subsequent processes, whereas replenishment is concerned in the more operative task how to fulfil the demands within the network with

J. T. Dickersbach, *Supply Chain Management with APO*,
DOI: 10.1007/978-3-540-92942-0_2, © Springer-Verlag Berlin Heidelberg 2009

a given supply quantity (which might be often a shortage). Figure 2.1 shows the processes in relation to the part of the supply chain.

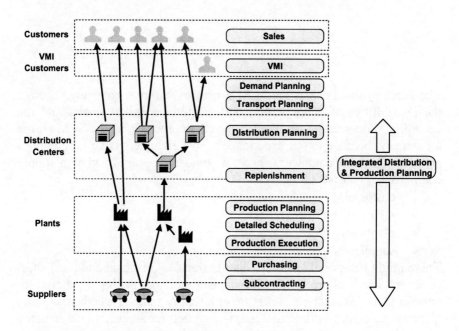

Fig. 2.1. Common supply chain processes

These processes differ both regarding their time horizon and their level of detail. A demand plan is usually established for 12 month to 5 years, whereas replenishment is carried out for some days to few weeks into the future. Regarding the level of detail, medium-term capacity planning is performed to get an overview of the monthly or weekly load on some bottleneck resources in contrast to a production schedule that contains the allocation of single operations with their exact duration to the resources. For additional information regarding the SCM processes Knolmayer et al. 2009 provides a good overview.

Figure 2.2 gives an indication about common time horizons of the respective processes.

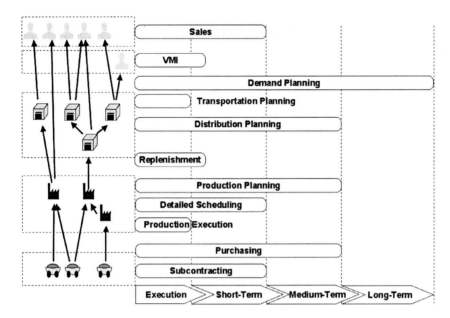

Fig. 2.2. Common time horizons for SCM processes

According to the different levels of the supply chain partners, time horizons and processes, SAP APO™ consists of different modules with different levels of detail. These modules are:

- Demand Planning (DP),
- Supply Network Planning (SNP) including deployment functionality,
- Production Planning & Detailed Scheduling (PP/DS),
- Available-to-Promise (ATP) and
- Transportation Planning & Vehicle Scheduling (TP/VS).

Figure 2.3 shows the positioning of these modules regarding the covered time horizon and the level of detail.

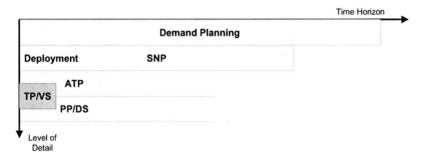

Fig. 2.3. Level of detail and time horizon for the SAP APO™ modules

Depending on the implementation, especially the time horizons for ATP and PP/DS might vary strongly. DP, SNP and ATP use time buckets:

- in DP usually months or weeks,
- in SNP usually months, weeks or days and
- in ATP days or parts of days.

PP/DS and TP/VS apply a time continuous calculation, so all orders are scheduled to hour, minute and second.

The supply chain processes identified above are generally modelled in the SAP APO™ modules as shown in figure 2.4. SNP provides three different methods for distribution planning resp. integrated distribution and production planning: the SNP heuristic which is not constrained by capacity restrictions, the SNP optimisation based on linear programming and the capable-to-match (CTM) heuristic which considers capacity constraints. Production planning and procurement – to a certain extent even distribution planning – are modelled either in both SNP and PP/DS, in SNP only or in PP/DS only, depending on the requirements for the process.

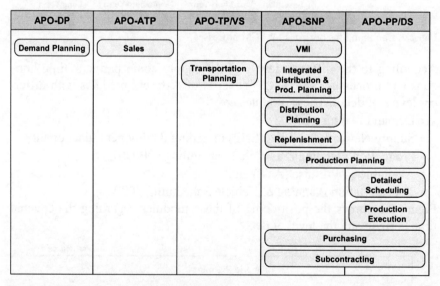

Fig. 2.4. SCM processes in SAP APO™ modules

From a supply chain project point of view figure 2.3 represents an implementation with the full scope of SAP APO™. There are some companies that apply SAP APO™ this way and even implement the complete scope at once. More often only a part of this scope is implemented – either as a first step or because this part is sufficient to satisfy the current needs. The advantage of keeping the scope of the implementation small lies in getting early results and having a shorter project duration.

Many implementations have only DP in scope since it is both technically and organisationally the part with the least complications. Especially in cases when the SAP APO™ implementation is done together with a change in the processes – e.g. from a (internal) customer – (internal) supplier relationship towards a supply chain planning in global or regional companies – the organisational aspects become most critical to the success of the project. Other common architectures are

- ATP for availability check across several plants,
- PP/DS for scheduling and sequence optimisation,
- PP/DS for finite production planning,
- DP & ATP for demand planning and availability check of allocations,
- DP & SNP for demand planning, distribution planning and replenishment,
- DP, PP/DS & ATP if there is no focus on distribution in the supply chain and sourcing decisions are either irrelevant or made using rules based ATP.

These are only some of the possible or even of the realised architectures. Dickersbach 2008 gives an indication about the incidence in implementations. Because of the multiple possibilities to model processes in SAP APO™ an experienced consultant should be involved at the definition of the architecture and the scope of an implementation.

Collaboration between companies has in some cases doubtlessly advantages, but for the vision of competing supply chains this is only one part. The other part is to establish SCM within a company – which is usually the more difficult part because it affects the organisation to a much higher degree. The change towards collaboration might be a bigger one looking from a change in the process but it affects only a small part of the organisation.

3 SAP APO™ Architecture

3.1 Technical Building Blocks of SAP APO™

SAP APO™ consists technically of three parts: the database, the SAP BI™ data mart and the live cache. The SAP BI™ data mart consists of info cubes. The live cache is basically a huge main memory where the planning and the scheduling relevant data are kept to increase the performance for complex calculations. Though there is technically only one live cache per installation, the data is stored in three different ways depending on the application:

- as a number per time period (month, week, day) and key figure (time series),
- as an order with a category, date and exact time (hour: minute: second) and
- as a quantity with a category and a date in the ATP time series.

We will refer to the according parts of the live cache as time series live cache, order live cache and ATP time series live cache.

Demand planning uses much of the SAP BI™ functionality and relies on the info cubes as data interface to any other system – SAP ERP™, SAP SEM™ or flat file. Therefore the historical data is always persistent in an info cube. For processing the data is written into the time series live cache. SNP and PP/DS use mainly the order live cache, though SNP is able to access the time series live cache as well, since there are many structural similarities between DP and SNP. ATP at last relies only on the ATP time series live cache. This way of data storage implies a certain redundancy, because orders are stored both in the order live cache and in the ATP time series live cache. The data model is however quite different, and the redundant data storage enables better performance for the applications. TP/VS finally uses the order live cache to reference other orders. Figure 3.1 shows how the SAP APO™ modules use the live cache and the data integration for transactional data from SAP ERP™ to SAP APO™.

J. T. Dickersbach, *Supply Chain Management with APO*,
DOI: 10.1007/978-3-540-92942-0_3, © Springer-Verlag Berlin Heidelberg 2009

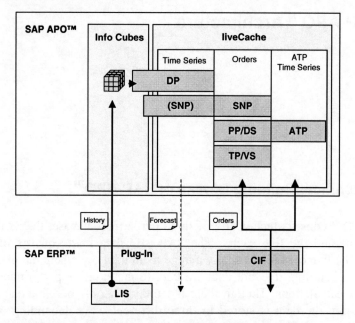

Fig. 3.1. SAP APO™ system structure and integration with SAP ERP™

The transactional data for planning purposes (e.g. planned orders and purchase requirements) are created in SAP APO™ and should preferably remain in SAP APO™ only to reduce the data load for the interface. In any case SAP APO™ should be the master for planning. For data with a close link to the execution (e.g. sales orders and purchase orders) SAP ERP™ is the master. Historical data finally is stored in SAP ERP™. For the integration of the transactional data and the data history with the plug-in SAP provides an interface from SAP ERP™ to SAP APO™. One part of the plug-in is the core interface (CIF), the other provides the interface to the SAP BI™ structures. The integration of SAP ERP™ to the SAP BI™ structures relies on the info structures of the logistics information system (LIS) in SAP ERP™, where transactional data is stored for reporting purposes according to the defined selection criteria. These data are uploaded into an info cube with periodic jobs.

In contrast to the periodic data upload to the SAP BI™ part the CIF provides an event triggered online integration. For example with the update of the goods receipt an event is created which triggers the transfer of the updated stock situation to SAP APO™. This information creates two entries in SAP APO™: one as an order in the order live cache and one as an element in the ATP time series live cache.

The release SAP SCM 2008 does not only contain SAP APO™ but other systems as well: the Supply Network Collaboration Hub (SAP SNC™), the Event Manager (SAP EM™), Forecasting and Replenishment (SAP F&R™), and the Extended Warehouse Management (SAP EWM™).

3.2 Master Data Overview

Like in SAP ERP™, master data plays an important role in SAP APO™ and controls many processes. Organisational entities like company codes or sales organisations on the other hand have no significance in SAP APO™. Some master data objects in SAP APO™ have analogies in SAP ERP™ like the product, and most of these are transferred from SAP ERP™. Others have to be maintained in SAP APO™.

● *Master Data and Applications*
Figure 3.2 provides an overview of the most important master data objects in SAP APO™ and in which application they are more or less mandatory (displayed in white) or only required for certain processes (displayed in grey).

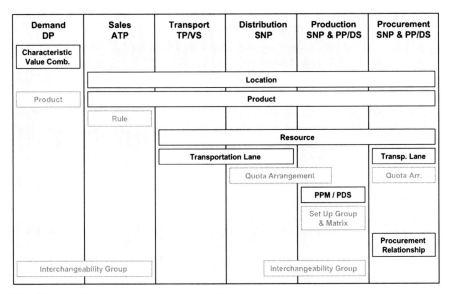

Fig. 3.2. Overview master data and applications

Most of these master data have correspondencies to the SAP ERP™ master data. Table 3.1 shows these correspondencies for those master data that are transferred from SAP ERP™ via CIF.

Table 3.1. Corresponding master data objects in SAP APO™ and SAP ERP™

SAP APO™ Master Data	SAP ERP™ Master Data
Location	Plant
Product	Material
Resource	Work Centre (PP) or Resource (PP-PI)
Transportation Lane	Info Record
Production Data Structure (PDS) Production Process Model (PPM)	Production Version (Combination of BOM and Routing/Recipe)
Procurement Relationship	Info Record, Scheduling Agreement, Contract

The characteristic value combinations (CVCs) can be generated from historical data that is transferred from SAP ERP™. If no historical data is available, e.g. for the demand planning of new products, the characteristic value combinations can be maintained in SAP APO™ as well.

The characteristics for planning in DP are chosen freely – the sales organisation for example will often be used for demand planning, but does not have any other correspondence. Only for the product and the location the characteristics 9AMATNR and 9ALOCNO are used per default as a link to the product master resp. the location master for the data integration between the time series live cache and the order live cache. Figure 3.3 visualises the master data integration from SAP ERP™ to SAP APO™. This is a one directional process, no master data information is transferred from SAP APO™ to SAP ERP™.

The integration of the characteristic value combinations (CVCs) is performed using the historical data that is transferred to the info cube. The CVCs are generated from the info cube.

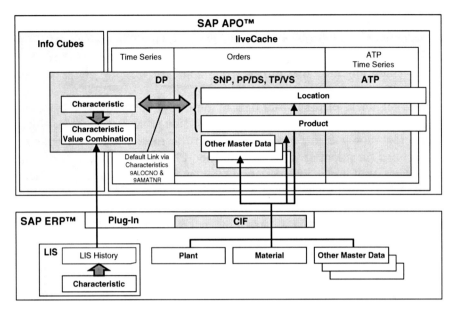

Fig. 3.3. Master data integration

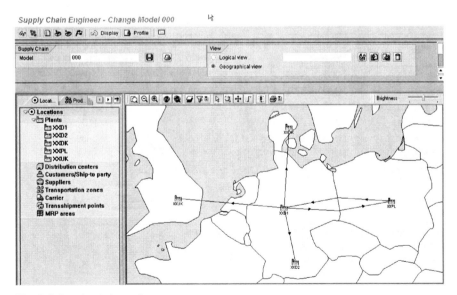

Fig. 3.4. Supply chain engineer

• *Supply Chain Engineer*

The supply chain is displayed and partially maintained in the supply chain engineer (SCE) with the transaction /SAPAPO/SCC07. Figure 3.4 shows the supply chain engineer that provides a graphical overview of the supply chain.

The assignment of the master data to the model or its removal from the model (see next chapter about model and version) is also carried out either within the supply chain engineer or from the master data maintenance transactions (location, product, resource and PPM). The model is a master data object itself and is locked during maintenance. Using the supply chain engineer for model maintenance has the disadvantage of locking the entire model for a comparatively long time. If master data is transferred from SAP ERP™, the assignment is carried out automatically.

• *Locations*

Differing from SAP ERP™ where a plant and a supplier are completely different objects, SAP APO™ uses only one object for a location. This object might have different types. The more common types, their usage and the corresponding SAP ERP™ entity is listed in table 3.2.

Table 3.2. Location types in SAP APO™

Location	Loc. Type	Application	Corresponding SAP ERP™ Entity
Plant	1001	SNP, PP/DS, TP/VS, ATP	Plant
DC	1002	SNP, PP/DS, TP/VS, ATP	Plant[1]
Transport. Zone	1005	TP/VS	Transp. Zone[2]
MRP Area	1007	PP/DS, SNP	MRP Area[3]
Customer	1010	TP/VS, SNP for VMI	Customer
Vendor	1011	SNP, PP/DS[4]	Vendor
Subcontractor	1050	SNP, PP/DS if one subcontractor is supplying more than one plant	-
Transport Service Provider	1020	TP/VS as carrier	Vendor[5]

[1]Assignment of the node type 'DC' in transaction DRP9
[2]Transferred implicitly via customer
[3]From material master, explicit transfer in CIF
[4]TP/VS possible but unusual
[5]Customising per account group

The transfer of customers to SAP APO™ is only recommended for VMI customers, for customers with consignment processes or for TP/VS. In SAP APO™ only the name of the location is used as the key, not the location type. Make sure that there are no code clashes due to plants, suppliers or customers with the same names – either by using different naming conventions in SAP ERP™ or by concatenation of a suffix or a prefix in the user exit APOCF001. The location object in SAP APO™ contains some additional information which is not transferred from SAP ERP™, such as calendars, storage and handling resources and planning relevant categories – namely the stock categories for SNP and the receipts and issues for deployment. These entries are maintained in the location master with the transaction /SAPAPO/LOC3.

• *Resource*
Resources are used in the applications SNP, PP/DS and TP/VS for many different purposes. The resources have a category (production, transport, handling or storage location) and a type (single, multi, bucket etc.). Category and type are selected when creating a resource with transaction /SAPAPO/RES01, figure 3.5.

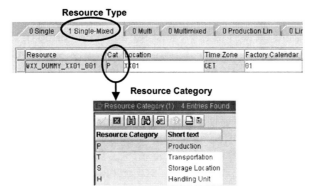

Fig. 3.5. Resource type and category

The usual combinations of type and category and their usage in the applications are shown in figure 3.6. The resource category is in brackets.

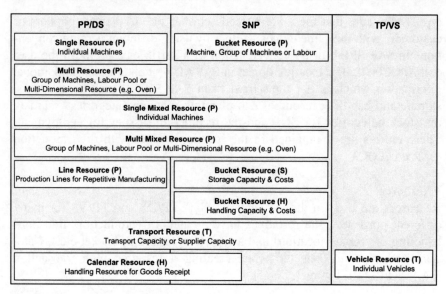

Fig. 3.6. Resource types and their usage

The resources are created in live cache for a certain period of time. Therefore it is important to run the report /SAPAPO/CRES_CREATE_LC_RES to extend the availability of the resource in live cache periodically as described in note 550330.

• *Overview about PDS, PPM and Resource Types*
In SNP and in PP/DS different objects are used for the resources and for the PDS resp. the PPM. To reduce the efforts for this manual master data maintenance it is possible to create mixed resources during the data upload from SAP ERP™. It is possible to define the resource type for SAP APO™ and maintain the SAP APO™ specific entries in the work center on the SAP ERP™ side. These settings are made in the capacity header using the button 'APO resource'.

Figure 3.7 provides a short overview about the different PDS resp. PPM types, the different resource types and how they are used within the PDS resp. PPM.

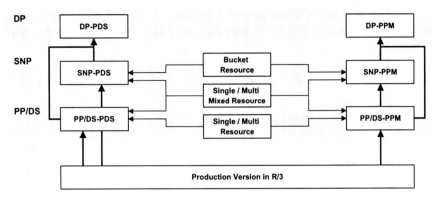

Fig. 3.7. PDS/PPM and resources

The SNP-PDS is directly transferred from SAP ERP™, while the SNP-PPM is generated from the PP/DS-PPM. For the generation of the PP/DS-PPMs to SNP-PPMs the use of mixed resources is a prerequisite.

3.3 Model and Version

The two basic structural elements in SAP APO™ are the model and the version. Several versions can be assigned to one model. The general idea is that the model contains the master data and the version the transactional data, but for some of the master data (location, product and resource) it is also the possible to make some version dependent changes. Figure 3.8 shows the assignment of the main master data to the model and the version. Possible scenarios for the use of version dependent master data are simulations of different shift models or lot sizes.

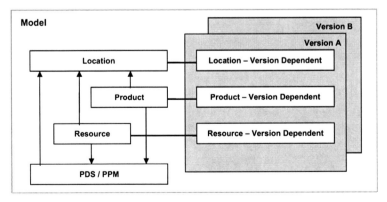

Fig. 3.8. Master data, model and version

By design it is possible to use different versions of any model for simulation purposes. Note that the ATP time series are only available for the active version and the according flag has to be set when creating the version. Before copying a version, it must be considered that the data volume is usually rather huge. Note 519014 describes some of the problems which might occur during the copying of a version and how to solve them. Versions are copied and maintained with the transaction /SAPAPO/MVM. For the copying it is recommended to select only the orders and use the transaction /SAPAPO/TSCOPY to copy the time series. The requirement for the live cache memory increases nearly linear with the number of simulation versions, which is in many cases already a knock-out criterion for the use of multiple versions.

If the master data is transferred from SAP ERP™ via CIF, it is automatically assigned to the active model 000. But if the master data is created in SAP APO™ it has to be assigned manually to the model. This can be done either in the SCE or from the maintenance of the respective objects:

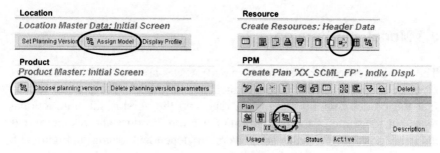

Fig. 3.9. Assignment of master data objects to the model

For the resource and the PPM you have to enter the object to assign it, for the location and the product master only the object name has to be filled in.

3.4 Planners

In SAP APO™ most of the organisational entities from SAP ERP™ are unknown. Since SAP APO™ is only concerned with the logistics and not with the accounting and costing, there are no company codes and no cost centers. Except for the locations – which is unlike in SAP ERP™ a master data in SAP APO™ – the main organisational entity is the planner. There are different planners for DP, SNP, PP/DS and TP/VS.

The planner is an organisational entity which is used within and across application for selection purposes (e.g. in some standard reports, the work area or in the alert monitor). The planner is defined with the transaction /SAPAPO/PLANNER for one or more of the applications PP/DS, SNP, DP and TP/VS. The key for the planner has to be unique – a planner XYZ is defined and the attributes 'production', 'SNP', ... are assigned to the planner. The planners are assigned to the product master in SAP APO™ and are not transferred from SAP ERP™. For a matching between the MRP controller in SAP ERP™ and the production planner in SAP APO™ it must be considered that in contrast to the MRP controller in SAP ERP™ the planner in SAP APO™ is not location dependent. If two MRP controllers exist in SAP ERP™ with the same key in different plants, a matching is not possible.

3.5 Order Categories

Each order in the live cache has a category which is used to select the orders for display (e.g. for the assignment to the key figures in the SNP planning book) or for functions as the forecast consumption. The categories are defined with the transaction /SAPAPO/ATPC03. There are four types of categories – stock, receipts, requirements and forecast – which determine the possibilities of their usage. The mapping of the transferred SAP ERP™ orders to the SAP APO™ categories is maintained within the category itself (for SAP categories). It is possible to define own categories (Non-SAP categories), but it is not possible to map them in customising with the SAP ERP™ elements. One or more categories are combined into a category group for display and copy purposes (e.g. SNP key figure, location master, copy jobs) with the transaction /SAPAPO/SNPCG.

3.6 Pegging

Pegging describes the attachment of supply nodes to demand nodes within the order live cache. A helpful notation for explaining planning situations in the order network is shown in figure 3.10. The basis of this figure is pegging areas that represent a product in a location (a locationproduct). Orders have a supply node in one pegging area for their output and demand nodes for their dependent demands in other pegging areas. The supply nodes and the demand nodes of one pegging area are connected by pegging.

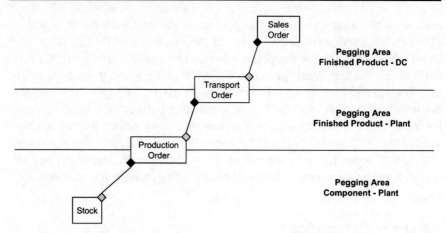

Fig. 3.10. Pegging and order network

Some elements as stock, purchase requisitions or purchase orders may have only a supply node while demand elements as forecast (planned independent requirement) or sales orders have only demand nodes. The pegging between these elements is calculated online in the live cache according to the settings in the 'demand'-view of the product master. The two strategies FIFO (first in, first out) and 'use latest receipt' are available. Figure 3.11 visualises the difference.

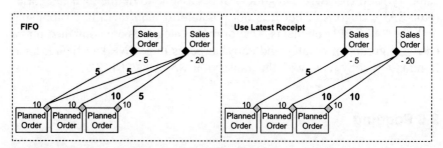

Fig. 3.11. FIFO and 'use latest receipt' pegging

In case of excess supply, with the FIFO strategy the unpegged quantity is available at the last order. If the option 'use latest receipt' is chosen, quantities of the first order are not pegged. An exception in the 'use latest receipt' procedure is the treatment of stocks. Pegging causes the excess coverage alerts which suggest the deletion of the respective elements. Since stock can not be deleted, with both strategies first the stock is pegged, figure 3.12.

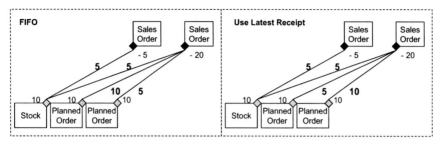

Fig. 3.12. FIFO and 'use latest receipt' pegging with stock

Since pegging is calculated online in the live cache, the pegging arcs change with the insertion of another demand or supply node. To avoid this dynamic change of the pegging it is possible to fix the pegging arcs, which fixes the receipt elements for the production planning heuristics and thus has severe implications for production planning. While dynamic pegging arcs do not cross by definition, fixed pegging can lead to rather confusing dependencies, as figure 3.13 shows.

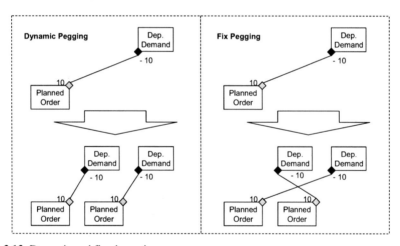

Fig. 3.13. Dynamic and fixed pegging

Pegging can be fixed either manually for the individual order or using a heuristic. The properties and limitations for fixed pegging are described in detail in section 19.4.

Pegging arcs can be used to model production constraints by specifying the maximum time between supply and demand nodes in the product master ('maximum earliest receipt'). An example for the application of such a constraint is the metal manufacturing where the products have to be processed before solidification. Pegging constraints however are not considered in production planning.

By default pegging is only created if the supply node is in time for the demand node. Lateness is tolerated if the setting for backward pegging in the demand view of the product master ('maximum latest receipt') is maintained. Having backward pegging instead of no pegging at all has advantages regarding the alerts, see section 24.2. The alert setting ('alert if receipt ... after demand') in the product master defines how late a receipt has to be to cause an alert. Since the lateness alerts do require a pegging relationship, the value for the alert threshold has to be smaller than the value for the backward pegging. Figure 3.14 visualises the impact of these settings.

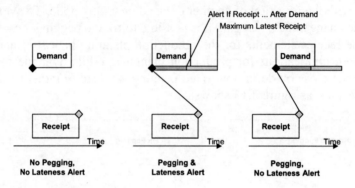

Fig. 3.14. Lateness settings for pegging

Independent of whether backward pegging is used or not, production planning tries to create the receipts in time.

Generally all the orders in the live cache are subject to pegging as long as they are not explicitly modelled as pegging irrelevant – like sales orders in combination with the production strategy 'make-to-stock' for example. SNP orders participate in pegging as well, but due to the bucket approach pegging is not that informative here, see section 14.6.

3.7 Data Locking

Depending on the module, the system behaves differently regarding the locking of data.

● *Data Locking in DP*
It is possible to choose on planning area level whether a conservative or a detailed locking logic is applied. If the conservative straightforward locking logic is applied, a set of data can only be processed by one person – or

one planning job – at a time. Using the detailed locking logic it is possible to limit the lock to the key figures in the planning book and to exclude read-only key figures from locking.

• Transactional Simulations
In PP/DS it is more often the case that different people want to access an object at the same time – for example for detailed scheduling and for production confirmation. To avoid unnecessary obstructions, PP/DS applies transactional simulations, which copy the objects of the work area and the connected objects from the current version. This way changes are performed independently.

• Order Merge
When saving the planning result the changes are written from the transactional simulation back to the current version ('merged').

There are certain rules for this merge – generally if one object is changed in different transactional simulations, the last save wins. Exceptions to this are e.g. production confirmations, because the reality overrules the planning.

• Temporary Quantity Assignments
In ATP there is no locking since it would not be acceptable if sales orders for overlapping materials could be checked only sequentially. However it is customisable whether sales orders that call the ATP check at the same time may be confirmed using the same element (see section 7.4 for temporary quantity assignments).

Part II – Demand Planning

4 Demand Planning

4.1 Demand Planning Overview

4.1.1 Demand Planning Process

The result of the demand planning process is the establishment of independent requirements which will trigger the planning activities as distribution, production and procurement planning. Usually a sales forecast is the key input to the demand plan. This sales forecast is consolidated and checked regarding plausibility, probably checked against a statistical forecast and corrected according to the experience of the planner before releasing it for the subsequent planning steps. Figure 4.1 shows this process for a very simple supply chain with global demand planning and single-sourcing local production planning:

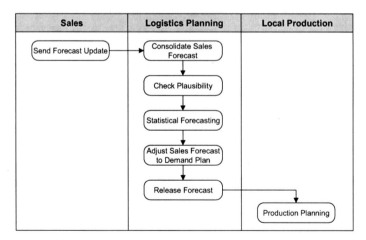

Fig. 4.1. Example process chain for demand planning

J. T. Dickersbach, *Supply Chain Management with APO*,
DOI: 10.1007/978-3-540-92942-0_4, © Springer-Verlag Berlin Heidelberg 2009

The monitoring of the forecast accuracy and a feasibility check against the planning constraints (e.g. capacity) are further common process steps. Depending on the business requirements there might be many others.

4.1.2 Planning Levels and Consistent Planning

The most elementary questions in demand planning are
- on which levels (product, product group, ...) is planning performed?
- in which granularity of time (weeks, months) is planning performed?
- which data is needed?

These questions have to be answered from a business point of view before starting with any kind of implementation.

In SAP APO™ the planning levels are represented by 'characteristics', the granularity of time corresponds to the 'time bucket' and the data is linked to 'key figures'. Aggregation and disaggregation (drill down) of the data according to the characteristics is one of the main advantages to a simple spreadsheet. Though data is maintained on all levels, it is always stored on the most detailed level (i.e. with each characteristic having a value) to keep the planning consistent. In this case the disaggregation is usually done either according to a different key figure or – as default – pro rata (the difference between the new and the preceding value is distributed according to the existing ratios). If the preceding value is zero, the new value is distributed equally. For the description of all disaggregation possibilities have a look into the system documentation.

Both the characteristics and the key figures are technically represented by 'info objects'. These info objects contain the data format definition, the name of the characteristic resp. the key figure plus other properties. The info objects are maintained with the transaction RSD1.

Since the understanding of the principle of aggregation and disaggregation with all its implications is the most important part to understand DP, this topic is stretched a little more by an example: planning shall take place on the levels sales organisation, location, product and product group. Each product is member of only one product group and each sales organisation is linked to only one location (this structure is represented by the characteristic combinations). The relevant data (i.e. key figures) are sales forecast and demand plan (the demand plan becomes relevant for planning), figure 4.2.

	Characteristics				Key Figure	
Sales Org	**Location**	**Product**	**Product Group**	**Month**	**Forecast**	**Demand Plan**
XXDE	XXD1	HEAVY_250	HEAVY	10-2003	100	
XXDE	XXD1	HEAVY_500	HEAVY	10-2003	200	
XXAT	XXD2	HEAVY_250	HEAVY	10-2003	50	
XXAT	XXD2	HEAVY_500	HEAVY	10-2003	150	
XXCH	XXD2	HEAVY_250	HEAVY	10-2003	80	
XXCH	XXD2	HEAVY_500	HEAVY	10-2003	20	
XXDE	XXD1	EXTRA_250	EXTRA	10-2003	160	
XXDE	XXD1	EXTRA_500	EXTRA	10-2003	140	

Fig. 4.2. Data structure for the planning example

In this example the forecast is maintained for each product by each sales organisation. To display the total forecast of all sales organisations for the product group HEAVY (i.e. 600 units) it is sufficient to make a selection for the product group only. The forecast values of the other characteristic combinations containing the product group HEAVY are cumulated. This way aggregation is carried out comfortably.

In the next step the planner enters the demand plan on the level 'location' and 'product group', figure 4.3.

Location	**Product Group**	**Month**	**Forecast**	**Demand Plan**
XXD1	HEAVY	10-2003	300	400
XXD2	HEAVY	10-2003	300	400
XXD1	EXTRA	10-2003	300	400

Fig. 4.3. Planning on aggregated level

Since there is no preceding value, the demand plan is disaggregated equally when saving the data, figure 4.4.

Sales Org	Location	Product	Product Group	Month	Forecast	Demand Plan
XXDE	XXD1	HEAVY_250	HEAVY	10-2003	100	200
XXDE	XXD1	HEAVY_500	HEAVY	10-2003	200	200
XXAT	XXD2	HEAVY_250	HEAVY	10-2003	50	100
XXAT	XXD2	HEAVY_500	HEAVY	10-2003	150	100
XXCH	XXD2	HEAVY_250	HEAVY	10-2003	80	100
XXCH	XXD2	HEAVY_500	HEAVY	10-2003	20	100
XXDE	XXD1	EXTRA_250	EXTRA	10-2003	160	200
XXDE	XXD1	EXTRA_500	EXTRA	10-2003	140	200

Fig. 4.4. Automatic disaggregation when saving the data

4.2 Data Structure for Demand Planning

4.2.1 Characteristics, Key Figures and Structure Overview

The display of the data and the planning is performed in planning books. To help understanding the technical configuration and the prerequisites for the planning book, figure 4.5 gives an overview about the structure of the relevant entities.

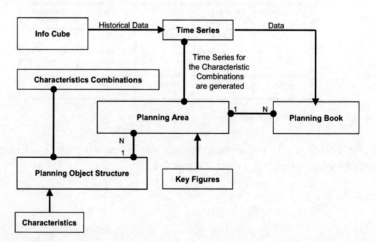

Fig. 4.5. Demand planning structure overview

A planning book is always linked to a planning area which contains the key figures. For the defined characteristic value combinations time series are created for each key figure of the planning area. These time series contain the actual data, so that the planning book serves to display the time series of the planning area. The planning area itself is connected to a basic planning object structure, where the relevant characteristics are defined. Historical data is stored in an info cube (which contains both characteristics and key figures) and loaded into the planning area. In the following chapter these entities will be described. Though an info cube is necessary for any realistic demand planning scenario, the set up of info cubes is not described here but in chapter 26.

4.2.2 Planning Object Structure and Planning Area

The basic settings for DP are the planning object structure and the planning area. The maintenance of these basic settings is accessed with the transaction /SAPAPO/MSDP_ADMIN, figure 4.6. Though it is possible to make some adjustments in these settings even after the characteristic combinations are generated and the planning books are designed, it is recommended not to change them too often to avoid inconsistencies. Note 332812 mentions reports to repair selections and documents.

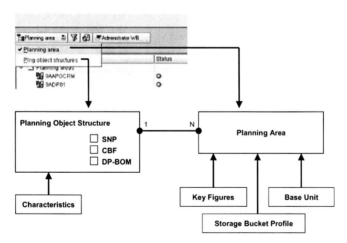

Fig. 4.6. Planning object structure and planning area

4.2.3 Configuration of the Planning Object Structure

The characteristics for planning are defined in the planning object structure. The characteristics 9ALOCNO and 9AMATNR correspond per default to the location and the product master in SNP and PP/DS, which are used for the release of the demand plan. It is nevertheless possible to use other characteristics for the correspondence. These have to be declared in the planning object structure using the menu '*Extra → Assign Prod./Loc.*'. For the release of the demand the correspondence has to be specified in the release configuration as well.

• *Navigation Attributes*
If there are levels on which the data is only displayed and no planning is performed, it is possible to model these levels as navigation attributes instead of characteristics. Navigation attributes have a reference to a characteristic, but are not part of the key for the data access. The advantage of the navigation attributes is that their values can be changed very easily without the necessity to realign any data. They are therefore very well suited to represent entities with more frequently changing values, for example the assignment of a planner to a product. Until only a handful of attributes are used, no performance restrictions are expected. Note that for the standard characteristics 9ALOCNO and 9AMATNR it is not allowed to use navigation attributes. Note 413526 compares the properties of navigation attributes and characteristics.

• *Aggregates*
An aggregate defines a higher level to store the data additionally. If data is often accessed on the specified aggregate, the advantage of the redundant data storage is that many calculations to aggregate the data are avoided. Note 503363 provides further recommendations for the use of aggregates.

Aggregates are defined for a planning object structure by right mouse click on the planning object structure → *Create Aggregate*. The characteristics for the aggregate are selected here. To have the data stored on this aggregate, it is necessary to select it in the planning area as well.

• *SNP, DP-BOMs and CBF*
If the planning book shall be used for SNP as well or the planning scenario includes DP-BOM (bills of material) or CBF (characteristic based forecasting) functionality, these functions are selected in the planning object structure. DP-BOMs are described in section 4.6, CBF in Dickersbach 2005.

• *Consistency Check*
The consistency of the planning object structure is checked with the transaction /SAPAPO/PSTRUCONS). For this purpose, see also reports /SAPAPO/TS_PSTRU_TOOLS and /SAPAPO/TS_PSTRU_GEN. If DP-BOMs are used (cf. section 4.6) the transaction /SAPAPO/PSTRU_PPM_MERGE checks whether the BOM information is consistent in the planning area.

4.2.4 Configuration of the Planning Area

The planning area contains the key figures and the settings regarding their properties, the time bucket profile, the unit of measure and the assignment of key figures to the aggregates for better performance.

The data is always stored in the unit of measure which is specified in the planning area. Nevertheless it is possible to switch between units of measures – as long as they are maintained in the product master as alternatives – with right mouse click on the top left cell of the table in the planning book. By setting the flag in the column 'UoM' of the key figure details (see figure 4.8) it is also possible to specify that always the base unit of the product master is used. The prerequisite for this is that the product characteristic is available in the planning object structure (see also note 429131).

During planning it might be desired to distinguish between initial values and periods where the planned value is zero. This distinction is controlled in the key figure details as well.

• *Time Buckets*
In DP there are two kinds of time bucket profiles, the storage bucket profile (transaction /SAPAPO/TR32) and the planning bucket profile (transaction /SAPAPO/TR30). The data of the planning area is always stored in the storage bucket profile, while the planning bucket profile is used to display the data in the planning book. In the planning bucket profile only those period types (month, week, ...) can be used that are also defined in the storage bucket profile. Therefore, if more than one time bucket is going to be used for planning or display, these have to be included in the storage bucket profile. For consistency reasons the data is stored on the lowest level. If the time buckets do not match – as with months and weeks – the periods are divided as shown in figure 4.7.

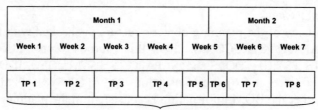

Fig. 4.7. Data storage in storage bucket profiles

In this case the data is saved on the lowest level, which is the technical period.

● *Activation of the Planning Area*
After saving the planning area remains inactive until time series are created. But before doing so, the characteristic combinations have to be created.

● *Number of Planning Areas*
One of the major decisions regarding the design of an implementation is whether to use one or more planning areas. Except for extreme cases it is not possible to give any recommendations. The use of several planning areas has some advantages:

- The maintenance and the repair of problems like corrupted time series is more flexible and less users are affected by the problems.
- There are less characteristic combinations in the planning area, which reduces the live cache requirements and improves the performance in principle. However no experiences regarding the limit of characteristic combinations and the potential performance improvement are known to us.
- If different business units use mostly different key figures, a large reduction of the live cache requirements can be achieved by avoiding empty time series for the disjunctive characteristic combinations and key figures.

On the other hand, different planning areas imply

- an increase of the complexity for the implementation,
- a risk of data errors and additional system load, if data has to be copied between planning areas,
- a multiplication of planning books and selections and
- a multiplication of the background jobs.

• *Locking*

In planning area it is also decided whether a detailed locking logic is used, see section 3.8.

• *Load Data from an Info Cube*

In the key figure details within the planning area additional properties are assigned to the key figures as shown in figure 4.8.

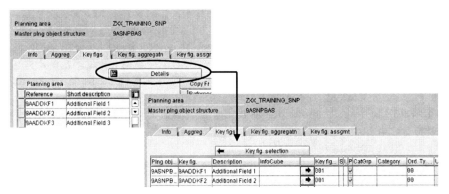

Fig. 4.8. Key figure settings in the planning area

One of these options is the assignment of an info cube to the key figure, which triggers that the data from the same key figure of the assigned info cube is read automatically. This data is displayed, but not stored in the planning area. There are three implications using this functionality:

- This key figure is not any more ready for manual input.
- If the planning object structure of the planning area contains more characteristics than the info cube, the data is not disaggregated to the additional characteristics of the planning object structure. Since the data is not saved in the planning area, no disaggregation takes place.
- It is not possible to use that key figure as a basis to disaggregate other key figures.

An alternative way of loading data from an info cube into the planning area to avoid these implication is to copy the data from the info cube to the planning area with the transaction /SAPAPO/TSCUBE. In this case the names of the key figures do not have to match anymore. The definition of these settings can be saved as a variant and used for background jobs. The disadvantage of this procedure however is that there is additional data stored in the live cache which increases the hardware requirements.

4.2.5 Disaggregation

The disaggregation options for the key figures are specified in the planning area as well. By default the disaggregation is pro rata, but disaggregation according to another key figure, no disaggregation or aggregation as an average are possible as well. For the disaggregation according to another key figure any key figure can be used, though the standard key figure for this is APODPDANT. It is possible to calculate the disaggregation ratio on the basis of any key figure and any time horizon of any info cube or planning area and save it into APODPDANT with transaction /SAPAPO/MC8V. Another option is to copy the disaggregation factor from an info cube to APODPDANT with transaction /SAPAPO/TSCUBE applying additional settings. These ratios are calculated either as a constant value for all periods or period dependent. If manual changes of the ratios in APODPDANT are allowed, the flag 'manual proportion maintenance' in the planning book must be set to include this key figure. There are five standard ways of key figure disaggregation:

- pro rata ('S'): data is disaggregated according to the proportions of the current data. If there is no current data, it is equally disaggregated.
- based on a different key figure ('P'): data is disaggregated according to the proportions of a different key figure. If there is no data in this key figure, no disaggregation is performed.
- average ('A'): there is no disaggregation downwards, upwards the average is aggregated.
- no calculation ('N'): neither aggregation nor disaggregation takes place
- pro rata, if initial based on a different key figure ('I'): as described in the name.

Figure 4.9 visualises the principle of the disaggregation types for an example where a value of 60 is entered on an intermediate level.

Additionally to this structural disaggregation there might be a time-based disaggregation, e.g. if the data is stored in weeks and planning is performed in months. Time-based disaggregation is carried out before structural disaggregation.

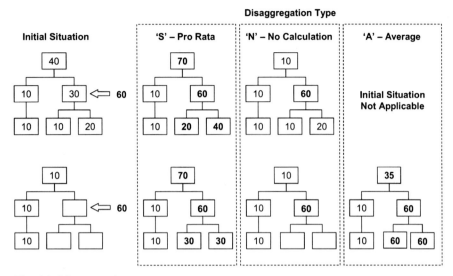

Fig. 4.9. Disaggregation types

4.2.6 Organisation of Characteristic Value Combinations

The data in DP is always related to the characteristic value combinations (CVC) which are used to access the data. Since the number of the characteristic combinations actually needed for planning is usually smaller than the product of possible characteristic values, the characteristic value combinations have to be created explicitly with the transaction /SAPAPO/MC62. In the example in figure 4.2, 8 CVCs are used, but the product of characteristic values is 24 (3 sales organisations times 2 locations times 2 products times 2 product groups), figure 4.10.

		Sales Org XXDE		Sales Org XXAT		Sales Org XXCH	
		XXD1	XXD2	XXD1	XXD2	XXD1	XXD2
Product Group HEAVY	HEAVY_250		X	X		X	
	HEAVY_500		X	X		X	
Product Group EXTRA	EXTRA_250		X	X	X	X	X
	EXTRA_500		X	X	X	X	X

Fig. 4.10. Active characteristic value combinations (white cells)

The characteristic value combinations are related to a planning object structure. When a planning object structure is activated, an info cube is created with the same name. The characteristic value combinations are stored in this info cube. If a planning object structure is deactivated, consequently all assigned characteristic value combinations are deleted.

• *Create Characteristic Value Combinations*
Characteristic value combinations are created with the transaction /SAPAPO/MC62. There are basically two ways to create the CVCs, either interactively or automatically from an info cube. Since manual creation of CVCs is rather tiresome, the generation of the CVCs from an info cube is the usual way. If an according entry exists in the info cube (e.g. a sales order), the CVC is created. Though it is possible to create the according time series in the live cache immediately, it is recommended in note 573127 not to set this flag but to create the time series in a separate step.

As a third option it is possible to load characteristic combinations from a file into a work list and modify these by replacing values for selected characteristics. This is especially helpful for the planning of new products – when a new product is launched, planning is required for a CVC before any historical data is available. If the CVC contains a new characteristic value, the according text is maintained as the master data for the info object with the transaction RSDMD.

Obsolete CVCs, i.e. CVCs that have no planning data (or which correspond to product or location master data which are marked for deletion), can be identified and deleted.

• *Change Characteristic Value Combinations/Realignment*
Another scenario is the change of CVCs – for example due to a change in the product group structure. In this case the data for the old CVC is copied to the new CVC with the transaction /SAPAPO/RLGCOPY. With this realignment it is also possible to copy the data of a CVC into a new CVC without deleting the source CVC. Figure 4.11 shows the settings for this transaction.

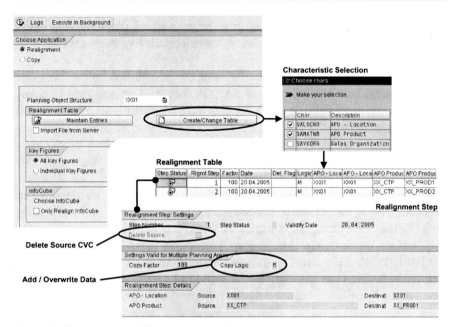

Fig. 4.11. Copy data to a different characteristic combination

To copy data from a CVC to a new CVC, first the relevant characteristics (i.e. the characteristics that will change) have to be selected. The value for not selected characteristics is simply copied to the new CVC. For the selected characteristics the realignment table is created, which contains one or more realignment steps. For each realignment step the source and the target characteristic values are defined, whether the data is added or overwritten (if the target CVC already exists) and whether the source CVC has to be deleted.

If there are characteristic values which change rather often, an alternative modelling using navigation attributes should be considered. In any case a close integration to the master data processes is necessary to determine the dependencies, the triggers and the responsibilities.

4.2.7 Time Series

As a last preparation step, the time series for the characteristic value combinations and the planning area are created with right mouse click on the planning area with the transaction /SAPAPO/MSDP_ADMIN. For automation the report /SAPAPO/TS_PAREA_INITIALIZE for the initialisation of the planning area or the report /SAPAPO/TS_LCM_DELTA_SYNC can be used as well – since time series have to be created regularly for new characteristic combinations. Now the data is written into live cache, therefore a version has to be specified.

The active version 000 is created with the installation, other versions can be created with the transaction /SAPAPO/MVM. As mentioned in chapter 3, the active version is used for the integration of SAP ERP™ to the order live cache and the ATP time series. Though it is possible to use version 000 for DP as well, the use of a different version has only advantages – e.g. the de-coupling of effects resulting from version copies and live cache initialisations.

The time series are created for a certain time period – by default the same horizon is taken as used for the storage bucket profile of the planning area. Instead of creating a very large horizon, it is possible to model a rolling horizon by moving the horizon period by period using the report /SAPAPO/TS_PAREA_INITIALIZE on a regular basis. It is also possible to define the horizon per key figure in order to reduce the allocated live cache space. This is especially rewarding for historical key figures.

4.3 Planning Book, Macros and Interactive Planning

4.3.1 Planning Book

The planning book is the main screen for DP where the data is displayed, entered and processed – where interactive planning takes place. With the transaction /SAPAPO/SDP94 the planning book is accessed. Figure 4.12 shows a planning book with the main features, that is
- the key figures displaying the data and
- the characteristics to select the data (planning levels).

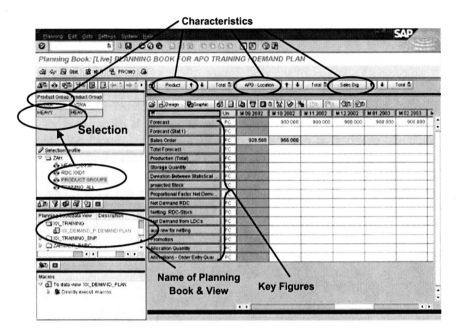

Fig. 4.12. Planning book in interactive planning

The four windows on the left side are called the 'shuffler'. They contain

- the data selections,
- the selection profiles,
- the planning books and the data views and
- the macros.

● *Data Selection*

To access the data in the planning book, the data has to be loaded explicitly from the live cache according to the ranges for the characteristic values specified in the selection. A selection contains the version, the characteristic for display and the restriction of the data. These selections are saved with a unique name and assigned to the data view and the user. These assignments are managed in the selection organisation with the transaction /SAPAPO/MC77.

Since the data in DP can be changed only by one person – or one background job – at a time, the locking logic restricts selections with overlapping criteria to display only. Via the path '*Goto → Lock Entries*' in the menu the current locks are displayed.

• *Create Planning Book*

Planning books are created with a reference to a planning area using the transaction /SAPAPO/SDP8B. In the planning book the desired key figures are selected as a subset of the key figures of the assigned planning area as well as the desired characteristics as a subset of the characteristics of the linked planning object structure. Each planning book contains one or more data views, where the key figures of the planning book are further restricted and grouped. If many key figures are in the planning book for different planning tasks, the data views help to keep a clear view of the key figures. In the data views the sequence of the key figures, the time buckets and the time horizons into the future and into the past are defined, figure 4.13.

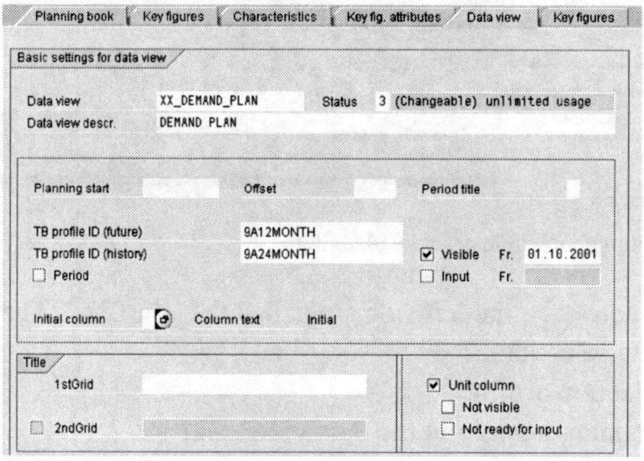

Fig. 4.13. Definition of the data view

For the future periods input is allowed (unless changed to display using a macro), the input possibility for past periods can be defined freely.

It is possible to use temporary key figures – e.g. for macro calculations – by defining them in the 'key figure attributes' of the planning book. These key figures do not have to be assigned to the planning area nor do they have to be defined as info objects. The value of the key figure is not saved to the live cache and is lost after finishing the planning session. Nevertheless these temporary key figures might be handy, especially for macro calculations.

• *Copy Planning Book*

To copy a planning book, an existing data view has to be used as a reference – copying takes place on data view level. If the reference data view belongs to a different planning area, the correct planning area has to be assigned afterwards, figure 4.14.

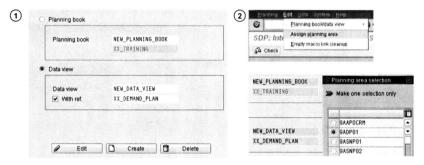

Fig. 4.14. Planning book copy

• *Aggregation and Disaggregation Using Header Bars*
Within the planning book in the interactive planning (transaction /SAPAPO/SDP94), header bars for the data aggregation and disaggregation can be set user specifically. To use the benefits of navigating this way in the data the selection has to be on a sufficiently high level. The header bar is set using the 'header'-button, figure 4.15.

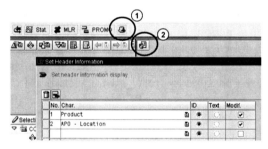

Fig. 4.15. Setting the header bar

Using the disaggregation buttons in the header bar, it is also possible to display all characteristic values for a characteristic by selecting 'details all'. In this case the key figure values are displayed in different rows as totals and per characteristic value.

• *Tool Bar*
The tool bar in the planning book provides some further customising and adjustment possibilities as well as some additional functionality, figure 4.16.

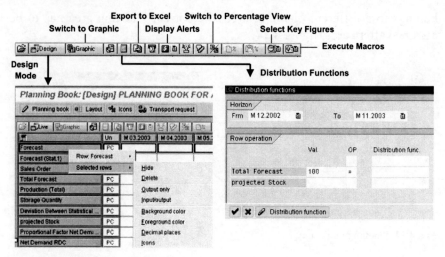

Fig. 4.16. Functions from the tool bar

Among these are:

- switch to design mode: in the design mode the possibilities exist to restrict key figures to output only or to hide key figures which are used rather for technical reasons than for planning (e.g. for calculation steps in macros). By switching to the design mode and using the right mouse click these options are displayed.
- switch between table oriented view and graphical view.
- distribution function: this function enables to process complete key figures, e.g. set the total forecast to 100 for the complete planning horizon. There are a couple of other functions available as a standard and it is even possible to create own distribution functions in a time series, e.g. a seasonal curve. In this case the specified value is distributed accordingly.
- download the current data selection to an excel sheet.
- display alerts (see chapter 24).
- switch to percentage view: this function makes only sense if 'details all' is selected first. The total of the row is fixed to 100%, and the proportions of the characteristic values are changed.
- select key figures: one, more or all key figures of the data view are selected display.
- execute macros.

● *Planning Book Handling*

Within the interactive planning there are some additional useful features as the display of row totals in a new column – either for the complete horizon, for the future, for the past or for selected columns only. The row totals are switched on and off via the menu path '*Settings → Row Totals*'.

Via the menu path '*Goto → Key Figure Overview*' it is possible to compare the periods of a key figure for different years. To enlarge the portion of the planning book on the screen it is possible to hide the shuffler.

Another feature is the assignment of notes to a cell using right mouse click → *Display Notes*. Though cells with attached notes remain marked after a level change, the note exists only and is displayed only for the characteristic combination it was created for. Using the button 'display note hierarchy' in the note display editor the according characteristic combinations are listed. A change of the cell value does not affect the note.

● *Assign User to Planning Book*
The planning book which is displayed when calling the transaction /SAPAPO/SDP94 is set for each user in the customising transaction /SAPAPO/SDPPLBK ('assign user to planning book'). The specified entry is valid for the first access to interactive planning and afterwards it is overwritten by the last selection. For the user maintenance process this should be considered.

● *User Specific Settings*
Per planning book it is possible to define user specific settings that contain e.g. the standard data selection, whether the data selection shall be loaded when opening the planning book and the key figures to be displayed.

● *Offline Planning*
Demand Planning is not only used by planners that have permanent access to the SAP APO™ system. For example, a sales person might want to update his forecast during a business trip. For this case it is possible to download the data from the planning book to a CSV-file, modify the data (e.g. in Excel) and upload the modified data to DP. In the download settings the flag 'prepare file for upload at later time' has to be set. This controls that additional source information is downloaded to the file. However, no aggregation and disaggregation may take place during offline planning.

Several options are available for the upload – execution of default macros and overwriting fixed values are some of them. During upload several consistency checks are applied and a log is written.

• *Authorisations*

Authorisations are checked on different levels. Note 400434 explains the authorisation concept in DP.

4.3.2 Macros

Macro functionality supports the data processing from a simple addition of rows to more complex calculations using a set of standard functions. Even own coding can be applied in user function macros. The macros are defined specifically for each data view in the macro builder. The macro builder is accessed via the macro workbench with the transaction /SAPAPO/ADVM. To avoid the tedious search for one's own data views, we recommend assigning a group to the data views and set a filter for this group as shown in figure 4.17.

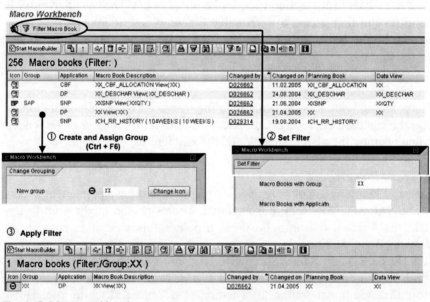

Fig. 4.17. Macro workbench

The macro builder is called for the data view by double-click onto the row of the according data view, figure 4.18.

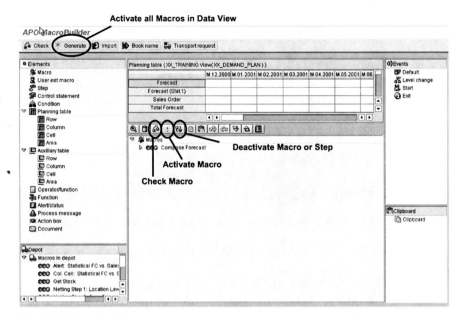

Fig. 4.18. Macro builder

To create a macro the according elements (starting with the 'macro' ele-ment) are picked from the 'element'-window on the top left by drag and drop. After the macro is defined, it has to be activated before it can be used. By default a macro is directly executable, unless it is inhibited within the macro definition. If the macro shall be executed at start of the interac-tive planning, when the level changes or after exit, the macro has to be assigned to the according events in the top right window by drag & drop. Note that frequent execution of macros affects the performance, therefore the assignment as default macro (i.e. the macro is executed after each 'enter') should not be generous.

Since the macro syntax is a bit tricky in some cases, the use of the 'check' function is helpful. Especially for the search of errors the deactiva-tion of macro steps is another useful feature.

● *Macro Syntax*
A macro consists of one or more steps with one ore more operations each. The processing area (total, future, past or freely defined) is selected on step level. The basic syntax – row 1 equals row 2 operator row 3 – is shown in figure 4.19. In this example the forecast is calculated as an average bet-ween sales forecast and statistical forecast.

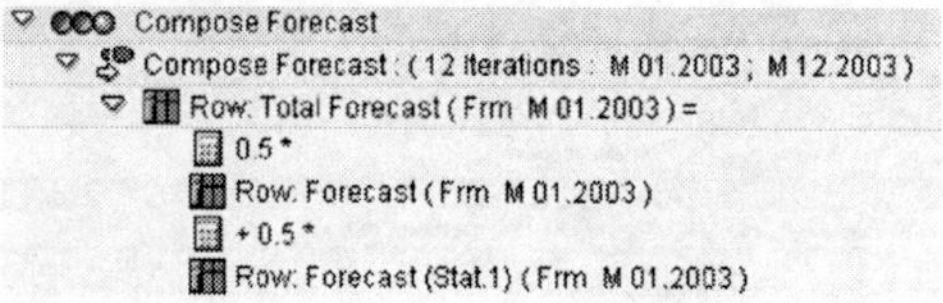

```
▽ ●●● Compose Forecast
   ▽ ₰ Compose Forecast : ( 12 Iterations :  M 01.2003 ;  M 12.2003 )
      ▽ ▦ Row: Total Forecast ( Frm  M 01.2003 ) =
            ▦ 0.5 *
            ▦ Row: Forecast ( Frm  M 01.2003 )
            ▦ + 0.5 *
            ▦ Row: Forecast (Stat.1) ( Frm  M 01.2003 )
```

Fig. 4.19. Macro – basic syntax

An IF-statement requires a control statement and a condition, figure 4.20.
The only difficulty is to place the objects into the right level.

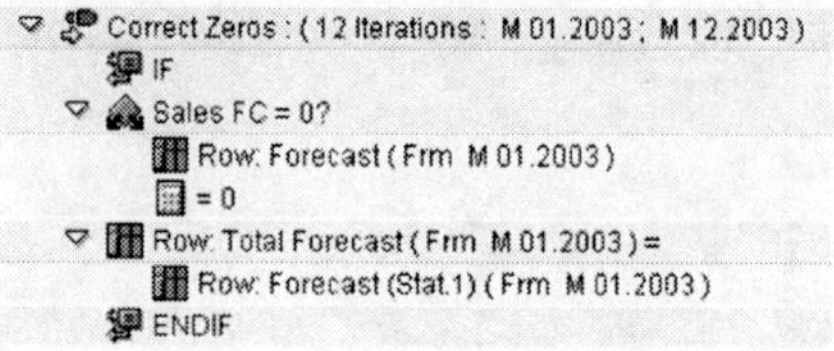

```
▽ ₰ Correct Zeros : ( 12 Iterations :  M 01.2003 ;  M 12.2003 )
   ₰ IF
   ▽ ▲ Sales FC = 0?
         ▦ Row: Forecast ( Frm  M 01.2003 )
         ▦ = 0
   ▽ ▦ Row: Total Forecast ( Frm  M 01.2003 ) =
         ▦ Row: Forecast (Stat.1) ( Frm  M 01.2003 )
   ₰ ENDIF
```

Fig. 4.20. Macro – syntax for IF-statement

Another peculiarity is the use of brackets for the functions, figure 4.21.
Make sure there is always a space between the expressions.

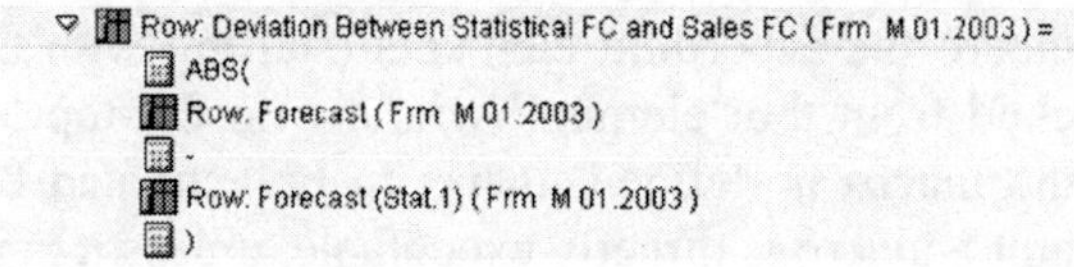

```
▽ ▦ Row: Deviation Between Statistical FC and Sales FC ( Frm  M 01.2003 ) =
      ▦ ABS(
      ▦ Row: Forecast ( Frm  M 01.2003 )
      ▦ -
      ▦ Row: Forecast (Stat.1) ( Frm  M 01.2003 )
      ▦ )
```

Fig. 4.21. Macro – syntax for functions

The available macro functions are described in the SAP standard help and
in the notes 403072, 438766, 433166 and others. There is also a collective
consulting note for macro functionality, note 539797.

For intermediate calculation steps an auxiliary key figure can be used
(as row, element or column). There is only one auxiliary key figure available which is used by all macros. In the macro it is specified whether this
auxiliary key figure is cleared before start.

• *Format Macros*
Macros help not only to calculate values but can also be used to change the
property of cells like their colour, the displayed symbols, their ability for
input or display only and other. To change these properties the change
scope of the according rows has to be switched to 'attributes' (this option
is selected by double clicking on the row in the macro builder).

Macros can be carried out depending whether rows or columns in interactive planning are marked, e.g. using the function ROW_MARKED (row

number). The corresponding function to find out the row number is MARKED_ROW.

● *User Function Macros*
With user function macros it is possible to include own routines and tables into the macro application. User function macros are defined in the menu path '*Edit → User Function*' within the macro builder. The naming convention for the user function macros is Z_[MACRONAME]. The interface parameters VALUE_TAB (TABLE, LIKE /SAPAPO/VALUE_TAB), F_ARGUMENT (CHANGING, TYPE /SAPAPO/MXVAL) and F_CALC_ERROR (CHANGING, TYPE C) are set by default. Additional interface parameters can be selected – mainly regarding the format of the planning book – when creating the user function in the macro builder. The parameter VALUE_TAB is used to transfer numbers and strings from the planning book to the function, whereas the parameter F_ARGUMENT is used to transfer the result of the user function back to the planning book. The user function with the according interface definitions has still to be created with the transaction SE37.

● *User Exit Macros*
User exit macros allow processing a whole grid – in contrast to user function macro. The BAdI /SAPAPO/ADV allows also to access the layout attributes.

● *Effects of Changes in the Data View*
If a change in the sequence of the key figures in the data view occurs, the logic of the macros is not affected. Before SAP APO™ 4.1 the selected rows in the macro were related to their position in the data view and not to the name of the key figure. Therefore a change in the sequence of the key figures – e.g. the insertion of a new key figure – had led to a disturbance of the macro semantic. The deletion of a key figure results in a deactivation of all concerned macros.

● *Copy Macros*
There are two ways to copy macros. To copy a macro within a data view, select the macro and use right mouse → *Copy to Clipboard*, then select any macro and use right mouse → *Insert From Clipboard*. The other way is to import the macros from another data view via the menu path '*Edit → Import Macro*'.

• Macro Execution

Macros are used in interactive planning by manual request from the planning book or triggered by events like the start of the interactive planning, the change of the level, the end of interactive planning or as a default macro. In these cases the macros process only the selected data on the current aggregation level. In general there is a difference whether a calculation is performed on the most disaggregated level or on an aggregated level and the result is disaggregated afterwards. In DP macros are used for alert calculation as well. This is described in chapter 24.

4.3.3 Fixing of Values

If planning is performed on mixed levels – e.g. on product level for key products and on product group level to include all products – it might not be desired that the data on the more detailed level is changed by the proportional automated disaggregation from entries of an aggregated level. To inhibit the overwriting of the value it is possible to fix key figure values by right mouse click on the cell. If fixing is performed on aggregated level, changes of the values on detailed level do not change the fixed value, but are counterbalanced by adjusting the other values on the same detailed level. A fixing on detailed level on the other hand represents a hard constraint, and a change on aggregated level has to take this constraint into account and to counterbalance the other figures on the detailed level.

To enable the fixing within a key figure, two key figures are required: one for the value of the key figure, the other for the fixing information. Within the info object of the key figure, the key figure for the fixing information has to be assigned as shown in figure 4.22.

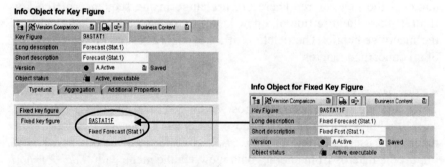

Fig. 4.22. Key figure configuration for fixing

The fixed key figure does not need to be assigned explicitly to the planning area. The notes 409181, 410680, 643517, 681451 and 687074 provide additional information on fixing values.

It is not necessary any more to have a persistent aggregate to fix key figure values on aggregated level. With this solution the fixing information on aggregated level exists only at the detailed level, and the fixed values of the underlying details are aggregated and are displayed as fixed values at aggregate level.

4.4 Statistical Forecasting

4.4.1 Basics of Forecasting

Statistical forecasting is usually applied to forecast sales quantities. There are two major groups of statistical methods, one is the univariate methods, where a key figure is forecasted from its past values, and the other group is causal models, where a key figure is forecasted according to the history of other key figures (also referred to as 'causal factors'). For the latter the most common method is multiple linear regression (MLR).

The quality of the statistical forecast depends on the successful approximation of inherent regularities. One basic fact of statistics is that these regularities are analysed and predicted the better the larger the number of data is. At the determination of the forecast level the trade off between increasing the data amount by using a higher aggregation level (e.g. product group instead of product) and losing information due to an inadequate disaggregation has to be considered. A common scenario is to carry out the forecast on the more detailed level (e.g. product level) for the short term horizon, when the accuracy of the operative data has priority, and to switch to forecasting on an aggregated level for longer horizons. This of course depends on what the data is used for. The logical prerequisite to forecast on aggregated level is that the history of each product is similar.

A further issue is the selection of the data basis. A typical question is whether to use third party sales or a mix of third party and inter-company sales for forecasting. To avoid a distortion of the inherent regularities of the sales data by local inventory policies and lot sizes, only third party sales should be used as data history whenever possible.

Another crucial factor is the application of the appropriate forecast model, which is able to model these regularities. Therefore an analysis of the data history has a significant impact on the accuracy of the statistical forecast. In some cases the data history is already examined and appropriate forecast models determined, but in many cases the effort of data analysis is regarded as too high for the expected benefit in accuracy.

4.4.2 Data History

Typical data history patterns are constant, trend, seasonal, trend and seasonal and sporadic history, figure 4.23.

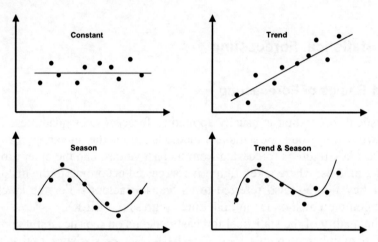

Fig. 4.23. Categories of data history patterns

4.4.3 Univariate Forecast Models

Depending on the pattern of the data history the forecast models are selected. Table 4.1 shows the most common models for these types of history.

Table 4.1. Forecast models

History Pattern				
Constant	Trend	Season	Trend & Season	Sporadic
• 1st Order Exponential Smoothing	• 2nd Order Exponential Smoothing (Holt)	• 2nd Order Exponential Smoothing (Winters)	• 2nd Order Exponential Smoothing	• Croston
• Weighted Average	• Linear Regression	• Seasonal Linear Regression		

• *Constant Models*

The most common forecast model is 1st order exponential smoothing. With 1st order exponential smoothing recent data has a bigger influence than data farther in the past. The weights of the recent data increase with the parameter α according to the formula

$$FC\ [t+1] = \alpha\ SH\ [t] + (1-\alpha)\ FC[t] \tag{4.1}$$

where FC is the forecast, SH the sales history and t the time period. Common values for α are between 0.1 and 0.3. Since 1st order exponential smoothing is an iterative approach, one period is needed for model initialisation. The value for FC of the current period is taken as constant value.

A more simple way to calculate a constant forecast value is by calculating the weighted average

$$FC = \Sigma\ w[t]\ SH\ [t]\ /\ \Sigma\ w[t] \tag{4.2}$$

where w[t] represent weights that are defined in the weighting group of the univariate forecast profile.

Constant models might provide appropriate forecasts usually for the next one to three periods.

• *Trend, Seasonal and Trend and Seasonal Models*

For data history that show either a trend or a seasonal pattern the most common forecast model is 2nd order exponential smoothing. The formula contains several intermediate variables – G as a base value, T for trend and S for season – that are strongly linked by recursions.

$$FC\ [t+i] = \{\ G\ [t] + i * T\ [t]\ \} * S\ [t-L+i] \tag{4.3}$$

$$G\ [t] = G\ [t-1] + T\ [t-1] + \alpha\ \{\ SH\ [t]\ /\ S\ [t-L] - G\ [t-1] - T\ [t-1]\ \} \tag{4.4}$$

$$T\ [t] = T\ [t-1] + \beta\ \{\ G\ [t] - G\ [t-1] + T\ [t-1]\ \} \tag{4.5}$$

$$S\ [t] = S\ [t-L] + \gamma\ \{\ SH\ [t]\ /\ G\ [t] - S\ [t-L]\ \} \tag{4.6}$$

Analogous to the factor α in 1st order exponential smoothing β weights the influence of recent trends and γ the influence of recent periods. Typical values for β and γ are 0.3. If there is only a trend pattern in the data history, γ is zero. The 2nd order exponential smoothing model without seasonal terms is also called Holt model. In the opposite case – only seasonality, no trend – β is zero and the model is named after Winters.

• *Model Initialisation*

Since most of the forecast models use a recursive approach, depending on the forecast model one or more periods are needed for model initialisation.

Table 4.2. Required periods for model initialisation

Forecast Model Type	Periods for Model Initialisation
Constant	1
Trend	3
Seasonal	one season
Trend & Seasonal	3 + one season

The number of required periods for model initialisation has an impact on the history horizon. If there is seasonality with a period of one year, at least 15 periods of history have to be available to use a trend & season model.

4.4.4 Multiple Linear Regression (MLR)

The general idea of MLR is to find a correlation between the dependent key figure (the one you want to forecast, usually sales history) and other key figures using historical data. The formula for this correlation is

$$SH[t] = \beta_0 + \beta_1 KF1[t] + \beta_2 KF2[t] + \ldots + \beta_n KFn[t] \qquad (4.7)$$

for periods in the past. The factors β_0, β_1, ..., β_n are calculated using ordinary least squares. Obviously there have to be at least as many periods of data history as independent variables. The forecast is calculated analogously

$$FC[t] = \beta_0 + \beta_1 KF1[t] + \beta_2 KF2[t] + \ldots + \beta_n KFn[t] \qquad (4.8)$$

for periods in the future. This requires forecasted values for the independent key figures KF1 to KFn. The key figures KF1 to KFn are maintained in the MLR profile.

4.4.5 Forecast Execution

Forecast is carried out from interactive planning using the buttons for univariate forecast respective MLR. A prerequisite for this is that the navigation to univariate forecast and MLR are flagged in the planning book. The forecast models and their parameters can be changed interactively and saved for the current selection either by overwriting the old profile or by creating a new profile.

The last forecast results are compared regarding their forecast errors with the path '*Goto* → *Forecast Comparison*' in the interactive planning. With the transaction for the forecast comparison it is possible to choose the forecast which fits best and create automatically a forecast profile.

• *Ex-Post Forecast*
The ex-post forecast is the result of the forecast model (after its initialisation) applied for past periods where historical data is available, figure 4.24.

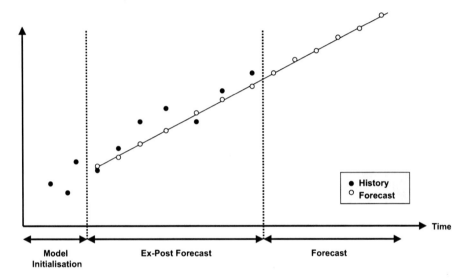

Fig. 4.24. Horizons for forecasting

The ex-post forecast is used to determine the accuracy of the forecast model.

• *Corrected History*
To reduce the impact of an outlier in the data history, which was due to an exceptional event, a correction of the historical data can be appropriate. There are two ways of data correction, either manually or automatically using the function 'outlier correction'.

Automatic outlier correction is performed if selected in the forecast profile. There are two options for outlier correction, the ex-post method or the median method. For outlier correction with the ex-post method a tolerance lane is calculated with a width related to the standard deviation:

$$\text{Tolerance Lane [t]} = \text{Ex-Post FC [t]} \pm \sigma * \text{MAD} * 1.25 \qquad (4.9)$$

Any historical value which exceeds this lane is corrected to the border value. With the customising path *APO → Supply Chain Planning → Demand Planning → Basic Settings → Maintain Customer Specific Settings for Outlier Correction* it is possible to define an initialisation period. Based on the corrected history the ex-post forecast and the forecast are carried out again. Since the outlier contributes to the calculation of the ex-post forecast, subsequent historic values might be 'corrected' to an undesired value as shown in figure 4.25.

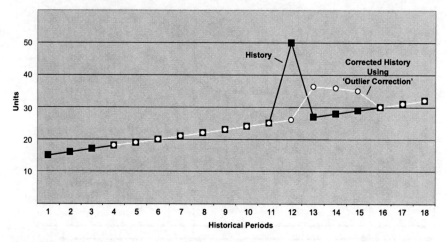

Fig. 4.25. Outlier correction

Nevertheless the impact on the forecast might be still an improvement, but the corrected history using outlier correction differs from what most people would expect.

The median method determines the basic value, the trend value and the seasonal indices as median values and calculates an expected value for the history. If the historical value is outside the tolerance lane it is corrected to the expected value.

An alternative way to identify outliers is to use the alert functionality and to correct the data manually. In this case the checkbox 'read corrected history data' has to be set in the forecast profile. By setting the flag 'corrected history' in the univariate forecast profile, corrected history is also calculated by subtracting promotions in the past from the original history.

● *Corrected Forecast*

Corrected forecast is used to adjust the forecast to the number of working days of the respective period, because the original forecast is calculated without taking any calendars into account. To use this functionality the assumed number of working days has to be maintained as 'days in period' in the univariate forecast profile.

• *Assignment of Key Figures*
Depending on the functionality to be used (e.g. outlier correction, forecast correction etc.) it is necessary to save the data into a key figure. Therefore the according key figures have to be assigned to the planning area for

- the forecast,
- the corrected history,
- the corrected forecast,
- the ex-post forecast and
- the ex-post forecast MLR.

The assignment of the semantic to the key figure is done within the planning area (transaction /SAPAPO/MSDP_ADMIN) with the path *Extras → Forecast Settings.*

• *Forecast Profile*
The settings for the statistical forecast are maintained in the forecast profiles with the transaction /SAPAPO/MC96B. Figure 4.26 provides an overview of the entries and the structure of the forecast profile.

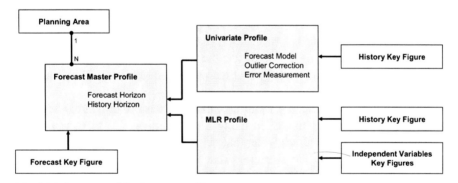

Fig. 4.26. Structure of the forecast profile

The forecast master profile is assigned unequivocally to one planning area only. The key figure to save the forecast has to be assigned explicitly. If it differs from the forecast key figure assigned in the planning area, the forecast is saved to the key figure assigned in the master profile. The horizons for reading the history as well as carrying out the forecast are specified here too. Especially for the background planning (see chapter 4.9) it is important that the time horizons in the forecast master profile match the time horizons in the data view. The forecast model specific parameters are defined in the univariate profile resp. in the MLR profile, and the profiles are assigned to the forecast master profile.

The forecast master profile contains additionally a composite forecast profile. The composite forecast is calculated based on multiple forecast models, and the one with the lowest error is selected.

If statistical forecasting is carried out in the interactive planning it is possible to change the forecast parameters. If in the user master the parameter /SAPAPO/FCST_GUID is set, these changes in the forecast profile are saved for the selection as a new forecast profile. The name of the forecast profile is created by the system as a GUID. The assignment of the forecast profiles to certain selections is displayed via '*Goto → Assignment*'. In the following sessions as well as in background jobs the system will first look for a selection specific forecast profile assignment and will apply the standard forecast profile only when no selection specific setting is found. Some additional user parameters are explained in note 350065. User-Exits and BAdIs for forecasting are explained in note 394076.

4.4.6 Life Cycle Planning

Life cycle planning is integrated into the statistical forecast functionality. To perform a statistical forecast for a new product which has no own sales history, some other data has to be used instead. The basic idea of the life cycle planning in SAP APO™ is to use the history of another product or a mix of other products instead. This is represented in SAP APO™ by the like profile.

New products often need a ramp-up time to reach the targeted sales figures. This behaviour is modelled using a phase-in profile to dampen the statistical forecast. Analogous to that a phase-out profile is applied to model decreasing sales for a product that is going to be substituted soon. Figure 4.27 shows the connection between these entities.

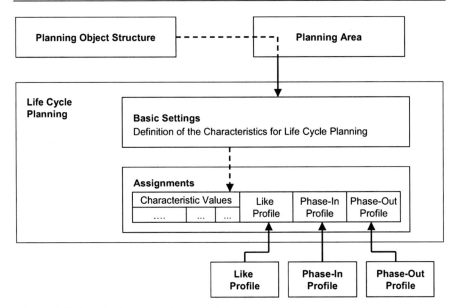

Fig. 4.27. Entities for the life cycle modelling

The configuration of the life cycle planning takes place with the transaction /SAPAPO/MSDP_FCST1. The settings are planning area specific. The like profile, the phase-in profile and the phase-out profile are created from the entry screen of the life cycle planning as well. In the profiles the reference characteristic values are maintained (i.e. the characteristics where the history or the forecast is taken from), and in the 'assignments' the profiles are linked to the characteristic values that have to be planned with the life cycle functions. Another prerequisite to carry out life cycle planning during the statistical forecast is to select the flag 'life cycle planning active' in the master forecast profile.

In former releases characteristic 9AMATNR had to be included, but with SAP APO™ 4.1 life cycle planning is also possible on other characteristic levels. Up to six characteristics are assigned to the basic life cycle. If during the execution of the basic assignment the warning 'output length of characteristic 9AMATNR is larger than 30' is given, it is possible to adjust the output length with the transaction /SAPAPO/OMSL. This enables the assignment of products in 'assign life cycle', but the warning will continue. The next step is to define the like profile.

• *Like Profile*
The like profile is simply a list of products with weight factors that determine the impact of the products' histories. The sum of the weights may be above 100% - with increasing weights the forecasted values will increase

as well. In the like profile the source characteristic value combinations are maintained – i.e. where the history is taken from. The like profile is assigned to the target CVCs (for the characteristics defined in the basic life cycle) with the radio button 'assign life cycle' in the transaction /SAPAPO/MSDP_FCST1.

For the phase-in and the phase-out a new time series is defined in which a percentage value for the specified number of periods is maintained. The forecast is decreased to that percentage value. The phase-in and the phase-out profiles are defined with absolute dates, i.e. it is not possible to have one phase-in profile relative to the start of the forecast horizon.

Figure 4.28 shows the life cycle planning settings for the substitution of PROD_OLD by PROD_NEW, where the history for the statistical forecast of the new product is composed to 80% of the old product and to 20% of the current product.

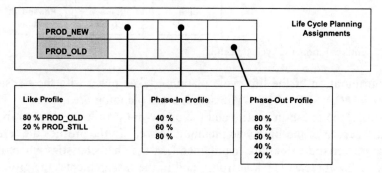

Fig. 4.28. Settings for a life cycle planning example

Figure 4.29 shows the result of the statistical forecast (applying a constant model):

		Past Periods			Future Periods					
Sales	PROD_NEW									
	PROD_OLD	100	100	100						
	PROD_STILL	200	200	200						
Forecast	PROD_NEW				48	72	96	120	120	120
	PROD_OLD				80	60	50	40	20	-
	PROD_STILL				200	200	200	200	200	200

Fig. 4.29. Example for the life cycle planning

The reduced forecast for the new product in the first three periods is due to the phase-in profile, analogously the fading of the forecast for the old product is due to the phase-out profile. By defining offsets in the profiles it is possible to postpone the beginning of the phase-in and the phase-out. Note 642593 provides additional information about life cycle planning.

4.5 Promotion Planning

Promotions are mainly used in consumer goods industries to create or keep customer awareness for a product or a brand. The effect of the promotion might be an increase of the demand or just a shift of the demand profile in time. Promotion planning functionality of SAP APO™ supports planning the impact of promotions on the demand. Promotions are modelled as an absolute or a relative increase in sales. Unlike life cycle planning, promotion planning is not integrated into the statistical forecasting process. The promotional demands are saved in the promotion object and written to the live cache for the promotion key figure. Per planning area one promotion key figure is defined with the transaction /SAPAPO/MP33. Figure 4.30 shows the structure of a promotion.

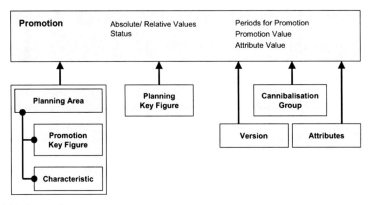

Fig. 4.30. Promotion

Promotion planning in SAP APO™ requires three steps:
1. create the promotion,
2. specify levels and characteristic values for the promotion ('assign objects') and
3. activate the promotion.

The promotional demand might be either used as a delta or as the total demand. This has an impact on the cannibalisation and on the history cleansing.

● *Create Promotion*

Promotions are created with the transaction /SAPAPO/MP34. The main information that are specified at the creation of promotions are

- the period type (week, month), the number of periods and the start date,
- whether absolute or relative values are used and
- the cannibalisation group.

Since the maintenance of the promotion is not necessarily intuitive, figure 4.31 shows the steps to create a promotion.

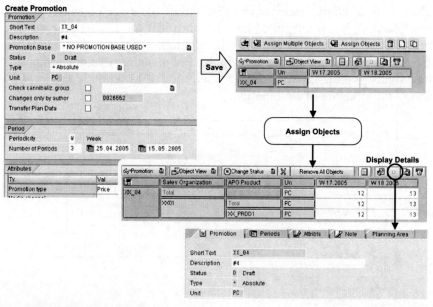

Fig. 4.31. Creation of a promotion

The promotional demands are maintained as absolute or as relative values. If relative values are chosen, these relate to the key figure defined as 'planning key figure', so that the promotional demand equals the defined percentage times the planning key figure. The total demand is calculated by the addition of the promoted and the non-promoted demand.

The choice whether absolute or relative values are used and whether and which cannibalisation group is assigned is irreversible.

• *Object Assignment*

In the next step the levels are defined, for which the promotion is valid. Using the button 'assign multiple objects' it is possible to assign the characteristic value combination more comfortably (than one by one like in former releases). Figure 4.32 shows the procedure to assign the CVC to the promotion.

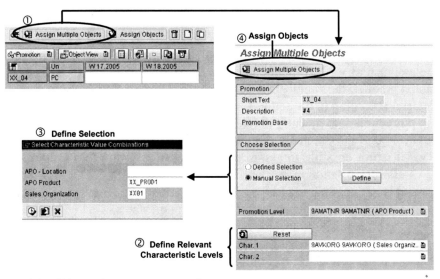

Fig. 4.32. Object assignment to a promotion

If promotions are maintained on different levels – e.g. on product level for one promotion and product group level for a different promotion – it is helpful to use the entity 'promotion base'. The promotion base is created with the transaction /SAPAPO/MP40 and allows to assign the promotional quantity to a different key figure than specified for the planning area with the transaction /SAPAPO/MP33. This is a helpful feature to increase the transparency for promotion planning.

After at least one object is assigned, the status of the promotion can be changed from 'D' (draft) to 'P' (planned in future). This status triggers the promotional demand to be written into the promotion key figure in live cache.

• *Promotion Attributes*

Attributes are defined with the transaction /SAPAPO/MP31 and help to select the promotions. The values of the attributes are maintained in the promotions. These attributes are used analogous to characteristics in the selection profile (the characteristic for display is 'promotion').

• *Cannibalisation*

In some cases a promotion for a product causes a decrease in the demand for a related product, which is substituted by the promoted product. This effect is called the cannibalisation of the demand.

The cannibalisation is modelled in SAP APO™ using a cannibalisation group (transaction /SAPAPO/MP32), which contains the promoted and the substituted products with the relation of their quantities (e.g. an increase of 50 units of the promoted product leads to a decrease of 20 units of the substituted product) with inverse signs. If an object is assigned to the promotion that is a member of the cannibalisation group, the other members are assigned as well. If the promotional demand is maintained for the promoted product, the cannibalisation is calculated as a negative promotional demand according to the settings in the cannibalisation group. The total of the promotion consequently diminishes. There are some restrictions in the application of the cannibalisation groups:

- it is not possible to assign a cannibalisation group to a promotion after the promotion is saved,
- it is not possible to change a cannibalisation group that is assigned to a promotion – independent of the promotion status,
- the prerequisite to include a product into the cannibalisation group is that the product master exists.

Another restriction exists regarding the assigned objects to the promotions. Cannibalisation works only if all products of the cannibalisation group have the same characteristic values for the additional characteristics, i.e. they belong all to the same part of the hierarchy (at least within the promotion). Figure 4.33 visualises this by an example:

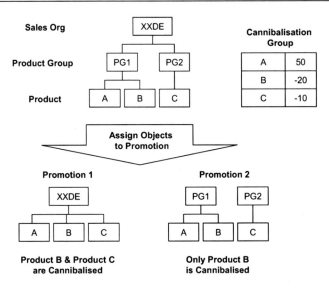

Fig. 4.33. Cannibalisation group and promotion

- *Update Promotions*

In case the planning area has been de-initialised and the time series have been created again, an update of the promotions is carried out to write the active promotional demand to the promotional key figure using the transaction /SAPAPO/MP38 (see also note 367031). For the update and copy of promotions transaction /SAPAPO/MP42 provides additionally a promotion management screen.

- *Reporting*

There are two options for promotions reporting. One is with the transaction /SAPAPO/MP39 that allows selecting promotions according to characteristic values, promotion name, start date, user or status and type, and display the promotions according to characteristic values, cannibalisation groups and promotion properties as status and type. The promotion attributes are not used as a selection criterion in promotion reporting. Note 384550 provides more information about promotion reporting.

The other option is to define promotion reports with the transaction /SAPAPO/MP41A as shown in Figure 4.34. Using the selections, it is possible to select promotions by their attributes.

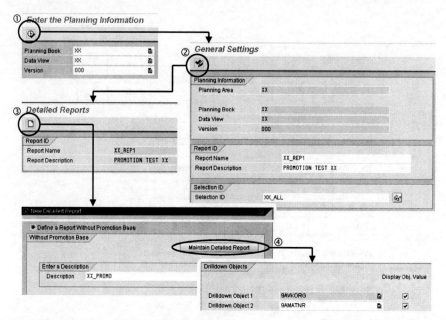

Fig. 4.34. Maintenance of the promotion report

The reporting itself is performed with the transaction /SAPAPO/MP41B.

The collective consulting note 540282 refers to further notes regarding promotions.

• *Alternative Ways to Plan Promotions*
An alternative and more easy way to plan promotions is by maintaining the promotion demand directly in a key figure in the interactive planning instead of in a promotion. In this case however, cannibalisation and promotion reporting can not be used.

4.6 Dependent Demand in Demand Planning

There are mainly two common scenarios where dependent demand provides a valuable input for demand planning. In one scenario a key component limits the supply quantity. If many products contain this key component – probably even with different quantities, as the active ingredients in the pharmaceutical industry – a rough feasibility check by a rule of thumb is not any more possible.

The other scenario is the use of sets (mainly in consumer goods industries), where different products are combined to a new product number. Here the information of the dependent demand is needed to achieve an overview of the total demand for some products, especially to analyse and to plan cannibalisation effects.

The information about the dependent demand is provided either by uploading this information from SNP (or from PP/DS) or by calculating it within DP.

• Uploading Dependent Demand from Order Live Cache
Assuming that SNP or PP/DS is used, the dependent demand is calculated during MRP anyhow. It is possible to transfer the categories for dependent demand (EL for SNP orders, AY for PP/DS orders) for the relevant products to DP as described in chapter 4.9.3.

The advantage of this solution is that no additional master data has to be maintained and no additional planning steps are required. The disadvantage however is that the dependent demand is available in DP only after the release of the independent demand and the MRP run. This implies that it is not possible to have a short cycle time to receive the results and that it is not possible to check the feasibility in DP before the disposition is triggered. Another disadvantage of this approach is that there might be time lags and lot size effects due to transports in the supply chain.

• DP-BOMs
Another approach to have the dependent demand available in DP is to calculate it within DP using DP-BOMs. The DP-BOM is either a PDS or a PPM – the properties of these master data types are described in chapter 13.4. The PDS for DP is generated from the PDS for SNP or PP/DS with the transaction /SAPAPO/CURTO_GEN_DP and contains the components of multiple BOM levels. The PPM for DP is created with the transaction /SAPAPO/SCC05 for the usage 'D' or generated with the transaction /SAPAPO/MC62. SNP-PPMs (usage type 'S') can be used as well, but are more complicated to create.

To use DP-BOMs the according flag has to be set in the planning object structure. This causes the selection of the standard characteristics 9APPMNAME and 9ABOMIO for DP-BOMs. In the planning area the key figures for the independent demand and the dependent demand require entries in the key figure semantic (401/501, 402/502, …). In the data view of the planning book at last the option 'second grid' has to be selected to display the dependent demand for a header product.

Finally the characteristic value combinations have to be complemented with the PPM values in the characteristic combination maintenance (transaction /SAPAPO/MC62) using the button 'add BOM information' (this button is only available for planning object structures with DP-BOM relevance). Adding the characteristic value for the DP-BOM to the characteristic value combination triggers the deletion of the characteristic combination for the header product (if the flag 'delete input product records' is set) and the creation of new characteristic combinations including the DP-BOM name. Regarding the components the procedure is as follows: if the characteristic combinations with the same values as for the header product exist, these are deleted as well and new ones are created instead. Else characteristic value combinations are created for the components with the same characteristic values as the header product. Figure 4.35 illustrates this principle.

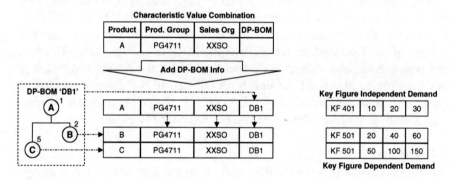

Fig. 4.35. Characteristic value combinations for the use of DP-BOMs

The implication of this characteristic value combinations handling is that all the data of the deleted characteristic value combinations will be lost.

The information about the BOM ratio is read from the DP-BOM at the point in time when the BOM information is added, i.e. the new characteristic value combinations are created, and is stored in the background. CVC-relevant changes in the DP-BOM are updated to the characteristic combinations unless the flag 'only edit combinations without PDS/PPM' is selected. If the BOM ratio is changed, the time series have to be updated either using a consistency check or via new initialisation.

Another issue might be that all the other characteristic values of the header product are copied for the components, in case the component belongs e.g. to a different product group. Realignment is not a possible solution for this, because the BOM information is lost during the realignment procedure.

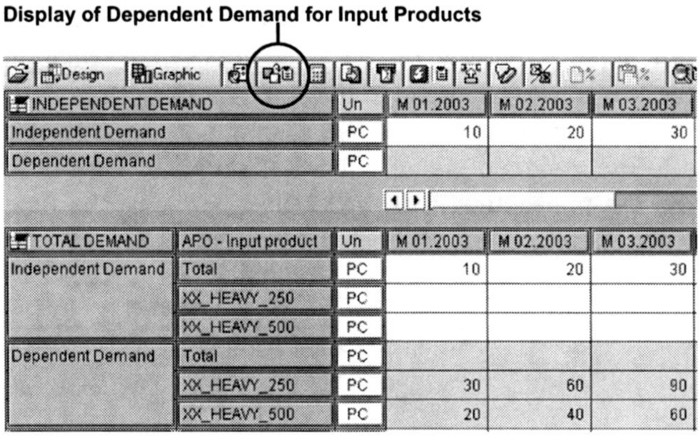

Display of Dependent Demand for Input Products

INDEPENDENT DEMAND		Un	M 01.2003	M 02.2003	M 03.2003
Independent Demand		PC	10	20	30
Dependent Demand		PC			

TOTAL DEMAND	APO - Input product	Un	M 01.2003	M 02.2003	M 03.2003
Independent Demand	Total	PC	10	20	30
	XX_HEAVY_250	PC			
	XX_HEAVY_500	PC			
Dependent Demand	Total	PC			
	XX_HEAVY_250	PC	30	60	90
	XX_HEAVY_500	PC	20	40	60

Fig. 4.36. Dependent demand in demand planning with DP-BOMs

As an example, demand planning is performed for a set containing three 250 ml bottles and two 500 ml bottles of beverage as shown in figure 4.36. In the first grid the demand is planned for the set. When saving this independent demand the BOM explosion is carried out. It is possible to display the dependent demand for the input products in the second grid using the button marked in figure 4.36.

4.7 Collaborative Forecasting

Collaborative forecasting provides the possibility to interact with customers – internal or external – in the demand planning process via internet. For the external partners no SAP APO™ installation is required. Some of the possible scenarios are

- demand transmission, where the customers maintain their forecasts themselves via internet in an SAP APO™ planning book. This procedure has the disadvantage that it is not possible to load data – e.g. from an excel sheet – into the planning book.
- demand confirmation, where forecasting for the customers is done by the central planning and the customer just confirms it, e.g. by copying it into another key figure using a macro and changing it if necessary.
- sales exception reporting, where major deviations from the previously forecasted values – e.g. due to unpredicted changes in the market demand – are reported as soon as possible. This information is

entered without any delay into SAP APO™ via Internet and evaluated by the planner, probably supported by alerts.

From a technical point of view, the basic concept of the collaborative forecasting is to allow a user to log on to the SAP APO™ system and to provide access to the planning book via Internet. The authorisation for the customer is restricted to the assigned planning book. Compared with the interactive planning with the transaction /SAPAPO/SDP94, some of the functionalities require user parameters as listed in table 4.3.

Table 4.3. User parameters for collaborative planning

Function	User Parameter
View Graph	/SAPAPO/CLP_WEBGRAPH
Maintain Notes	/SAPAPO/CLP_WEBNOTE
Download Grid Values	/SAPAPO/WEBDOWN
Use Pivot	/SAPAPO/WEBPIVOT

The planning book, the data view and the selections have to be assigned to the collaborative user as for any other user. Additionally the user has to be assigned to the entity 'collaboration partner' with the transaction /SAPAPO/CLP_SETTINGS. The entity 'collaboration partner' controls and enables the functionality for collaborative planning via Internet. Figure 4.37 provides an overview about these settings.

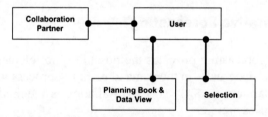

Fig. 4.37. Entities for the collaborative forecasting

User specific settings are maintained with the transaction /SAPAPO/CLP_SDP_USER. Note 564186 lists the restrictions using the planning book via Internet.

As a technical prerequisite a web server has to be installed. The address of the server has to be entered into the table TWPURLSVR in the transaction SM30.

4.8 Background Planning

A major part of the planning activities is carried out in the background, either because they are periodic activities or they are used for mass processing. The activities described in this chapter are
- copying data between key figures,
- macro calculation and
- forecasting.

The release of the demand plan is described in section 4.9.

● Data Copy
Data in DP is copied from one key figure to another using the transaction /SAPAPO/TSCOPY. It does not matter whether these key figures are within one planning area or not, but there has to be at least one common time characteristic in the source and the target planning area.

● Forecast and Macro Execution
The background planning for the macro and the forecast execution has a similar structure, as shown in figure 4.38.

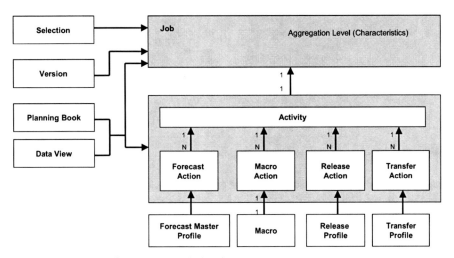

Fig. 4.38. Structure of the background planning

The first step for the background planning is to create a planning activity with the transaction /SAPAPO/MC8T. The activity contains one or more actions, where each action might be of the type 'forecast' or 'macro'. The planning job itself is created with the transaction /SAPAPO/MC8D (and changed with /SAPAPO/MC8E) and contains only one activity. On job level the selection and the aggregation level (planning level) for the activity is specified. Depending on the aggregation level the results might differ significantly – e.g. due to disaggregation (see chapter 4.2.5). For performance reasons the same characteristics should be used in the selection as in the aggregation level. Again for performance reasons a separate planning book should be used for the background planning with macros and for the background forecasting. These planning books should be restricted only to the necessary key figures. Notes 39876 and 546079 provide some recommendations to configure the background jobs.

• *Scheduling of Jobs*
Planning jobs are scheduled in DP with the transaction /SAPAPO/MC8G with a similar functionality as in the standard job definition with the transaction SM36. To include the DP background jobs into the standard basis job procedure the report /SAPAPO/TS_BATCH_RUN can be applied, which uses the name of the DP background job as only input. By saving this as a variant, all parameters for the standard job scheduling are available, figure 4.39.

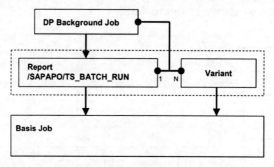

Fig. 4.39. DP background job and basis background job

Logging is nevertheless related to the name of the DP background job. To get the relation between these two kinds of jobs the complete assignments has to be checked.

• *Logging*
The log file of the background planning job is displayed with the transaction /SAPAPO/MC8K. The log helps to identify possible mistakes. Some of the most common problems are

  •   the selection is locked, for example if interactive planning is carried out for overlapping selections at the same time and

- the macro wants to write in periods for which no writing is allowed in the data view.

4.9 Release of the Demand Plan

4.9.1 Topics for the Demand Plan Release

To make the established demand plan relevant for disposition, it has to be released as an independent demand to an application where requirement planning takes place, that is either to the order live cache as 'release to SNP' (in this case both SNP and PP/DS have the information) or to SAP ERP™, figure 4.40.

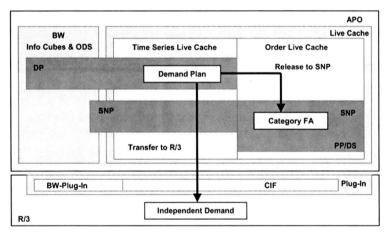

Fig. 4.40. Release of the demand plan

4.9.2 Forecast Release

There are two possibilities to release the forecast, either via the transaction /SAPAPO/MC90 or using a release profile (transaction /SAPAPO/MC8S) in the background planning (see chapter 4.8). For mass data the release via background planning is recommended (note 403050). In both cases a number from a key figure in the time series live cache is transferred as an order category into the order live cache. It is possible to specify the key figure to be released as well as the category for the orders to be created. The standard category for independent requirements is FA, and forecast consumption and many applications in SNP are based on this category. In the order live cache the demand is linked to a product and a location. By default the characteristics 9AMATNR and 9ALOCNO are used to match product and location, but they can be overwritten in the release profile.

If demand planning is carried out without any location characteristic (for example only on sales levels) the locations for the demand in SNP are determined according to the location split which is defined in the transaction /SAPAPO/MC7A (proportions ≤ 1, use comma as decimal point). The advantage of using the location split lies in the reduction of the number of characteristic combinations. Its disadvantage is the loss of flexibility to change the proportions in a different way than by changing the table. Another implication is that if different countries plan the same product number for their own market, it is not possible to use a country specific location split. Analogous to the location split it is possible to define a product split with the transaction /SAPAPO/MC7B.

• *Time Granularity*
At the release to SNP two bucket profiles can be assigned, the planning bucket profile and the daily bucket profile (if the release is performed using a release profile, the planning bucket profile is taken from the planning book). For the selection of the planning bucket profile the same restrictions apply as in the data view. The daily bucket profile is technically a planning profile as well, but contains only days as periods. Time buckets in SNP are usually either of the same or of smaller size. In case of a smaller size the question of disaggregation (regarding the time buckets) arises. Figure 4.41 shows the system behaviour in this case without the use of daily bucket profiles. The demand is released to the first bucket possible.

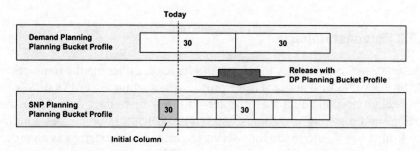

Fig. 4.41. Disaggregation of time buckets

Using the daily bucket profile and the setting 'period split' in the 'SNP2'-view of the product master, it is possible to achieve a more even disaggregation as shown in figure 4.42:

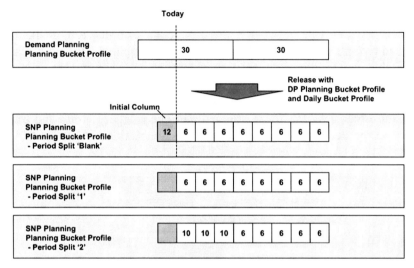

Fig. 4.42. Disaggregation of time buckets using the daily bucket profile

The period split profile offers enhanced disaggregation options, e.g. applying ratios for the disaggregation. For even more complex disaggregation there is always the possibility to set up a second planning area in demand planning with smaller periods and perform the disaggregation there with macros or to use the BAdIs /SAPAPO/SDP_RELDATA (see also note 403050) or /SAPAPO/SDP_REL_SNP.

• *Direct Release from Info Cube*
In some cases demand planning is performed in an external system but nevertheless SNP or PP/DS shall be used. In this case the forecast can be loaded into an info cube, and the data can be released directly from the info cube without using DP.

4.9.3 Forecast After Constraints

To get the opportunity to react as early as possible to potential constraints (e.g. capacity), a feasibility check of the demand plan is helpful. In case of non feasibility, decisions for capacity expansion, allocation or focussing on key markets can be triggered in advance.

With a simple production structure a rough feasibility check might be performed by an aggregation of the demand according to a certain characteristic (e.g. capacity group). If this is possible – that is the assumptions to simplify are sufficiently correct – it is the easiest way to get this information. In many cases however this is not sufficient. There a loop with production planning can help: After the production planning run, stocks and planned receipts are transferred back to DP and compared with the demand. A suitable way to interpret these figures is by cumulating the demands and the supplies and check – probably supported by alerts – whether the cumulated supply falls below the cumulated demand. To transfer the receipts all relevant categories have to be grouped into one category group and transferred to the dedicated key figure with the transaction /SAPAPO/LCOUT, figure 4.43.

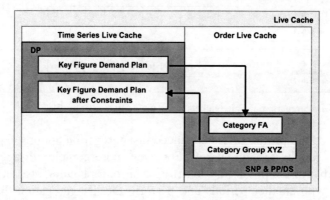

Fig. 4.43. Data exchange between the time series live cache and the order live cache

It is possible to influence the transfer from SNP to DP – e.g. in order to split values to different DP characteristic combinations or to change the unit of measure – via the BAdI /SAPAPO/SDP_REL_SNP.

The process chain for forecasting with capacity check is shown in figure 4.44:

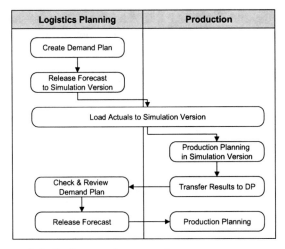

Fig. 4.44. Process chain for forecast after constraints

4.9.4 Transfer to SAP ERP™

If SNP or PP/DS are not used, the independent requirements are transferred directly to SAP ERP™.

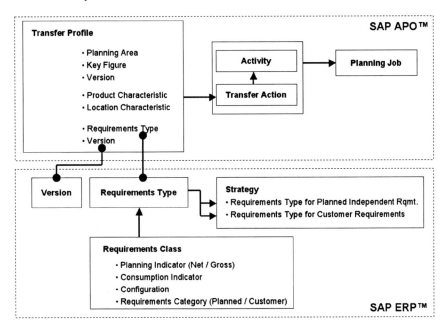

Fig. 4.45. Structure of the transfer to SAP ERP™

The structure for the transfer to SAP ERP™ has similarities to the background planning as shown in figure 4.38. In this case instead of the forecast profiles and the macros a transfer profile (maintained with the transaction /SAPAPO/MC8U) is assigned to the activity, figure 4.45.

The entries for the requirements type and the version in the transfer profile relate to the settings in SAP ERP™ as shown in figure 4.45. The transactions in the SAP ERP™ customising for these entities are grouped in table 4.4.

Table 4.4. SAP ERP™ transactions for independent demand configuration

Entity	Transaction
Requirements Strategy	OPPS
Requirements Type	OMP1
Requirements Class	OMPO
Version	OMP2

The transferred independent requirements are displayed in SAP ERP™ with the transaction MD63. A prerequisite for the data transfer from SAP APO™ to SAP ERP™ is that the distribution definitions for the publication type 'planned independent requirements' with the transaction /SAPAPO/CP1 are maintained (see also chapter 25).

5 Forecast Consumption and Planning Strategies

5.1 Make-to-Stock

The classical production strategies are make-to-stock and make-to-order, which determine the planning to great extent. In a typical make-to-stock environment planning is triggered only by independent requirements and therefore demand planning has a great significance. Sales orders provide merely information to monitor whether the forecasted quantities are appropriate. Typical industries where make-to-stock strategy is applied are commodities and consumer goods, since the same products are usually sold to many customers and the lead time of the sales orders is usually very short. The order life cycle for make-to-stock is shown in figure 5.1. There is always a balanced situation shown as a result of a production planning run.

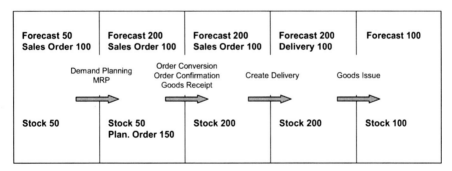

Fig. 5.1. Order life cycle for make-to-stock

The initial situation – the demand of the sales order exceeds the forecast – is a result of inappropriate planning and should not occur. This situation is chosen for this example because it helps to clarify that sales orders are not relevant for production planning in a make-to-stock environment. The sales orders are excluded from pegging as well (and the sales order receives the flag 'pegging irrelevant').

J. T. Dickersbach, *Supply Chain Management with APO,*
DOI: 10.1007/978-3-540-92942-0_5, © Springer-Verlag Berlin Heidelberg 2009

The forecast is reduced by goods issue according to the forecast consumption settings in the product master.

To exclude obsolete forecast from planning – e.g. because the requirement date of the forecast slips into the past without the sales having met the forecasted quantities – it has to be deleted in a reorganisation run. The reorganisation is defined per locationproduct and time horizon (past and future) with the transaction /SAPAPO/MD74.

Make-to-stock is often used for consumer products, where mainly an anonymous market is served and for industries with a long lead time for production, e.g. the chemical industry, where the possibility to react to changes in the customer demand are limited.

5.2 Make-to-Order

The opposite strategy to make-to-stock is make-to-order. In this scenario there is no demand planning, because the complete production planning process starts with the entry of a sales order and independent requirements are not relevant for planning. Choosing the planning strategy 'make-to-order' causes the creation of separate planning segments for each sales order (the business event AE is transferred from SAP ERP™, see chapter 7). Therefore production planning is performed for each sales order individually, figure 5.2. Since production is only triggered by a sales order, there is usually no stock (except due to lot sizes or reduced scrap).

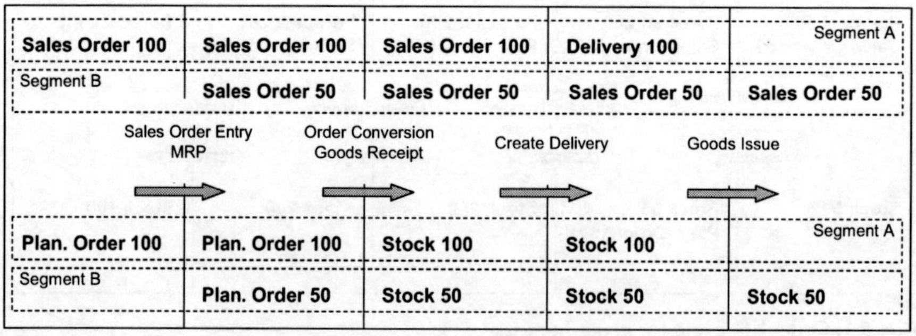

| Sales Order 100 | Sales Order 100 | Sales Order 100 | Delivery 100 | Segment A |
| Segment B | Sales Order 50 | Sales Order 50 | Sales Order 50 | Sales Order 50 |

| Sales Order Entry / MRP | Order Conversion / Goods Receipt | Create Delivery | Goods Issue |

| Plan. Order 100 | Plan. Order 100 | Stock 100 | Stock 100 | Segment A |
| Segment B | Plan. Order 50 | Stock 50 | Stock 50 | Stock 50 |

Fig. 5.2. Order life cycle for make-to-order

Typically this scenario is used for more complex products which often have customer specific features and for highly configured materials.

Technically it is possible to create forecasts in demand planning, release them into a 'make-to-stock' planning segment and perform production planning for these. The sense of demand planning combined with the use of the requirements strategy 'make-to-order' is however questionable.

5.3 Planning with Final Assembly

Many companies apply a mixed scenario, that is they perform demand planning and use the independent requirements for planning, but take sales orders into account as well. In these cases the forecast is consumed by the sales order, figure 5.3. Production planning takes the total of the forecast and the sales order into account. This strategy is called 'planning with final assembly' in SAP.

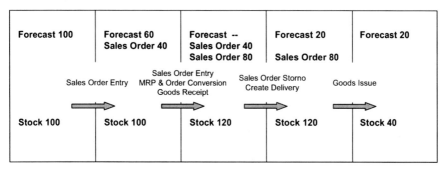

Fig. 5.3. Order life cycle for planning with final assembly

The horizons for the forecast consumption (number of days forwards or backwards from the confirmed date of the sales order) and the consumption mode (backwards only, forwards only or both) are maintained in the product master. Figure 5.4 shows an example for a product with a backward consumption of 4 days and a forward consumption of 3 days.

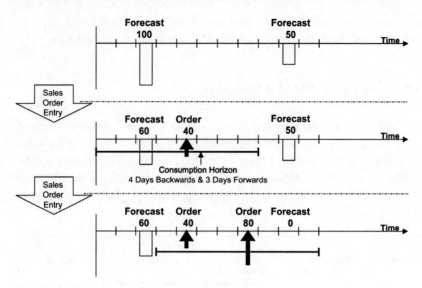

Fig. 5.4. Forecast consumption mode and horizons

The forecast consumption is calculated each time a sales order is created, changed or deleted and each time the forecast is released from DP. The necessary background information to keep the forecast consumption consistent for these changes is the consumed forecast and the sales order entry. Table 5.1 lists these entries for the example shown in figure 5.4.

Table 5.1. Forecast and consumed quantity

Step	Info	Periods							
Initial	Planned Quantity	100							50
	Allocated								
	Plan Remainder								
	Sales Order								
1st Sales Order	Planned Quantity	100							50
	Allocated	40							
	Plan Remainder	60							
	Sales Order			40					
2nd Sales Order	Planned Quantity	100							50
	Allocated	40							50
	Plan Remainder	60							0
	Sales Order			40			50		

The reorganisation of the forecast with the transaction /SAPAPO/MD74 deletes only the remaining quantity of the forecast, i.e. the planned quantity is reduced to the allocated quantity. The current forecast consumption situation is displayed with the transaction /SAPAPO/DMP1 and contains the same information as listed in table 5.1.

A setting in the category group controls per category (e.g. sales order) whether the requested quantity or the confirmed quantity is consumed.

The consumed quantity remains allocated after goods issue as well. The prerequisite for this is that the order category AT for goods issue is included into the category group of the strategy and that the CIF-model for manual reservations is active.

The consumption of the forecast is performed on the locationproduct level. It is possible though to restrict the forecast consumption to a more detailed level (e.g. locationproduct and customer group) using descriptive characteristics. The use of descriptive characteristics is explained in Dickersbach 2005.

5.4 Planning Without Final Assembly

In some cases planning is performed to 'prepare' the production – e.g. to procure components or to produce assembly groups – but the finished product is only produced if an according sales order exists. This might be the case in a business area with diverging material flow or if the added value of the last production step is very high.

With the strategy 'planning without final assembly' planned orders are created in a separate planning segment. Though these planned orders use the same master data, they can not be converted into production orders. Since the master data is the same, they create the according dependent demand and they use the capacity as well. After the sales order entry and the forecast consumption the production planning run deletes the order in the planning segment and creates an according one in the make-to-stock segment, figure 5.5.

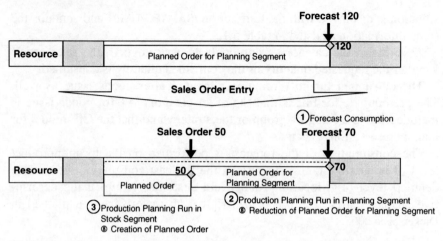

Fig. 5.5. Forecast consumption and production planning without final assembly

A hard restriction for the use of this strategy is that no finite planning may be applied. The dependency on the right planning sequence – i.e. that the planned order in the planning segment is reduced before the planned order in the stock segment is created – contains risks which might destabilise the planning, e.g. by an unfortunate scheduling result. For the same reasons this strategy may not be used in combination with CTP (see section 16.2). If fix lot sizes are used, finite planning is not possible anyhow and the planned orders will usually exceed the demand, see figure 5.6.

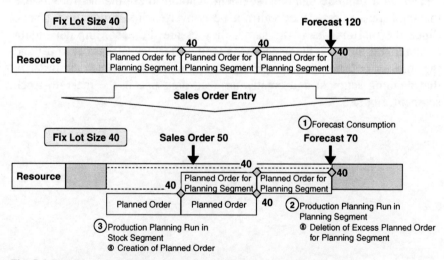

Fig. 5.6. Problems of planning without final assembly and with fix lot size

The main objective of this strategy should be the procurement (internally or externally) of the components. Still some other restrictions apply, for example that stock is not used to cover the forecast, since forecast and stock are in different planning segments. Note 519070 describes these problems in detail.

• *Planning with Planning Product*
The planning with a planning product is basically similar, the only difference is that the master data is different. For the use of a planning product a hierarchy for the planning product has to be created for the object type product with the transaction /SAPAPO/SCCRELSHOW. Several products can use the same planning product.

Both strategies – planning without final assembly and planning with planning product – are only used in PP/DS, neither in SNP nor in CTM. The ATP product check does neither take receipts in the planning segments nor for the planning product into account.

If the main purpose is to create planned orders which are only converted into production orders if a sales order exists for them, an alternative approach might be to use a conversion rule as described in note 519070. In this case an order – whether planned order or purchase requisition – is only converted if it is pegged to a sales order.

5.5 Planning for Assembly Groups

In cases of divergent material flow (i.e. many finished products are produced out of few key components) or make-to-order where assembly groups or components have to be procured in advance, planning on assembly level is common.

In this case the dependent demand of the planned order resp. the order reservation of the production order for the finished product consumes the forecast for the assembly group. Therefore the category AY for the dependent demand and the categories AU, AV and AX for the order reservation have to be included into the category group of the strategy.

Another prerequisite is that the checkbox 'assembly group' is flagged on the demand view of the assembly product. Note 159937 provides additional information.

5.6 Technical Settings for the Requirements Strategies

The requirements strategies define whether and how the forecast is consumed by a sales order or another requirement. Additionally to the definition of the ATP categories in scope – the forecast category and the category group for the requirements – the requirements strategy defines the type of the assignment between forecast and requirement and the planning segment for the forecast. Figure 5.7 gives an overview about the requirements strategy settings for the forecast consumption.

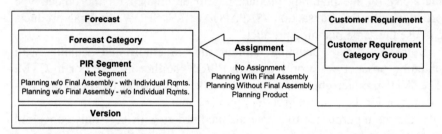

Fig. 5.7. Requirements strategy settings

The customising path for the requirements strategy is *APO → Master Data → Product → Specify Requirements Strategies*. The required settings for an assignment between make-to-stock and make-to-order on requirement side and net segment and planning segment on forecast side are listed in table 5.2.

Table 5.2. Requirements strategy settings for forecast consumption

Sales Order and Forecast		Check Mode	Requirements Strategy Settings	
Sales Order	Forecasts Segment	Assignment	Assignment	PIR Segment
Make-to-Stock	Net Segment	1	1	0
Make-to-Order	Planning Segment	2	2	1
Make-to-Stock	Planning Segment	2	2	2
Make-to-Order	Net Segment	1	2	0

Assignment:	1: Customer Requirement to Planning With Assembly
	2: Customer Requirement to Planning Without Assembly
PIR Segment:	0: Net Segment
	1: Planning Without Final Assembly, With Individual Requirements
	2: Planning Without Final Assembly, Without Individual Requirements

The assignment type of the check mode is not always the same as the assignment type of the requirements strategy.

• *Standard Requirements Strategies*

The requirements strategy is assigned to the product in the product master. The strategies that are transferred from SAP ERP™ are listed in table 5.3.

Table 5.3. Strategies in SAP ERP™ and SAP APO™

Strategy in SAP ERP™	Strategy in SAP APO™	
10	10	Make-to-Stock
20	-	Make-to-Order
40	20	Planning With Final Assembly
50	30	Planning Without Final Assembly
60	40	Planning With Planning Product

Note that the check mode for the customer orders (transaction /SAPAPO/ATPC06, see chapter 7) must contain the same assignment mode as the applied requirements strategy.

• *Category Group*

The category group defines which categories consume the forecast. Another setting in the category group controls whether the requested or the confirmed date and quantity is used for consumption. Note that all relevant categories have to be included into the category group for a consistent forecast consumption (including the creation of deliveries and goods issue) and that the CIF model for manual reservations has to be active. Notes 159937 and 421940 provide additional information.

• *Offline Consumption*

As a standard the forecast consumption is carried out online with each sales order transfer and each demand release. Note 609883 describes the possibilities of a forecast consumption in the background.

• *Requirements Strategies in SNP*

SNP supports only the strategies make-to-stock (10) and planning with final assembly (20). Planning without final assembly and make-to-order are not supported by SNP.

• *Calculation of the Relevant Demand in SNP*

Within SNP the possibility exists to simulate a kind of forecast consumption for planning purposes using the 'forecast horizon' setting in the product master. This kind of consumption is part of the calculation of the total demand (using the macro function DEMAND_CALC). Within the forecast horizon only the sales order categories are relevant for the total demand, outside the forecast horizon the maximum of the customer orders and the forecast is used for the total demand. The prerequisite for this is that no strategy is maintained in the product master.

Part III – Order Fulfilment

6 Order Fulfilment Overview

Order fulfilment contains the processes related to the customer from order taking, availability check and confirmation to the shipment to the customer. Figure 6.1 shows the process chain for order fulfilment processes.

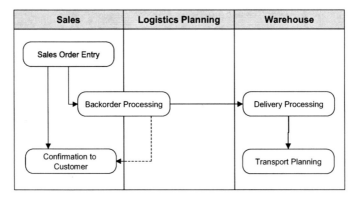

Fig. 6.1. Process chain for order fulfilment

The sales order entry is performed in SAP ERP™ either manually (i.e. for order per telephone) or per EDI, but the ATP check is carried out in SAP APO™ during the sales order entry. Backorder processing is performed in SAP ERP™ – usually as a background job – and the results are sent as updates to the sales order to SAP ERP™. The deliveries are created in SAP ERP™ – usually as a background job as well – several times per day. Again the ATP check is performed in SAP APO™. The transportation planning is performed in SAP APO™ based on the deliveries and sends shipments back to SAP ERP™.

There is an alternative way to perform transportation planning first based on sales orders and trigger the delivery creation from SAP APO™, but this has disadvantages in the flexibility if the ATP check turns out to be unsuccessful.

The main object for the order fulfilment is the sales order and the main functionality from a logistical point of view is the ATP check. The ATP check is performed during sales order entry, backorder processing and delivery processing to determine the available quantity. The criteria for the

ATP check for the delivery is usually more restrictive than for the sales order since it is closer to execution. The order life cycle of a sales order is displayed in figure 6.2:

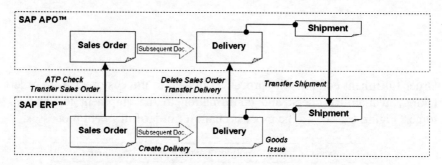

Fig. 6.2. Order life cycle

The sales order represents the contract with the customer. Depending on the business, the sales order might be placed a considerable time before the requested delivery date or with the request for immediate delivery. The sales order might be changed several times before it is actually delivered. Each time the sales order is changed, a new ATP check is carried out.

The first step towards execution is the creation of the delivery. At this point in time an ATP check is carried out again – usually with a much more restricted scope (i.e. stock only). The deliveries (and other documents as stock transfer orders and returns) are grouped to a shipment for transport. The shipment is a document which has a link to the included documents but does not replace them. Accordingly a shipment is not relevant for planning and does not show up in the product view.

7 Sales

7.1 Sales Order Entry

The sales order is the key document in the sales process. The sales order is created in SAP ERP™, and only the transportation and shipment scheduling and the ATP check are performed in SAP APO™. Other tasks during the sales order entry process that are performed on SAP ERP™-side is pricing and credit limit check. Figure 7.1 shows this process between SAP ERP™ and SAP APO™:

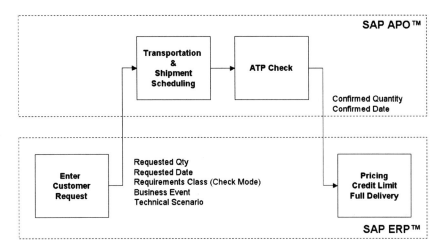

Fig. 7.1. Tasks during sales order entry

The decision whether the ATP check is carried out in SAP ERP™ or in SAP APO™ is made per material and plant by activation of the appropriate CIF model (see chapter 25).

Though the ATP check is usually triggered by SAP ERP™ (with the exception of backorder processing, see section 7.9), it is possible to carry out a simulative ATP check within SAP APO™ with the transaction /SAPAPO/AC04. The purpose of this simulation is to check the ATP settings which can be rather complex. The ATP check in SAP APO™ might be triggered as well from SAP CRM™ or from a BAPI. For the use of the BAPI from a legacy system there are however severe limitations.

● *ATP Check*

The ATP check is called with a requested date and quantity for a location-product (i.e. a product at a location) and returns with confirmed dates and quantities. The availability check is a basic functionality mainly for sales, but also for distribution and production. ATP is carried out for the objects sales order, scheduling agreement, delivery, stock transfer order and production order. Though ATP has especially in distribution and production a rather close connection to the execution, there are many possibilities and opportunities to influence the planning processes by appropriate ATP settings. With the ATP module SAP APO™ offers an alternative to the ATP check in SAP ERP™ with enhanced functionality. Some of the main advantages are

- global rules-based ATP with substitution of locations and products,
- global allocations planning through integrated functionality with demand planning,
- check against the current planning result (if planning is performed in SAP APO™),
- trigger production if there is no availability with CTP (capable-to-promise) functionality,
- check of component availability using multi-level ATP,
- enhanced check options in backorder processing and
- high performance using the time series live cache.

The properties and limitations of triggering production from ATP are described in chapter 16.

7.2 Availability Check Overview

7.2.1 Functionality Overview for the Availability Check

The general idea of the ATP check is to calculate whether or when a requirement can be met and confirm the respective dates and quantities. SAP offers three basis methods for availability check – product check, allocation check and forecast check – and the advanced methods of rules-based ATP and to trigger production. The availability check is triggered by a requested date and quantity and returns with one or more confirmed dates and quantities. The requested date for the ATP check is calculated from the customer request in the transportation scheduling.

• *Basis Methods*
There are different ways to determine whether a request is or will be available at a certain point in time. The basis methods within SAP APO™ are the

- product availability check, where the request is checked against receipt elements (e.g. stock or planned orders), the
- product allocation check where the request is checked against a defined allocation quantity and the
- forecast check, where the request is checked against existing forecast as another demand element. The assumption here is that everything that has been forecasted can actually be procured.

These methods can be combined. Additionally there is the advanced method of rules-based ATP (RBATP) and two methods to trigger production from the ATP check: Capable-to-promise (CTP) and multi-level ATP (ML-ATP). The check instruction determine which of the basis methods are used.

7.2.2 ATP Functionality for Document Types

The ATP check in SAP APO™ is used by sales documents, by stock transfer orders and by production orders (for their input components). Sales orders are the most complex objects that trigger ATP, they have the biggest impact on planning and offer the most possibilities for ATP configuration. Therefore most of the following examples relate to sales orders.

• *Sales Order*
The sales order in SAP ERP™ consists of one or more items (i.e. materials) with a requested date and quantity. Each item has one or more schedule lines, depending whether the full quantity is confirmed for the requested date. For each (partial) confirmation a schedule line is created. Figure 7.2 shows a sales order in SAP APO™ with three schedule lines in the product view.

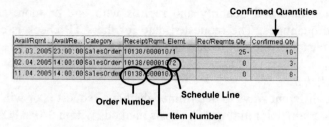

Fig. 7.2. Sales order in SAP APO™

The time of the requirement is set by default to noon. Note 443500 describes these details regarding the scheduling of sales orders. Another information that is transferred from the SAP ERP™ sales order item is the delivery priority from the shipping view, which is used as the order priority in SAP APO™, e.g. for optimisation.

The idea of the sales process is that all conditions including the quantities and dates are determined in the sales order. Therefore any complex ATP logic relevant for planning should take place when checking the sales order. Since the sales order represents a contract with the customer, the delivery merely executes the agreed conditions. The purpose of the ATP check in this case is therefore to make sure that the planned quantities are really available on stock. Accordingly only the product check is to be used in these cases and the scope of check is limited to certain stock categories.

● *Scheduling Agreement*

Scheduling agreements are used if a customer orders a product on a regular basis with a high frequency. In this case the individual customer requests are modelled as schedule lines to the scheduling agreement.

A sales order type is defined as a scheduling agreement in the transaction type of the order type (transaction VOV8). The scheduling agreement is created in SAP ERP™ with the transaction VA31. The standard scheduling agreement type with delivery schedules is BL. Figure 7.3 shows a scheduling agreement in SAP ERP™ and in SAP APO™.

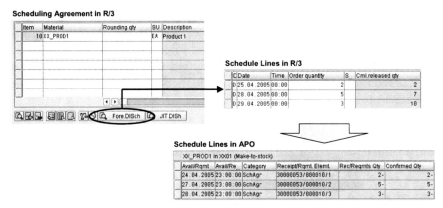

Fig. 7.3. Scheduling agreement and schedule lines in SAP ERP™ and in APO

In SNP the demand of the schedule lines is considered in an aggregated way, but any additional information – e.g. the customer or the individual scheduling agreement – is not available in SNP.

● *Delivery*

Since the delivery is created very shortly before the goods are actually issued, the ATP check is usually limited to stock categories. Therefore only the product check is supported for the delivery.

● *Stock Transfer Order*

With the creation of the stock transfer order in SAP ERP™ an ATP check is triggered in the source location. Figure 7.4 shows a stock transfer order with a partial confirmation.

Stock Transfer Order in Target Location

XX_TRANSFER in XX02 (Make-to-stock)

Avail/Rqmt ...	Avail/Re...	Category	Receipt/Rqmt. Elemt.	Rec/Reqmts Qty	Confirmed Qty	Avail. Quantity	Surplus/Shortage	Destination / Source
05.05.2005	00:00:00	PchOrd	4500015866/000010/1	10	0	10	0	XX01

Stock Transfer Order in Source Location   **Confirmed Quantity** ↓

XX_TRANSFER in XX01 (Make-to-stock)

Avail/Rqmt ...	Avail/Re...	Category	Receipt/Rqmt. Elemt.	Rec/Reqmts Qty	Confirmed Qty	Avail. Quantity	Surplus/Shortage	Destination / Source
02.05.2005	00:00:00	ConRel	4500015866/000010/1	10-	6-	4-		4- XX02/

Fig. 7.4. Stock transfer order in SAP APO™

In this case only a partial confirmation about 6 units instead of the requested 10 units was possible. This information is only available at the source location and not at the target location – the stock transfer order is not automatically adjusted.

Up to SAP ERP™ 2005 the stock transfer object does not support sub-items, and therefore the rules-based ATP is not supported. With SAP ERP™ 2005 the stock transfer order object is enhanced rules based ATP is also available for stock transfer orders (in combination with SAP APO™ 5.0). The original request is also stored, so that stock transfer orders may take part in backorder processing (cf. section 7.9).

• *Functionality Matrix*
Not all document types are supported for the different types of check. Table 7.1 provides an overview.

Table 7.1. Overview about ATP-methods and document types

ATP Method	Sales Order	Delivery	Scheduling Agreement	Stock Transfer Order
Product ATP	Yes	Yes	Yes	Yes
Allocation	Yes	No	Yes	Yes
Forecast Check	Yes	No	No	Yes
Rules-Based	Yes	No	No	Yes[2]
CTP	Yes	No	No[1]	Yes
ML-ATP	Yes	No	No[1]	Yes

[1] technically possible but not recommended
[2] with SAP ERP™ 2005 and SAP APO™ 5.0

Note that for the forecast check the forecast consumption is mandatory. Therefore the appropriate ATP categories must be included in the requirements strategy, e.g. BI for stock transfer orders.

7.3 Master Data and Configuration

7.3.1 Master Data for ATP

The basic master data which is needed for the ATP check is the location-product since all quantities are stored on the locationproduct level. Additional levels might be batch (in ATP called 'version'), storage location and characteristic value (see Dickersbach 2005). If rules-based ATP is used additionally the rules have to be considered as another set of master data. The maintenance of the rules is a task which might require some effort. Customers are not required in SAP APO™ for the ATP check.

7.3.2 Basic ATP Configuration

The parameters mentioned above are maintained in the entities check mode, ATP group, check control and check instruction in the ATP customising. Figure 7.5 shows which entities contain these parameters and the interdependencies between these entities.

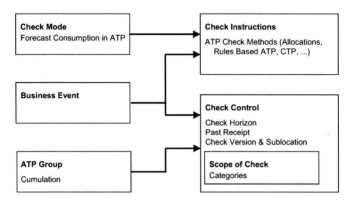

Fig. 7.5. ATP parameters and entities

Check mode and check instructions contain settings which control the ATP check independent of the method whereas the ATP group and the check control are only relevant for the product check. To help the overview they are depicted nevertheless but will be explained later. The check mode and the business event are provided as input parameters from the calling SAP ERP™ system, the ATP group (choose individual requirements) is maintained in the product master and transferred during the master data upload.

• *Business Events*
The business event is 'A' for sales orders ('AE' if the requirements strategy 'make-to-order' is used, see chapter 5) and 'B' for deliveries ('BE' for make-to-order) and can not be changed without modification. Table 7.2 shows the business events for the common order types.

For the stock transfer order the checking rule is used. The customising path in the SAP ERP™ system is *Materials Management → Purchasing → Purchase Order → Set-up Stock Transport Order → Create Checking Rule*. The business event for the production order is defined in the global parameters for PP/DS (transaction /SAPAPO/RRPCUST1).

Table 7.2. Business events

ATP Check Triggered By	Business Event
Sales Order	A
Sales Order (Make-to-Order)	AE
Schedule Line	A
Delivery	B
Delivery (Make-to-Order)	BE
Goods Receipt	01
Goods Issue	03
Stock Transfer Order	Checking Rule in R/3
Production Order	PP/DS Customising

• *Check Mode*

The check mode contains the assignment mode for forecast consumption (see chapter 5) and is used to determine the check instructions. Additionally it contains the setting for the production type – its significance is explained further on (see also chapter 16).

The check mode in SAP APO™ corresponds to the requirements type in SAP ERP™ and is transferred during the ATP check from the sales order. The check mode is maintained in the product master as well and is used for the ATP check in the case of rules-based ATP, multi-level ATP or backorder processing.

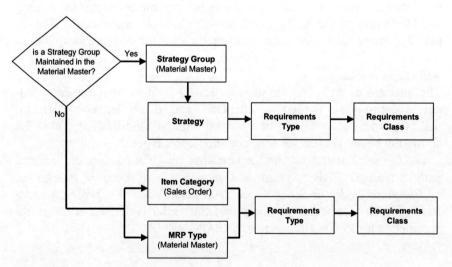

Fig.7.6. Determination of the requirements class

• Determination of the Check Mode

The check mode in ATP corresponds to the requirements class in SAP ERP™. The requirements class is determined in SAP ERP™ using the strategy group in the material master or – if no strategy group is maintained – the combination of item category in the sales order and MRP-type in the material master, figure 7.6.

The strategy is assigned to the requirements type with transaction OPPS and the requirements type to the requirements class with transaction OMP1. The default assignments for the most common strategies are listed in table 7.3.

Table 7.3. Default assignment of strategy to requirements class

Requirements Strategy	(Customer) Requirements Type	Requirements Class
10 (Make-to-Stock)	KSL	030
20 (Make-to-Order)	KE	040
40 (Planning With Final Assembly)	KSV	050
50 (Planning Without Final Assembly)	KEV	045

The item category and the MRP-type are assigned to the requirements type with the transaction OVZI. If no requirements class is found, the mode from the SAP APO™ product master is used to determine the check instructions. If rules-based ATP or multi-level ATP is used, the check mode of the product master is used for all products which are not transferred from the SAP ERP™ order – i.e. the substitution products resp. the components.

• Check Instructions

The check instructions mainly define which basis method – these are product check, allocation check (see chapter 7.5) and forecast check – are used for the ATP check. Additionally the use of the advanced methods – rules-based ATP and start of production – are selected here.

7.3.3 Time Buckets and Time Zones

For the demands and for the supplies different ATP time buckets are used. Per default the time buckets for supplies are switched by 12 hours in order to decrease the probability of an overconfirmation – e.g. if a production order is scheduled for 10 p.m., it is not desired to use it for the confirmation of a sales order that has to be delivered at 11 a.m. Figure 7.7 shows these two ATP time buckets and the way the orders are stored in the buckets.

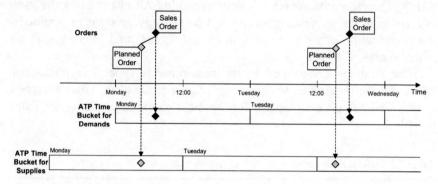

Fig. 7.7. ATP time buckets

Up to SAP APO™ 4.1 this was the only possibility, and the reference point between the time buckets for demands and supplies was 12:00 UTC – independent of the local time zone. With SAP APO™ 4.1 it is possible to use local time zones for the ATP time buckets and to use different time buckets than a day. However, if more than one bucket is used per day it might affect the performance. The use of local time zones and the size of the ATP time buckets is a setting on client level, figure 7.8.

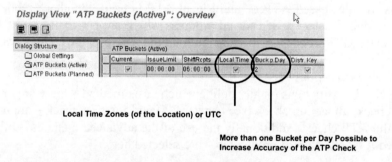

Fig. 7.8. Configuration of the time zones and the bucket sizes

The customising path is *APO → Global ATP → General Settings → Maintain Global Settings for ATP* and a change of the settings requires a live cache downtime. If local time zones are used, it is possible to select a time zone for display, figure 7.9.

Fig. 7.9. Display of ATP times in local time zone

The ATP time buckets are always displayed in UTC.

7.4 Product Availability Check

7.4.1 Product Availability Check Logic

The most obvious method of ATP is the product check, where the requirements are balanced with the receipts. This calculation is performed using time buckets of one day in the ATP time series live cache where all orders of a locationproduct are stored with their quantities and their order category. These time series are displayed with the transaction /SAPAPO/AC05, whereas transaction /SAPAPO/AC03 already applies the scope of check (described in the following paragraphs) and filters the customised categories.

Figure 7.10 shows an example for the product check, where a sales order with a requested date and quantity – represented by an arrow – is confirmed regarding both date and quantity. The confirmed order is represented by the rectangle. The sales order of 30 units is confirmed since there are 70 units of stock and only 40 units of requirements.

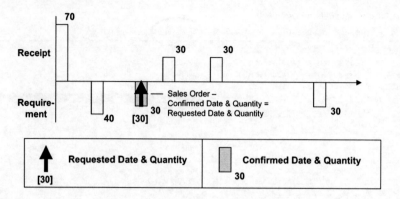

Fig. 7.10. Product check for sales order – full confirmation

The logic that is applied for the product check is explained in figure 7.11. A 'stack' is calculated containing the free receipts that can be used for confirmation by subtracting each (confirmed) requirement from the nearest receipt element starting from the far future moving towards today. The calculation of the stack is performed each time anew during the check. There is no fix correspondence between receipt and requirement elements.

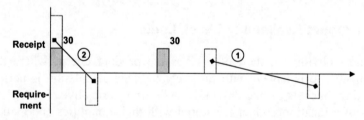

Fig. 7.11. Stack calculation

The stack calculation starts from the far future in order to ensure that the receipt element which is used to confirm a requirement that might be further in the future is not used again to confirm a new requirement that is closer to the receipt. This way an overconfirmation is avoided.

In the second scenario (figure 7.12) the sales order for 70 units is only partially confirmed with 30 units on the requested date and 30 units on the date of the first free receipt as calculated in the stack. An open quantity of 10 units remains.

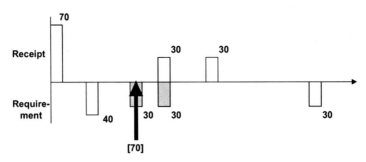

Fig. 7.12. Product check for sales order – partial confirmation

7.4.2 Product Availability Check Configuration

The parameters for the check are defined mainly in the check instruction, the ATP group and the check control resp. the scope of check. An additional control is whether a cumulation of the requests is used for the ATP check. These entities and their influence on the product availability check are described in the following.

• *Cumulation*
The system behaviour regarding confirmations in cases of shortages is defined by the parameter for cumulation, which is set in the ATP group. An example with a shortage situation as shown in figure 7.13 illustrates the effect of this parameter for the calculation of the stack and thus for the confirmation of orders.

A sales order of 80 units was fully confirmed, since there had been 80 units of stock. However 30 units had to be withdrawn for scrapping. To resolve the shortage a planned order about 30 units is created, but it will be available only after the confirmation date of the sales order. Now a new sales order with a requested quantity of 30 units is checked.

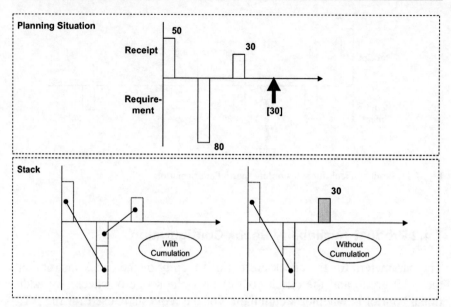

Fig. 7.13. Cumulation

With cumulation the shortage is taken into account so that the cumulated ATP quantity is zero at the requested date and subsequently the sales order is not confirmed.

Without cumulation however negative ATP quantities are not considered, which represents the idea that late receipt do not help to resolve the shortage and can therefore be used to confirm other orders. Accordingly the ATP quantity is 30 units and the sales order is confirmed.

● *Check Control*
The check control contains the settings about the scope of check, the consideration of past receipts, the consideration of the check horizon and whether the check considers sublocation and batches.

The key for the check control is the business event and the ATP group. Different document types can influence the scope of check via the business event, different materials via the ATP group.

● *Scope of Check*
The scope of check defines the categories for the receipt and the requirement elements which are used for stack calculation (stock, production orders,…) and thus determine the available quantities.

Depending on the document type – sales order, delivery, stock transfer or production order (for the input components) – the order type or the material, the scope for the ATP check might be different. For a sales order for example a planned receipt in the future might be sufficient for confirmation, whereas for the confirmation of a delivery usually stock on hand is required. The scope of check is assigned to the check control.

● *Consider Past Receipts*
The use of past receipts determines whether receipt elements with due dates in the past shall be used for confirmation or not. If e.g. a production order has its due date already in the past, this could mean either that it is going to become inventory soon or that there is something fundamentally wrong with it and should not be used therefore. This entry must be set according to the business context.

● *Checking Horizon*
The basic assumption for the use of checking horizons is that any amount can be procured (internally or externally) within a specified number of days, figure 7.14. This number of days is entered in the field 'checking horizon' in the ATP view of the product master.

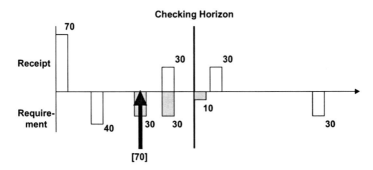

Fig. 7.14. Product check with checking horizon

Within the checking horizon the selected basis methods are used to check the availability. Any quantity that can not be confirmed by these methods is confirmed outside the checking horizon.

● *Check Sublocation and Version*
For one locationproduct data is stored within the ATP time series in hierarchical time series according to the sublocation (storage location) and the version (batch). If the request contains information regarding the batch or the storage location (latter usually from the production order), the request is checked both at detailed level (version or sublocation) and location

level. With the according setting in the check control the check at the version resp. the sublocation level can be prevented.

• *Time Series vs. Pegging Network*
Alternatively to the ATP check using the time series (as discussed until now) it is possible to carry out the ATP check using the pegging network in the order live cache. In this case the ABAP class for CTP of the planning procedure (a setting in the PP/DS view of the product master, see chapter 16 and 20) gains relevance. Since SAP APO™ 4.1 the check horizon and the scope of check are considered by the ATP check using the pegging network as well. To use the pegging network for the product check the setting for the production type in the check mode has to be 'characteristic evaluation'. The notes 574252 and 601813 provide additional information.

• *Temporary Quantity Assignments*
Confirmed sales order quantities reduce the available quantities only after saving. If several sales orders perform ATP checks for the same product at the same time, an overconfirmation may occur, since the same receipt element may be used several times for confirmation. To prevent this situation it is possible to write temporary quantity assignments which create a requirement element for the checked sales order until the order is either saved or cancelled. Whether temporary quantity assignments are used or not is defined in the global settings for ATP customising. Frequently asked questions are answered in note 488725. The temporary quantity assignments are displayed with the transaction /SAPAPO/AC06. It is recommended to work with temporary quantity assignments.

7.5 Allocations

7.5.1 Business Background and Implications

In cases where the demand or the expected demand exceeds the supply – e.g. because of production limitation due to very high investments – allocations help to ensure that customers or customer groups are provided with their share according to the sales and marketing policies. Generally allocations prevent 'first come first serve' behaviour.

It is most important to understand that allocations are not reservations for actual receipts (e.g. stock), but limitations for the sales order confirmation. The effect of reserving quantities for customer groups is achieved by hindrance of other customers to claim more than their share. If allocations are used for a product it is therefore necessary to allocate the complete scope of the product. If some orders are allowed to sneak through without allocation check, the use of allocations does not control the distribution of quantities to customers or customer groups anymore.

Another common motivation to apply allocations is to check against a limited production capacity, assuming that the according amount can still be produced if necessary. Combined with a product check using the check horizon even a rough estimation of the lead time can be taken into account.

A scenario that is often applied in supply chains with multiple (and hierarchical) distribution centres (DCs) focuses on the share of inventory. Figure 7.15 shows a supply chain where several sales organisations access the stock of one DC.

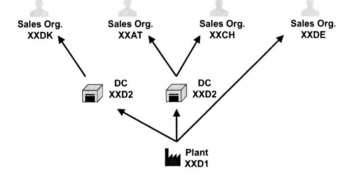

Fig. 7.15. Case for allocations in a supply chain

To meet the service level agreements and/or to motivate sales organisations to increase their forecast reliability, it is important to provide inventory for each sales organisation according to their forecasted amounts. Typically there are two conflict situations in this scenario, even if the provided quantities equal the sum of the forecasts.

One conflict arises if more than one sales organisation accesses the inventory of one DC – in this case the sales organisations XXAT and XXCH in DC XXD2. If for example XXAT sells the complete inventory, profits for XXAT rise while XXCH gets into serious trouble because of stock outs.

The second problem arises from a similar effect on a different level. Stock transfers from a plant or a central DC are increasingly often planned by a central sales organisation, which controls the distribution of the inventory and is responsible to meet the service levels for stock availability in the local DCs. If the plant or the central DC serves both other DCs and customers as in figure 7.15, the respective sales organisation – here XXDE – has the privilege to access freely a well stocked location. There is no possibility to detain sales organisation XXDE from selling the complete inventory of the plant respective the central warehouse XXD1 and thus causing problems for the supply chain organisation as well as for the other sales organisations. Since the supply chain organisation is responsible for the distribution of the inventories but has only no control of the consumption by the sales organisation XXDE, this set up has some potential for conflicts.

Using allocations – for example with an allocation quantity of the forecasted amount plus a tolerance of e.g. ten percent – inhibits in both cases the 'stealing' of inventories that are intended for other purposes.

7.5.2 Configuration of the Allocation Check

The level of allocation is modeled by characteristics (product, product group, customer, sales organisation, …) like in demand planning and can be quite freely defined. The only restriction is that the characteristic has to be in the table /SAPAPO/KOMGO – it is however possible to include any characteristic via user exit. From there it is possible to assign it to the field catalogue and use it in the product allocation group. The characteristic KONOB is mandatory. The allocation object is used as a value of the characteristic KONOB.

The allocation quantities are stored in the key figure KCQTY on the level of the allocation group. With the confirmation of a sales order the consumed quantity of this allocation increases and the open quantity decreases in the respective period. The consumed quantity is stored in the key figure AEMENGE while the open quantity is calculated. The appropriate period is determined either by the product availability date (MBDAT), the delivery date (LFDAT) or the goods issue date (WADAT). This date is chosen in the allocation group. Figure 7.16 provides an overview of the settings described above.

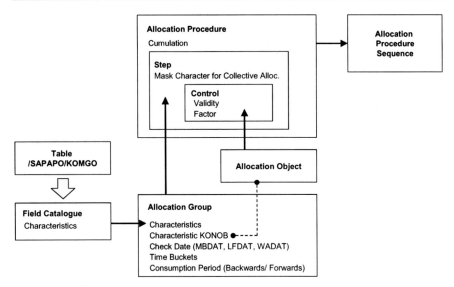

Fig. 7.16. Allocation settings

• Combination of Allocations

Within one allocation procedure several allocation groups can be checked in subsequent steps. The minimum of the available quantities of each allocation group is taken as confirmed quantity. The open quantities of each allocation group are reduced by the confirmed quantity.

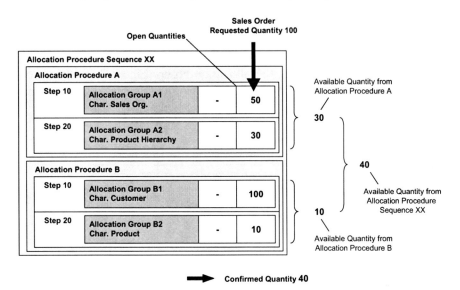

Fig. 7.17. Combined allocation check

If an allocation procedure sequence is used, the confirmed quantities of each allocation procedure are added. Figure 7.17 illustrates this logic with a probably rather far fetched example. In this example a sales order for customer XXCUST1 is entered by sales organisation XXDE with the request of 100 units of product HEAVY_250, which is part of the product hierarchy HEAVY. In the first allocation procedure A in step 10 50 units are available, in step 20 only 30 units. The minimum of the two – 30 units – is therefore available after checking allocation procedure A. Analogously 10 units are available after checking allocation procedure B. Since the available quantity of an allocation procedure sequence is the sum of all allocation procedures, 40 units are confirmed. The open quantities are reduced after saving the sales order as shown in table 7.4.

Table 7.4. Reduction of open quantities in combined allocation check

Allocation Group	Open Quantity Before Check	Open Quantity After Confirmation
A1	50	20
A2	30	-
B1	100	90
B2	10	-

• *Consumption of Open Quantities from Other Periods*
To allow the consumption of not used quantities from other periods the flag 'cumulation' in the allocation procedure is a prerequisite. If this flag is not set, only the period of the requested date is checked. The details regarding the period consumption – mainly the number of periods backwards and forwards – are defined in the allocation group. Figure 7.18 illustrates an example for the consumption of open quantities 2 periods backwards and 3 periods forwards.

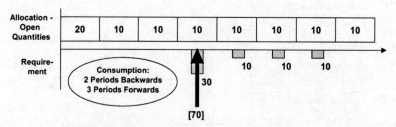

Fig. 7.18. Consumption of open quantities

The requested quantity of 70 units is partially confirmed with 30 units from the current and the two previous periods. Since only two periods are allowed for backwards consumption, the three periods towards the future are used for further confirmation. Still only 60 units of the requested 70 units are confirmed in this example.

Within backwards consumption the least recent period is consumed first. The sequence of the period check for this example is shown in figure 7.19.

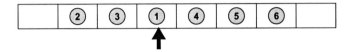

Fig. 7.19. Check sequence for consumption

Depending of the check sequence the result might differ.

● *Order Types*
Allocation Check is supported for sales orders, scheduling agreements and stock transfers. It is not supported for deliveries.

● *Prerequisites for Allocation Check*
A prerequisite for checking allocations is that an according check step is defined in the check instructions and that either an allocation procedure or an allocation procedure sequence is defined in the product master. Both allocation procedure and allocation procedure sequence can be maintained as a global or as a location dependent setting.

● *Switching Allocations On and Off*
The appropriate procedure to switch between allocated situations and non allocated situations is using the settings for validity and active or inactive for the allocation object in the control view of the allocation procedure. Other ways require changes in the master data – e.g. to remove the entry for allocation procedure or allocation procedure sequence – or in customising (changing the check instructions). Increasing the allocation quantity would be another alternative.

● *Combination of Allocations and Product Check*
In most cases a check of allocations does not substitute the product check but is used in addition. This means that an order is only confirmed if there are open quantities both in the allocation and the product check. The check instruction defines whether allocations and product availability are checked and in which sequence. Independent of the sequence the minimum of the two checks is used for confirmation. The sequence is only of

importance in combination with CTP, if 'start production after product check' is chosen.

7.5.3 Allocation Maintenance and Connection to DP

One of the advantages of the allocation solution in SAP APO™ is that the allocation quantities are planned in demand planning. To transfer the allocation quantities from DP to ATP it is necessary to connect the planning area to the allocation group in customising and assign the characteristics and the key figures as shown in figure 7.20.

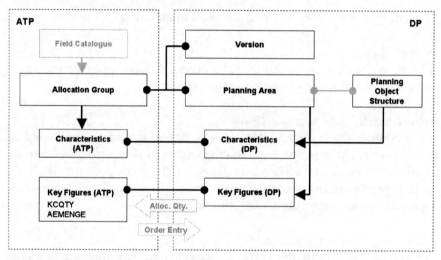

Fig. 7.20. Connection of the allocation group to the planning area

Before transferring the quantities, the characteristic value combinations (CVCs) have to be generated in ATP. The characteristic value combinations are copied with the transaction /SAPAPO/ATPQ_PAREA_K from DP according to the connected planning area. The CVCs in ATP are displayed with the transaction /SAPAPO/ATPQ_CHKCHAR. The allocation quantities are read from DP with the transaction /SAPAPO/ATPQ_PAREA_R, and the order quantities are transferred from ATP to DP with the transaction /SAPAPO/ATPQ_PAREA_W.

As an alternative to this periodic transfer in both ways it is possible to connect the planning area online to ATP by setting the flag 'check planning area' in the allocation group. The advantage of the online connection is given if the allocation quantities are changed frequently and an immediate impact is desired. Note that there might be some locking problems.

7.5.4 Collective Product Allocations

As mentioned before, if allocations are used, all orders have to be allocated. Often it is too much effort or even impossible to plan allocations for each characteristic value – for example if allocations are planned on customer level. In this case collective product allocations can be used to aggregate characteristics in a way that first the explicit characteristic value combinations are checked (in this example all allocations with characteristic values for 'real' customers) and afterwards the characteristic value combinations for collective product allocations. These CVCs for collective product allocations use mask characters as wildcards to match the characteristic value from the sales order with the characteristic for the allocation group and are created in transaction /SAPAPO/ATPQ_COLLECT. The mask character to be used as wildcard is defined in the allocation procedure.

The downside of collective allocations regarding the integration of the allocation quantities with DP is that there is no aggregation functionality. This means that the collective (masked) characteristic value combination must exist in DP as well and is planned explicitly like any other 'real' characteristic value combination.

If the level for collective allocations – that is the characteristic connected to the allocation group with masked values – is not used for other planning purposes, it does not cause too many problems to use an additional CVC value instead of a set of real CVCs. But if this is not the case, it has to be examined whether the introduction of an additional characteristic or the copying of the key figures between CVCs (e.g. via an info cube using routines in the update rules) is more appropriate to the given circumstances.

Creating CVCs using the mask characters might require an 'introduction' of these characters first with the transaction RSKC. Make sure that the characteristic from table /SAPAPO/KOMGO and the connected info object have the same data format.

7.6 Forecast Check

An alternative method of an availability check is a forecast check, which might be used in a make-to-stock environment assuming that the forecasted amounts are always procured. This method is – at least in SAP APO™ – not very often used since the check against stock and planned receipts is usually more accurate. However, one business case might be the planning for phantom assembly groups combined with a multi-level ATP check. The prerequisite for the confirmation in the forecast check is the correct configuration of the forecast consumption (see chapter 5).

7.7 Rules-Based ATP

An ordinary ATP check is restricted to the requested locationproduct and checks only the according time series. Using rules-based ATP it is possible to substitute both the location and the product.

In supply chains where mostly uniform products are kept at multiple locations – often in consumer goods industries – or where different product numbers are sold to customers as one product – either because these numbers represent only some internal information which is irrelevant to the customer (e.g. the production site) or because quality information are encoded in the product number and upgrades are allowed – rules-based ATP is a valuable tool to increase the delivery performance and to reduce the safety stock levels, since the inventories of the entire network can be used to cover unpredicted deviations in the demand. Therefore the safety stock at each location just has to cover the average deviations in the demand per location, which are less than the demand deviations which have to be taken into account for each location.

Rules-based ATP is called with a requested locationproduct from the item of the SAP ERP™ order and returns with the same or another locationproduct that is created as a sub-item. Therefore only those SAP ERP™ objects can use rules-based ATP that support the sub-item structure. These are sales orders, inquiries and quotations (and from SAP ERP™ 2005 on stock transfer orders). Another restriction exists regarding the use of sales BOMs, since these already use the sub-items and therefore no rules based ATP can be applied. To apply rules-based ATP first a rule is needed and a procedure to determine the rule.

* *Rules Maintenance*
The rules for rules-based ATP are maintained with the transaction /SAPAPO/RBA04. Note that these rules are master data and therefore have to be maintained in each system (development, quality assurance, productive) and are not transported like customising entries. The rule is a rather complex structure that contains several entities. These are mainly

- substitution lists ('procedures' for products, locations, locationproducts or PPMs, which define the substitute objects for the checks),
- a rule control, which determines the check sequence,

- a calculation profile and
- a location determination activity.

Figure 7.21 shows the structure of the entities within the rule.

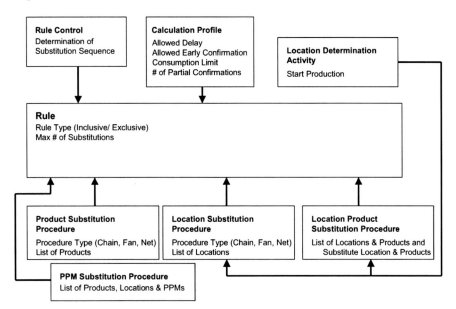

Fig. 7.21. Rule structure

To access the entities in figure 7.21 which are sketched above the rule (rule control, calculation profile and location determination activity) the button 'profile & parameter' has to be used, figure 7.22.

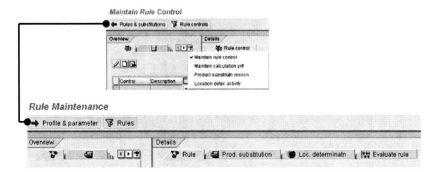

Fig. 7.22. Rule maintenance

• Triggering Rules-Based ATP from the Check Instructions
The prerequisite to trigger rules-based ATP is that the according checkbox is set in the check instructions. The basic methods used within the rules for

the availability check (product check, allocations, forecast) are defined in the check instructions. For the substituted locationproducts the check mode of the product master is taken.

If no full confirmation is possible, an additional sub item can be created with a requested quantity for the difference for either the originally requested locationproduct or for the locationproduct which was used last for substitution. If the checkbox 'requirements' is not selected, the originally requested quantity is not any more available in SAP APO™. In this case backorder processing will not be able to check the requested quantity (except if a re-evaluation of the rules is triggered).

• Substitution Procedures
The product substitution procedure is merely a list of products that can generally be used as substitutes. The location substitution procedure is similarly a list of locations which might be used as substitutes. Additionally to each location a location determination activity is assigned, which is able to trigger production and to overwrite the check mode and the business event. The location product substitution procedure finally contains a list of locationproducts with assigned location determination activities.

• Location Determination Activity and Calculation Profile
The location determination activity allows to trigger production with the same parameters as available in the check instructions. Both the check mode and the business event can be overwritten with the value maintained in the location determination activity. This way it is possible to modify the check instruction and the scope of check.

The calculation profile contains entries regarding the allowed delay or earliness between requirement and receipt. Using these entries it is e.g. possible to prevent the creation of schedule lines with unacceptable delays or – the other way round – it is possible to prevent that a sales order with a requested date in the far future blocks stock which should be used for the confirmation of immediate requests.

• Stock Transfer Creation for Location Substitution
From a business point of view two different kinds of behaviour might be desired in the case of a location substitution. As an example the customer places its order to location XX01 and rules-based ATP determines the available quantity in location XX02. The two options in this case are:
1. The sales order is passed on to location XX02. The customer is then delivered from location XX02 instead of XX01.

2. The sales order remains at location XX01 and a stock transfer requisition is created from location XX02 to XX01. The customer is delivered from location XX01.

Figure 7.23 visualises these two options:

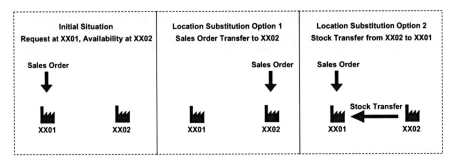

Fig. 7.23. Location substitution with and without stock transfer requisition

The first option is the standard behaviour for rules-based ATP. For the configuration of the second option the checkbox for 'stock transfer' has to be set in the location determination activity.

Technically the stock transfer requisition is created using the ATP tree structure as for multi-level ATP (see chapter 16). For this reason it is not possible to combine this option with the CTP functionality (see also chapter 16). Notes 510313 and 557559 provide additional information.

• *Rule Control*
The rule control determines in which order the substitutes for products and locations are checked. It contains the parameters
  • access strategy for substitution lists,
  • restriction of products and locations,
  • combination of lists as a union or as an intersection and
  • usage of the complement of the combined list.
These parameters are explained in the following examples.

• *Access Strategy*
The access strategy describes the sequence in which the substitute lists ('procedures') are evaluated. Figure 7.24 visualises the possible access strategies.

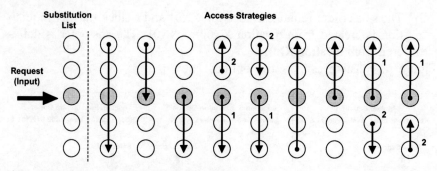

Fig. 7.24. Access strategies to the substitution list

For the product location substitution the two substitution lists for products and locations have to be combined to a substitution sequence. This is defined in the combination settings within the rule control. The impact of the radio buttons

- 'combine qualified product with all locations, then qualified location with all products' and
- 'combine qualified location with all products, then qualified product with all locations'

is shown in figure 7.25.

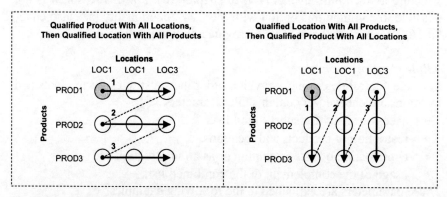

Fig. 7.25. Combination of substitution lists

In the rule control there are possibilities to restrict the product and/or the location substitution list to one entry only, either to the requested product or to the first product of the list. Figure 7.26 shows the result for restrictions in combination with the setting 'combine qualified product with all locations, then qualified location with all products' (checking products first).

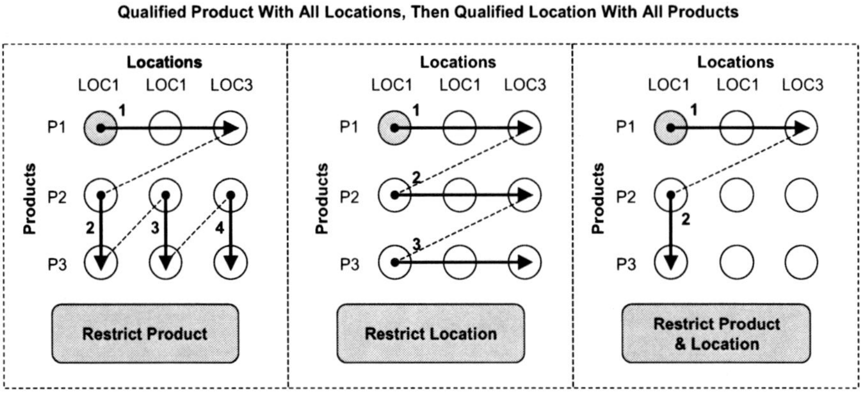

Qualified Product With All Locations, Then Qualified Location With All Products

Fig. 7.26. Restriction of the substitution list checking products first

Compared with an unrestricted check the restriction of the location has no influence when checking products first.

Figure 7.27 shows the impact of the restrictions when checking locations first. Analogously the restriction of the product has no influence.

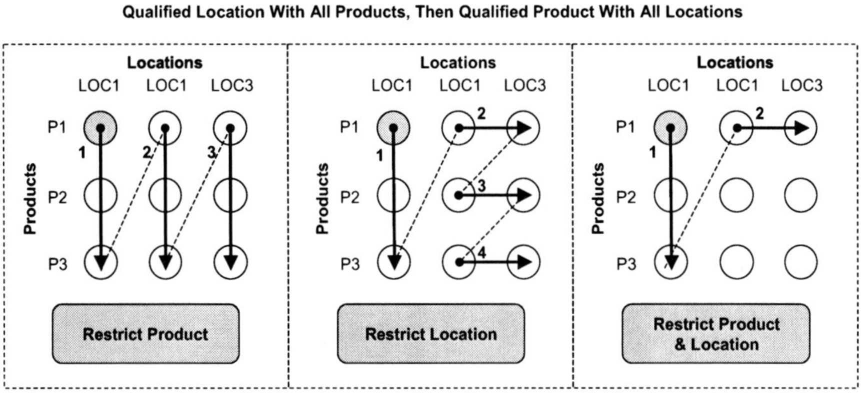

Qualified Location With All Products, Then Qualified Product With All Locations

Fig. 7.27. Restriction of the substitution list checking locations first

Per default the substitution lists are combined as a union. By changing the combination to 'intersection', further restrictions of the combined substitution can be achieved in combination with the restriction of products and locations, figure 7.28.

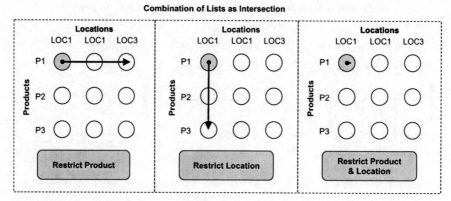

Fig. 7.28. Combination of lists as intersection

Setting the 'complement' checkbox causes the check to be carried out only on the excluded elements. Figure 7.29 illustrates this behaviour. If no elements are excluded, check is carried out only for the requested product and location.

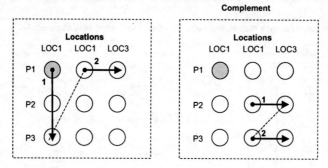

Fig. 7.29. Complement

● *Result Screen*

The result screen of the rules-based ATP differs from the result screen of the standard ATP check and is displayed in a separate window. Figure 7.30 gives an example for a check with the substitution of products as shown in the second picture of figure 7.28.

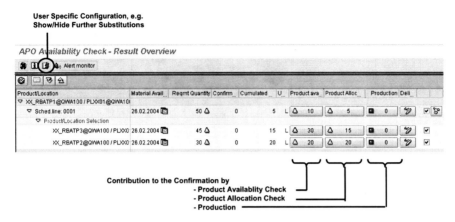

Fig. 7.30. Result screen for rules-based ATP

A yellow traffic light represents that the sales order is partially confirmed from that locationproduct. The green traffic light signals that the full requested quantity is confirmed, while a red light shows that this location-product does not contribute to the sales order confirmation. The blue information box notes that there are still locationproducts farther on the substitution list which have not yet been checked.

A small calendar indicates that there has been a late confirmation and schedule lines are created. Note that in the column 'confirmed quantity' only those quantities are listed, which are confirmed at the requested dates. Therefore it is possible to get a green traffic light and an entry of zero in the 'confirmed quantity' column.

A legend of the symbols is available using the white info box in the top left corner. It is possible to branch into the availability situation (transaction /SAPAPO/AC03) of the listed locationproduct using the button with the glasses.

● *Rule Determination*
The rule for the rules-based ATP check is determined using the condition technique, where characteristics from the sales order are chosen to determine via condition type a rule according to the values of the selected characteristics. Figure 7.31 gives an overview of the involved entities.

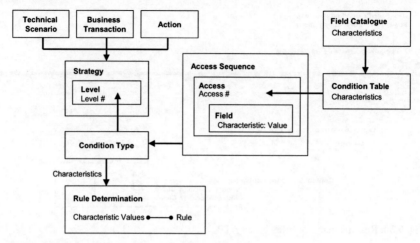

Fig. 7.31. Rule determination using condition technique

The parameters to determine the strategy – technical scenario, business transaction and action – are determined in SAP ERP™ and transferred to SAP APO™. The values for the technical scenario and the action type are set in the program RVDIREKT and can not be changed. The technical scenario is AA for dialogue, BB for batch input and DD for EDI, and the action type is A for create, B for change and C for copy. Only the values for the business transaction are customised in the ATP customising in SAP ERP™ *Logistics → Sales and Distribution → Basic Functions → Availability Check and Transfer of Requirements → Availability Check→ Rule-based Availability Check → Define Business Transaction* and assigned to the order type with the transaction VOV8.

The transactions to maintain the settings for the condition technique are listed in table 7.5.

Table 7.5. Transaction for condition technique

Entity	Transaction
Field Catalogue	/SAPCND/AO01
Condition Table	/SAPCND/AO03
Access Sequence	/SAPCND/AO07
Condition Type	/SAPCND/AO06
Strategy	/SAPCND/AO08
Rule Determination	/SAPCND/AO11

The most common cause for not finding any condition is an incorrect data format in the rule determination. If a customer is used as a characteristic for example, leading zeros have to be considered.

• *Exclusive Rules*

By default a rule has the type 'inclusive'. These are the ones which have been considered so far. Exclusive rules on the other hand exclude the locationproducts which are determined in the rule from the check. The use of exclusive rules can be valuable in combination with subsequent inclusive rules. Following example illustrates the use of exclusive rules.

To the customer group CUSTGR one hundred customers (CUST001 to CUST100) are assigned. These customers may be delivered from any of the five locations 1000, 1100, 2000, 3000 and 5000. Only for the customers CUST038 and CUST073 delivery is only allowed from the locations 1000, 1100 or 2000. Instead of creating 100 rule determinations (98 for the rule with five substitutions and 2 for the rule with three substitutions) an exclusive rule for the complementary two locations (3000 and 5000) is created and checked before the inclusive rule containing all five locations. The settings for this subsequent checking of an exclusive rule on more detailed level and of the inclusive rule in the next step is shown in figure 7.32.

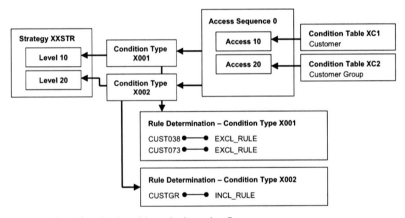

Fig. 7.32. Settings for checks with exclusive rules first

As mentioned above, rule INCL_RULE contains all locations, while rule EXCL_RULE contains only locations 3000 and 5000.

• *Rules-Based ATP and Make-to-Order*

Rules-based ATP in combination with a make-to-order strategy is not supported because rules-based ATP causes the creation of sub-items and the account assignment is not consistently copied to the sub-items.

• *Multi-item Single Delivery Location*

Location substitution is performed for each sales order item. Using the rule type for multi-item single delivery location (MISL) all items of the same delivery group are sourced from the same location. The standard logic is that as soon as an item is not fully confirmed the next location is checked for all items. Another feature in this context is the use of a consolidation location in order to consolidate partial confirmations from multiple location substitutions before sending them to the customer. The material availability date for the ATP check is determined in two steps – first from the customer to the consolidation location and then from the consolidation location to the sourcing locations.

7.8 Transportation and Shipment Scheduling

The transportation and shipment scheduling is an integral part of the availability check in SAP APO™, and calculates the time difference between the requested delivery date at the customer site and the required material availability date at the factory. If SAP APO™ is used for the ATP check, the transportation and shipment scheduling functionality of SAP APO™ has to be used instead of the analogous functionality in SAP ERP™.

In between the requested delivery date and the material availability date the goods issue date, the loading date and the transportation planning date are scheduled. Figure 7.33 visualises these dates.

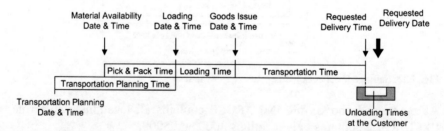

Fig. 7.33. Dates for transportation and shipment scheduling

The scheduling starts from the requested delivery date and time. The requested delivery date is meant as the date when the material has to arrive at the customer site. This is the date which is entered in the sales order. The requested delivery time is determined by the allowed unloading times at the customer site, which are maintained in the SAP ERP™ customer master. If for example the unloading times at the customer are from Monday to Friday from 6:00 to 18:00 and the requested delivery date is a Monday, the earliest unloading time is set as the requested delivery time (i.e. Monday 6:00). Both the requested delivery date and the requested delivery time are sent to SAP APO™.

The goods issue date and time are determined by subtraction of the transportation time. This is when for example the truck has to leave the factory towards the customer. If loading time is required at the factory, this has to be considered as well and leads to the loading date. At this date the loading of the truck has to start. The pick & pack time finally considers the preparation of the goods for the loading of the truck. The start of the picking and packing of the material is the material availability date. To avoid conflicts in the scheduling, the material availability time is set to 12:00 of the according day, because the ATP time series are in daily buckets.

Besides the packed material, it is necessary to have a free truck available at the shipping point to start with the loading. Therefore the transportation planning time is subtracted from the loading date as well. All scheduling steps are performed on the basis of hour and minute.

If the resulting material availability date and the transportation planning date are future dates, no further scheduling is necessary at this stage. If either of the two dates is in the past, the according date is set to today and forward scheduling is performed with the result of a later delivery date.

In a second step the system checks the availability of the material for the calculated material availability date. If the material is available at the requested date, no further scheduling is necessary. If the material is only available at a later date, a forward scheduling from the earliest material availability date is carried out.

The scheduling steps are modelled using the condition technique as described in the paragraph about the rule determination (cf. section 7.7). Though the structure is the same, a different type of objects is used. The transactions for the according entities for transportation and shipment scheduling are listed in table 7.6.

Table 7.6. Transactions for the condition technique for scheduling

Entity	Transaction
Field Catalogue	/SAPCND/AU01
Condition Table	/SAPCND/AU03
Access Sequence	/SAPCND/AU07
Condition Type	/SAPCND/AU06
Strategy	/SAPCND/AU08
Rule Determination	/SAPCND/AU11

The standard condition types for the scheduling are LEAD for the transportation planning time, PICK for the pick & pack time, LOAD for the loading time, TRAN for the transportation time and UNLD for the unloading time at the customer.

The related field catalogue contains a huge number of geographical characteristics and other customer related information. The condition types are named similar as the scheduling steps described above. The rather complex settings can be tested with the transaction /SAPAPO/SCHED_TEST in a simulation.

7.9 Backorder Processing

The current availability situation might differ significantly from the planned one at the time of the sales order confirmation – the more the confirmed date is in the future. This means that the currently confirmed dates would not be confirmed any more. One main motivation to resolve such a situation is to enhance or restore the visibility and the clearness of the planning situation. If it is clear that the confirmed dates can not be met, unrealistic requirement dates cause alerts for shortage or lateness and misleading goals for production planning and optimisation. Nevertheless the first attempt has to be to cover the confirmed demands. Therefore the backorder processing must only be performed after the production planning is finished. A second point is that in many cases, especially when the lead time between placing an order and the requirement date is large, the customer should be informed about the delay. Whether information to the customer - e.g. by sending a new confirmation – is required, depends on the sales order process. Sending confirmations to customers is a process entirely within sales in SAP ERP™, which can be triggered by a change in the confirmation. A new confirmation however does not necessarily cause a notification to the customer.

The basic idea of backorder processing is to carry out a new ATP check for a set of order items. This way backorder processing can also be used to distribute quantities in case of shortage (resp. lateness) according to priorities. Backorder processing is usually performed in the background, but interactive backorder processing is also possible. The settings for the background planning are described first.

The methods of the ATP check itself are the ones defined in the check instruction, with the only restriction that CTP can not be triggered (production planning has to be completed before backorder processing starts).

There are mainly three categories of settings which influence the backorder processing. These are
- the selection of the order items to be checked,
- the sequence in which the check for the selected order items is carried out and
- the parameters for the check itself.

The selection of the order items and their sequencing is the work-list.

● *Work-List*

The work-list for the backorder processing contains a set of order items, which are checked in the sequence of their listing. To create a work-list a filter type, a filter variant and a sort profile are required as shown in figure 7.34.

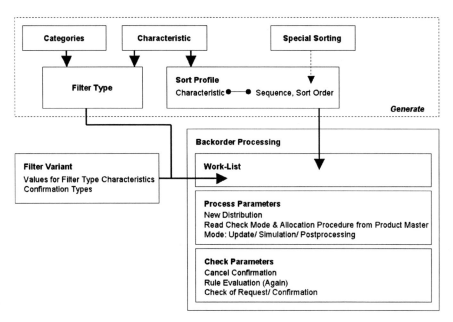

Fig. 7.34. Settings for the backorder processing

The order items are selected according to the filter type, which defines the characteristics for the selection, and the filter variant, which defines the value ranges for the characteristics of the filter type. An additional selection criterion is the confirmation type (confirmation on requested date, partial confirmation, no confirmation, …) of the order items. If these criteria should not be sufficient, it is possible to create own filter criteria per user exit as described in note 376773.

The sort profile defines the sequence in which the order items are checked one by one. Obviously the sequence of the check has a great impact on the result, since the first order item that is checked gets the available quantity or – in the case of shortage and backorder processing without new distribution – is the first to lose its confirmation. The sequence is determined by sorting the order items according to one ore more characteristics in ascending, descending or free order. A free order for sorting is helpful in cases where no other criterion makes sense, e.g. for customers. The characteristics of the sort profile do not have to match the characteristics of the filter type. Both the filter type and the sort profile have to be generated explicitly before they can be used.

When calling backorder processing with the transaction /SAPAPO/BOP the work-list is calculated using the assigned filter type and sort profile. The filter variant is either defined from there or assigned if variants already exist, figure 7.34. It is possible to display the work-list with the transaction /SAPAPO/BOP_WORKLIST. Table 7.7 lists the transactions for the entities that are shown in figure 7.34.

Table 7.7. Transaction for the backorder processing entities

Entity	Transaction
Filter Type	/SAPAPO/BOPC_FILTER
Sort Profile	/SAPAPO/BOPC_SORT
Special Sorting	/SAPAPO/ORDER
Backorder Processing	/SAPAPO/BOP

● *Parameters for Backorder Processing*
The main parameters for backorder processing are:
- New distribution: By selecting 'new distribution' all confirmations for the selected orders are released, so that the according receipts are all available for the following ATP checks. If 'new distribution' is not selected, only the current order (the one which is currently checked) releases its confirmation.
- Check of confirmation or request: This setting determines whether the check is done for the originally requested or for the confirmed date and quantity. Checking the request makes sense if a change towards the request is regarded as an improvement, whereas checking of the confirmations is rather used when keeping changes in the confirmations as less as possible is preferred, e.g. because the customer has already adjusted his processes to the confirmed date.

- Read check mode & allocation procedure from product master: Setting this flag, the check mode and the allocation procedure are read from the product master instead from the order item. This is especially helpful if allocation procedures have been assigned recently.
- Rule evaluation (again): If rules-based ATP is used, the substitution of product and location in backorder processing can differ from the previous result. Especially if the sales orders have a direct impact on production planning, a new change of the locationproduct might not be desired.
- Cancel confirmation: The idea of this setting is to select a couple of order items to provide their confirmed quantities to other orders. These order items are a subset of the selected order items in the work-list and can be further restricted with own filter types and filter variants. These order items do not take part in the succeeding checks.

To clarify the influence of the settings 'new distribution' and 'check of request or confirmation' figure 7.35 shows an example with four sales orders. Again the arrows represent the requested date and quantity while the rectangles represent the confirmed dates and quantities. In this example two sales orders are confirmed differing from their requests.

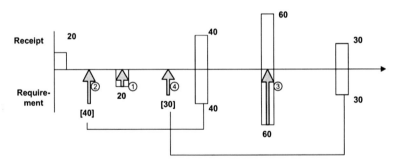

Fig. 7.35. BOP example – situation at the time of order confirmation

Now a change in the availability situation leads to a shortage for one sales order on one hand and allows improving the confirmation – i.e. to confirm closer to the requested date – on the other hand. Figure 7.36 shows this situation where an additional receipt element of 20 units is created (e.g. a manually created production order to utilise a free capacity slot) and an existing receipt element about 40 units is delayed (e.g. due to the scheduling of a prioritised order).

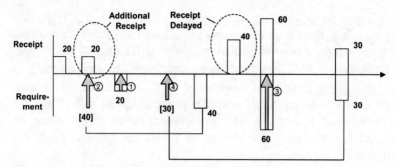

Fig. 7.36. BOP example – situation after changed availability

Starting from this unbalanced situation figures 7.37 to 7.40 visualise the impact of the two parameters 'new distribution' and 'check of request or confirmation'. The orders are checked one by one in the sequence order 2, order 3, order 1 and order 4.

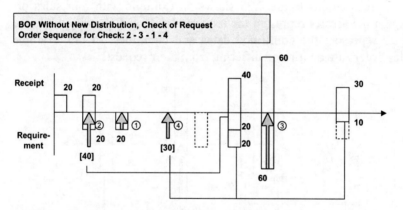

Fig. 7.37. BOP without new distribution and check of request

Order 2, which is checked first, is partially confirmed to its requested date with 20 units using the free additional quantity. Since the receipt which was originally used for its confirmation is delayed, the second schedule line of order 2 is confirmed with an even bigger difference to its requested date. Orders 3 and 1 are already confirmed at their requests and – since their receipts have not changed – receive no change in their confirmation. Order 4, which was confirmed far from its requested date, can now use the open quantity of 20 units of the receipt element that order 2 does not need any more to get a partial confirmation closer to its requested date.

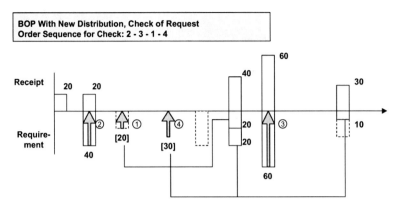

Fig. 7.38. BOP with new distribution and check of request

Using the setting 'new distribution', the same situation leads to quite different results. All receipt elements are now available for the ATP checks, so that order 2 gets confirmed according to its requests. For order 3 nothing changes, but order 1 is now confirmed differing from its request with the delayed order, since the stock is already used for order 2. Order 3 finally uses the remaining 20 units of the delayed receipt element to improve its confirmation like in the example before. Now backorder processing is carried out with check of confirmation. Without new distribution the only change is that order 2 gets another schedule line with a confirmation date even farther from its request date, figure 7.39.

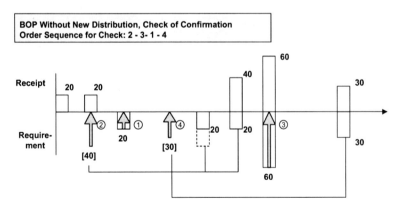

Fig. 7.39. BOP without new distribution and check of confirmation

Using 'new distribution' the confirmation situation for order 1 gets worse, since order 2 is the first to use the stock and the first receipt element for its confirmation.

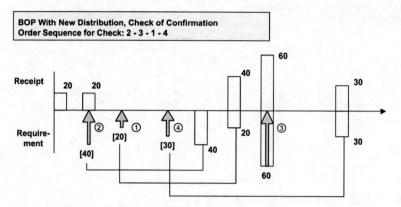

Fig. 7.40. BOP with new distribution and check of confirmation

If just a subset of orders is selected, both the not selected orders and the receipt elements for their confirmation are excluded from the check.

• *BOP and CTP*
It is not possible to trigger production via CTP in backorder processing. It is however possible to include products which are planned with CTP into the BOP and check these without creating new planned orders.

• *Check Level*
As a default the BOP is performed on item level, i.e. a list of all items is created and then the check is performed item by item, i.e. all schedule lines of an item are checked before the next item. In the case of scheduling agreements this implies that the whole scheduling agreement with all its future schedule lines is checked before the next schedule line. The implication of this is that in the case of a shortage some orders (i.e. schedule lines) for the near future will not be confirmed any more, though the shortage might be over long before the future schedule lines of the first scheduling agreement are due. To avoid this behaviour it is possible to configure the BOP for a check on schedule line level per ATP category with the customising path *APO → Global ATP → Tools → Define Check Level*.

• *BOP for Stock Transfer Orders*
It is possible to include stock transfer orders into the BOP. Up to SAP ERP™ 2005 the stock transfer order contains less information than the sales order, e.g. only the confirmed date and quantity is stored, not the requested date and quantity. This implies that the BOP is not able to check

the requested date and quantity for stock transfer orders. With SAP ERP™ 2005 and SAP APO™ 5.0 these limitations do not apply any more, see sections 7.2.2 and 7.4.

• *Integration to SAP ERP™*
The complete backorder process contains the steps:
- backorder processing in SAP APO™,
- transfer confirmation to SAP ERP™ and
- update order in SAP APO™,
as shown in figure 7.41.

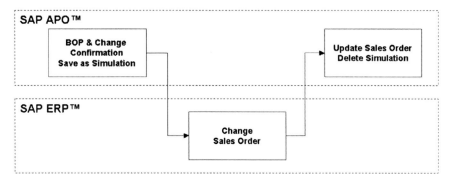

Fig. 7.41. Backorder processing - integration to SAP ERP™

• *Result Display*
It is possible to display the results of the BOP planning run with the transaction /SAPAPO/BOP_RESULT. To achieve an overview about the results it is possible to search for multiple criteria, e.g. for all orders that have received a worse confirmation situation.

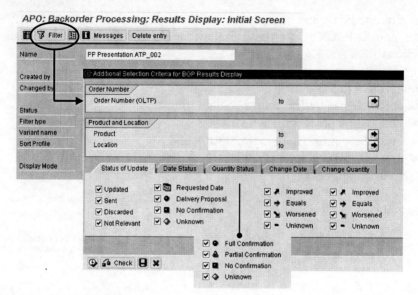

Fig. 7.42. BOP result display

• *Interactive Backorder Processing*

Interactive backorder processing (transaction /SAPAPO/BOPI) allows changing the confirmations of order items interactively. It is even possible to create overconfirmations, though the system gives a warning message for this. The only restriction is that it is not possible to exceed the requested quantity.

Interactive backorder processing is always carried out for one location-product at a time. Though it is possible to access interactive backorder processing using the work-list, interactive backorder processing branches to the locationproduct of the selected order item.

For interactive backorder processing the business event can be defined in the location master (general view) or entered explicitly for individual access. This way it is possible to change the check scope.

• *Performance*

With SAP APO™ 5.0 it is possible to parallelise the BOP check. Nevertheless, backorder processing for mass data should be carefully applied and sufficiently tested within the designed scenario.

• *Order Due List*

The order due list (ODL) is an alternative approach to deal with backorders. The basic idea is that only a few sales orders of high importance are included to this list manually at the beginning of the day and the change in the availability during the day – e.g. receipts from production or distribution – can be monitored and assigned. The default priority is 500.

Triggered by events like goods receipt an assignment of the available quantity to the orders in the ODL list is performed (EDQA). For additional information see Dickersbach 2007.

8 Transportation Planning

8.1 Transportation Planning Overview

Transportation planning is the second step for order fulfilment from a planning point of view. It is either executed in batch mode several times per day after creating the deliveries or in an interactive mode by the transportation planners. Both delivery creation and transport planning are usually responsibilities of the warehouse as explained in chapter 6, though there might be dedicated transportation planners for the latter task. The first step towards execution is the creation of the delivery. At this point in time an ATP check is carried out again – usually with a much more restricted scope (i.e. stock only). Optionally deliveries (and other documents as stock transfer orders and returns) might be grouped to a shipment to make the execution of the shipping easier.

The objective of transportation planning is to group the deliveries into shipments. The challenge in creating the shipments is to minimise the effort – i.e. the number of the shipments and the length of the shipments – while taking

- the due dates,
- the calendars of the customers (for loading and unloading),
- the capacity restriction of the vehicles (i.e. how much can be loaded into a vehicle),
- the vehicle availability (i.e. if there are not enough vehicles available) and
- incompatibilities (e.g. of the goods or locations)

into consideration. To solve this problem TP/VS offers an optimisation tool. The result of the transportation planning is the creation of a shipment in SAP APO™. The shipment is a document which has a link to the included documents but does not replace them. Accordingly a shipment is not relevant for requirements planning and does not show up in the product view.

Subsequent process steps are the selection of the carrier and the release of the shipment. The shipment is only transferred to SAP ERP™ after the release in SAP APO™. With the transfer a shipment is created in SAP ERP™. Figure 8.1 visualises the document flow.

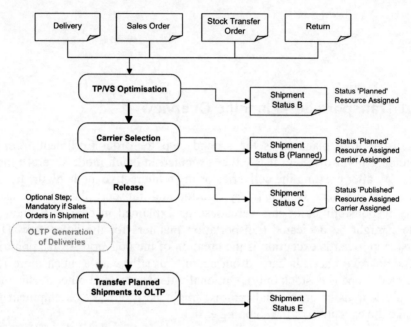

Fig. 8.1. Order life cycle for transportation planning

The common process flow is to create deliveries in SAP ERP™ first before running TP/VS. Alternatively it is also possible to trigger the delivery creation from TP/VS, i.e. to plan in TP/VS on sales order basis. The ATP check for deliveries has usually a more restricted scope than the sales order and therefore the delivery might not get the full confirmation. If shipments are created in SAP APO™ based on sales orders, they have to be corrected in that case. Depending on the probability of not getting the full confirmation for the deliveries, transportation planning based on sales orders might not be advisable.

It is important to note that TP/VS is designed for the transportation planning of a producing or trading company and not for a transport service provider since TP/VS does not cover some of their common functional requirements but do require the master data (products, locations and resources).

• *Input Documents for Shipments*
Planning in TP/VS is possible for orders which contain a start location
(LOCFROM) and a destination location (LOCTO) and are relevant for requirements planning. Though it is possible in SAP APO™ to combine inbound and outbound orders in one shipment, this is not supported by SAP
ERP™. Therefore it should not be done by SAP APO™ either. Outbound
documents are sales orders, stock transfer orders and returns, inbound orders are purchase orders. Triangular business is not supported.

Planning in TP/VS is usually performed on basis of deliveries, but it is
possible to plan for sales orders as well. If TP/VS plans for sales orders,
planning is performed either on basis of sales orders, sales order items or
schedule line. Which of these is used depends on the consolidation level
which is a setting on client level and is maintained with the customising
path *APO → TP/VS → Basic Settings → Basic Settings for Vehicle Scheduling.*
This setting is only relevant if TP/VS is performed on sales order level.

8.2 Master Data and Configuration

8.2.1 Master Data for TP/VS

The main master data for TP/VS reflect the geographical condition described by the locations – the own plants resp. DCs, the customers and
transportation zones – and the transportation conditions described by the
transportation lanes, means of transport, vehicle resources and the transport service providers. Figure 8.2 provides an overview about the settings.

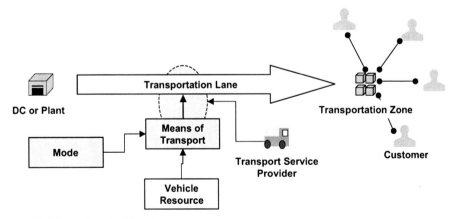

Fig. 8.2. Master data for TP/VS

The transportation zones and the transport service providers resp. carriers are locations of the type 1005 (transportation zone) resp. 1020 (transport service provider). Both are transferred from SAP ERP™ – the transportation zone implicitly with the customer and the transport service provider using vendors in SAP ERP™. To transfer the vendor as a transport service provider, it is necessary to assign the account group in the table CIFVENTYPE in SAP ERP™.

The means of transport is a customising setting. The products do not need to exist for customer locations or for the transportation zones. In order to have consistent documents in TP/VS the customer has to be transferred before the transactional data.

● *Transportation Zone*
Transfer of the transportation zone is done implicitly using the assignment from the customer. The prerequisite for this is the existence of hierarchy structure and the hierarchies for TP/VS in the SAP APO™ customising.

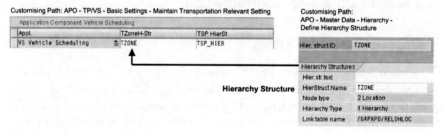

Fig. 8.3. Hierarchy structure for TP/VS

The hierarchy itself is displayed with transaction /SAPAPO/SCCRELSHOW, figure 8.4. The hierarchy has to be assigned to the active model 000.

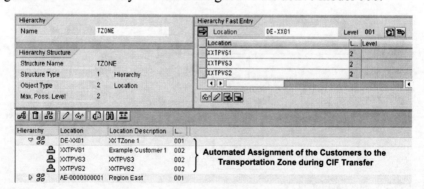

Fig. 8.4. Hierarchy for TP/VS

• *Transportation Lane*

Transportation lanes are required from plant resp. DC to the transportation zone and have to be created manually. The allowed carriers (transport service providers) have to be assigned per means of transport and transportation lane explicitly.

The restrictions of the validity of a transportation lane per products is ignored by TP/VS. The only possible restriction is using compatibilities.

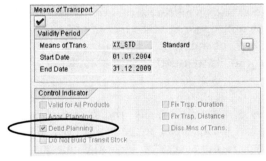

Mns Trans.	Mns/Transp	Start date	End Date	All Prods	Aggr. Plng	Detld Plng	Trsp. Cal.	Fix Duratn	Trsp. Dur.	Ret.Period	Fix Dist.
XX_STD	Standard	01.01.2004	31.12.2009	☐	☐	✔		☐	120:43		☐

TSP	Text	Mns Trans.	Start date	End Date	Trsp.Costs	Unit	TSP Costs	Priority	Share	Min. Numbr	Max. Numbr
XXCAR1	Carrier1	XX_STD	01.01.2004	31.12.2009	0,00		1,000	1	80,0	0	0
XXCAR2	Carrier 2	XX_STD	01.01.2004	31.12.2009	0,00		5,000	2	10,0	0	0

Fig. 8.5. Carrier assignment to the transportation lane

Another prerequisite for TP/VS is that the flag for detailed planning is set in the transportation lane, figure 8.6 (aggregated planning is used for SNP).

Means of Transport ✔

Validity Period
Means of Trans.   XX_STD   Standard
Start Date   01.01.2004
End Date   31.12.2009

Control Indicator
☐ Valid for All Products   ☐ Fix Trsp. Duration
☐ Aggr. Planning   ☐ Fix Trsp. Distance
✔ Detld Planning   ☐ Disc Mns of Trans.
☐ Do Not Build Transit Stock

Fig. 8.6. Detailed planning in the transportation lane

The end date of the validity should be less than 31.12.9999 due to possible time zone shifts which would lead to a date overflow. Note that the transport calendar of the transportation lane is not used (but of the vehicle resource).

• *Vehicle Modelling*

For the modelling of the vehicles the three entities mode, means of transport and vehicle resource are available. Figure 8.7 provides an overview about the information within the different entities.

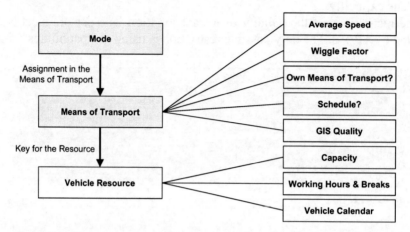

Fig. 8.7. Entities for vehicle modelling

The mode is maintained with the customising path *APO → TP/VS → Basic Settings → Maintain Transportation Mode* and is used only for grouping purposes. The means of transport is maintained with the customising path *APO → TP/VS → Basic Settings → Maintain Means of Transport* and contains the information shown in figure 8.7. The means of transport should correspond either to the type of transport vehicle (e.g. one for 20 t trucks, one for 40 t trucks etc.) or to the transport service provider.

• *Vehicle Resource*

For TP/VS resources of the type 'vehicle' and the category T are required. As a location the depot location can be maintained but this is – unlike for the other resource types – not necessary. Instead the means of transport has to be maintained as a key field. We recommend not to use free breaks. The header dimension is the alphabetically first one (usually AAAADL for pallets). Though up to 8 dimensions are supported, 3 dimensions are usual.

If the resource is created for a means of transport which contains the flag for an 'own means of transport' then there has to be either a transportation lane back from the transport zone to the plant resp. DC or the resource will only be loaded once per planning run.

Finite planning in TP/VS has the focus rather on the consideration of the capacity dimensions – e.g. that a resource with the capacity of 20 t is not

loaded with 22 t – than on the limitation by few vehicles. The consequence
is that enough vehicle resources have to be created.

If infinite planning is used (i.e. resources without load restriction) there
is a danger that too many activities are loaded onto them. A possible modelling for this is to have e.g. seven resources (one per day) and to load
them periodically.

● *Schedules*
Schedules represent vehicles that have a regular schedule – e.g. a ferry or a
regular truck. For the maintenance of a schedule first an itinerary has to be
created with transaction SAPAPO/TTW1 which lists the sequence of the locations. The locations have to be connected by transportation lanes.

Except for the schedule itself (transaction /SAPAPO/TTC1) a validity period has to be maintained with transaction /SAPAPO/TTV1. Figure 8.8 provides an overview about the required settings. Breaks for schedule vehicles
are ignored and might lead to errors.

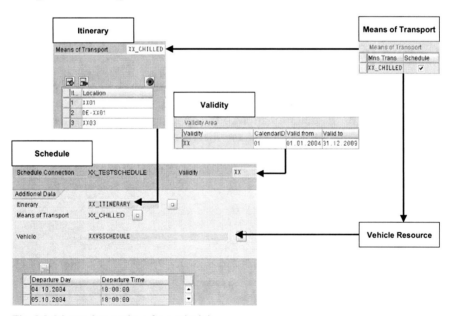

Fig. 8.8. Master data settings for a schedule

● *Hubs and Location Hierarchies*
For the optimiser there is the limitation that unloading is allowed maximum two times within a route. Parallel hubs should only be used after consultation with SAP.

• *Transport Groups*

Transport groups have to be customised in SAP APO™ analogous to SAP ERP™ with the customising path *APO → TP/VS → Basic Settings → Maintain Transport Group*. There is no automated synchronisation with SAP ERP™.

• *Number Ranges*

The internal number range for shipments in SAP APO™ must be the same as the external number range in SAP ERP™.

• *Shipment Simplification*

The normal functionality of TP/VS is that if a shipment is manually created or changed in SAP ERP™, the shipments and the orders are excluded from planning in SAP APO™ and from live cache. The table /SAPAPO/VS_E_DLV is used for this exclusion, where the shipment and the delivery numbers are used as keys. For the shipment simplification the deliveries of the published shipments are excluded from planning via the user exit APOCF029.

8.2.2 Geo-Coding

The calculation of the transport durations depends on the accuracy of the geo-coding of the locations and on the accuracy of the distance between the locations. For the determination of the geographical settings (longitude and latitude) of the locations and the distance of the transportation lanes SAP offers different options.

TA SZGEOCD_GEOCD2CLS

Mapping Table: Geocoder ID for ABAP OO Class

Source	Class	Function Module	Destination
5GLD	CL_GEOCODER_ZIP5GOLD		
CASO	CL_GEOCODER_CAS		
IGS	CL_GEOCODER_IGS	GEOCODE	GEO_IGS_RFC_DEST
RFCG	CL_GEOCODER_GENERIC..	RFC_GEOCODE	RFCGEOCODER_TEST_PTV
SAPO	CL_GEOCODER_SAPO		

TA SZGEOCD_GEOCODERS

Geo-Coder Maintenance

C..	Sequence	So..	Excl.
DE	5 000	SAPO	☐
DE	1 000	RFCG	☑

TA SZGEOCD_GEOCDRLFLD

Address Fields Relevant for a Geo-Coder

Source	Field
IGS	CITY1
IGS	COUNTRY
IGS	HOUSE_NUM1
IGS	POST_CODE1
IGS	REGION
IGS	STREET
RFCG	CITY1
RFCG	COUNTRY
RFCG	POST_CODE1
SAPO	COUNTRY
SAPO	REGION

Fig. 8.9. Customising for the determination of the geo-coder

The determination of the geographical settings of the locations is performed either based on country and region (the standard setting), based on postal code (requires geo-coding software which is included in the standard licence) or based on the address (requires geo-coding software and is not included in the standard licence). Figure 8.9 shows the customising settings for the geo-coding of the locations.

The determination of the distance for the transportation lanes is performed either as the air-line distance (standard) or as the actual distance between addresses using a route planning. The latter requires geo-coding software as well (which is linked via the IGS interface) and the exact longitude and latitude of the locations as input. The geo-coding software is not included in the standard licence. Figure 8.10 visualises the recommended combinations – note that it is not recommended to perform scheduling based on country and region.

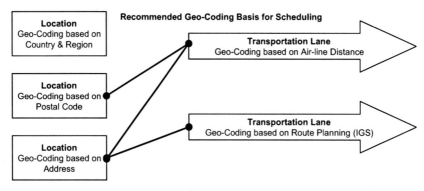

Fig. 8.10. Geo-coding combinations for scheduling

For the calculation of the geo-coding relevant data following transactions are helpful: The transportation zone coordinates are calculated with the transaction /SAPAPO/LOCTZCALC. The mass generation of GIS distances is possible with the transaction /SAPAPO/TR_IGS_BPSEL. Alternatively the duration of the transportation lanes is calculated within the supply chain engineer (transaction /SAPAPO/SCC07) with a right mouse click on the selected transportation lanes.

For an automated adjustment of the basic country settings between SAP APO™ and SAP ERP™ it is possible to choose the customising path *General Settings → Set Countries → Define Countries in mySAP.com Systems* in SAP APO™ and perform the adjustment with the menu path *Utilities → Adjustment* per RFC-connection. The address settings have to be synchronised in a similar way.

8.2.3 Configuration of the CIF

The publishing types for deliveries (type 340) and for shipments (type 330) have to be maintained on SAP APO™ side with the transaction /SAPAPO/CP1 (see chapter 25). These publishing types are maintained for the source location.

8.3 TP/VS Planning Board

The central tool for TP/VS planning is the TP/VS planning board that is called with transaction /SAPAPO/VS01 as shown in figure 8.11.

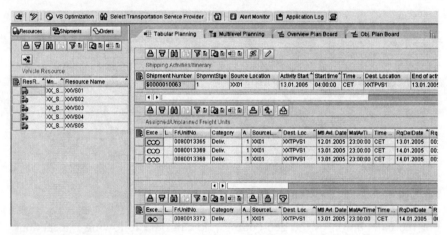

Fig. 8.11. TP/VS planning board

On the right side in the top row the shipments (as a result of the TP/VS planning) are displayed, below the assigned deliveries resp. orders and in the bottom line the unassigned deliveries resp. orders. With the navigation on the left side it is possible to select the shipments and the assigned freight units via resources, shipments or orders.

When calling the planning board an optimisation profile has to be entered. This optimisation profile contains restrictions regarding the resources, locations, compatibilities or order types (ATP categories) as shown in figure 8.12.

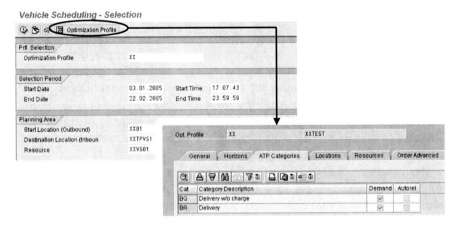

Fig. 8.12. Selection or order types for TP/VS planning board

● *Multi-level Planning*
Within the planning board it is possible to perform an interactive planning
of shipments. There is a consistency check when saving the shipments
(e.g. all relevant stages are assigned).

● *Heuristics*
Additionally to the manual planning of individual orders it is possible to
create and to use heuristics in the 'multi-level planning'-view of the planning board, figure 8.13.

Fig. 8.13. Heuristics in the TP/VS planning board

The prerequisites are that the heuristics are activated for TP/VS with the
customising path *APO → TP/VS → Interactive Planning → Activate Heuristics
Interface* and that the heuristic is maintained in the BAdI with the customising path *APO → TP/VS → BAdIs for TP/VS → BAdI: Heuristics Interfaces*
(definition /SAPAPO/VS_HEURISTIC).

8.4 TP/VS Optimisation

The task of the optimiser is to create a plan (i.e. shipments) which considers the constraints and produces a plan which is optimal in the sense of low penalty costs. Hard constraints for the optimiser are compatibilities, opening hours (modelled by the handling resource) and the finiteness (as in the resource master). Soft constraints are defined in the optimiser profile (e.g. earliness and lateness). However, using the pick-up and delivery window these might become hard constraints.

The procedure for the TP/VS optimisation is to begin with some start solutions and successively load further orders. After a while a few solutions out of these are taken into account for further processing and the final result is the best solution of the whole cycle. Both genetic algorithms and evolutionary algorithms are used – the TP/VS optimiser is a mixture of local search and evolutionary search.

The complexity of the problem is driven by number of means of transport, hubs, handling resources and orders. The TP/VS optimiser is called either from the planning board or with transaction /SAPAPO/VS05.

• *Optimiser Profile*
For the configuration of the optimiser the optimiser profile has to be created with transaction /SAPAPO/VS021. The optimiser profile contains e.g. the information about the horizons, the locations (source and target), the vehicle resources, the ATP categories and the cost profile.

Two horizons are relevant for the optimiser profile: the planning horizon controls the dates for the shipments to be created and the demand horizon the horizon for the deliveries resp. sales orders which are taken into account. The demand horizon should start in the past.

The parameters on the 'additional' tab are not used.

• *Cost Profile*
The cost profile is maintained with transaction /SAPAPO/CTRP. The maintenance of the cost profile is a little bit unusual because it is accessed by the model (i.e. for any operational relevance 000), where first the cost profile has to be defined and subsequently on each tab the cost profile and its entries have to be maintained, figure 8.14.

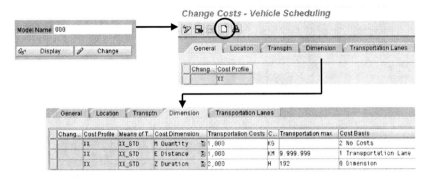

Fig. 8.14. Cost profile

On the tabs always the tuples for the cost profile and the entries have to be maintained. The costs which are maintained on the different tabs are listed in table 8.1.

Table 8.1. Costs within the cost profile

Cost Profile Tab	Cost
Location	Costs of delayed and premature delivery and pick per day.
Transportation	Costs per means of transport. This cost represents the cost per vehicle in a planning run (one time cost - not per day, not per tour) and can be used to diminish the number of vehicles to be used.
Dimension	Costs are maintained for up to four dimensions: • Distance per UoM. If a maximum distance is maintained, unit must be maintained (else error during pre-processing). • Quantity per UoM and km. The UoM has to be the LTL unit. • Duration. • Stop Offs.
Transportation Lane	Transportation cost per UoM (e.g. kg). Cost for means of transport per km. The tab 'transportation lane' is just a view on the same fields as maintained in the respective transportation lanes.

• *Conditions*

Conditions represent a restriction by one or more fields from the field catalogue for TP/VS and are used for the selection of the planning board, the compatibilities of delivery items and the pickup/delivery window for customers.

The conditions for the compatibilities are maintained from the 'order advanced'-view of the optimisation profile as shown in figure 8.15.

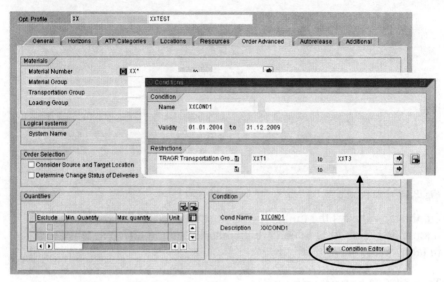

Fig. 8.15. Condition maintenance for compatibilities

- *Compatibilities*

Up to SAP APO™ 4.0 compatibilities could only be modelled using the transport group. From SAP APO™ 4.1 on a new and more flexible concept for the compatibilities exists. The idea is that compatibilities or incompatibilities may exist between any kinds of fields – not only transport groups. The relevant fields have to be included into the field catalogue and those two fields which are relevant for compatibility are selected and saved as a compatibility type with the customising path *APO → TP/VS → Optimiser → Define Compatibility Type.*

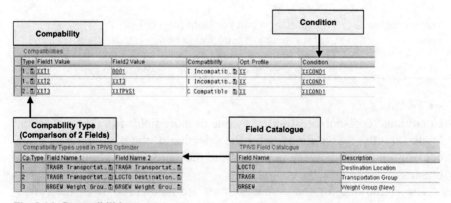

Fig. 8.16. Compatibilities

The compatibilities themselves (i.e. the values of the fields which are compatible or incompatible) are maintained with transaction /SAPAPO/VS12 and are related to the optimisation run by the condition as shown in figure 8.16.

• *Pick-up and Delivery Window*
With transaction /SAPAPO/VS11 it is possible to define the constraints for the optimiser regarding early and late pick-up resp. delivery. These constraints can be restricted by the conditions. These entries are created either manually or with the report /SAPAPO/VS_UPGRADE_SCM_41.

8.5 Scheduling

Depending on the precision of the information (GIS, geo-coding,..) that is used in SAP APO™ the quality of the transport distance differs.

The distance of the runtime lanes is either based on the GIS information or is calculated using the geo-coding distance and the wiggle factor from the means of transport, figure 8.17.

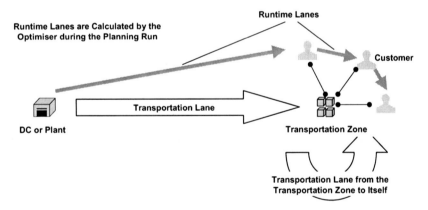

Fig. 8.17. Scheduling with runtime lanes

If the runtime lanes are based on the geo-coding information, the distances are calculated offline once resp. periodically and stored in the table /SAPAPO/TRDIDUPS. The calculation is triggered with transaction /SAPAPO/TR_IGS_BPSEL and might lead to very many entries. Therefore it is important to find an appropriate restriction for the calculation of the lanes. As a rule of thumb the transportation lanes within a transportation zone and their neighbours should be calculated.

If the more detailed distance calculation with geo-coding is used, the means of transport has to have 'GIS quality' flagged and average speed for city (low), country road (medium) and motorway (high) have to be maintained.

Using the geo-distance might provide significantly better results than using the airline distance from the transportation lanes because the real route is considered which might lead to a combination of transports e.g. if the same motorway is used to a big extent.

• *Transportation and Shipment Scheduling*
The transportation and shipment scheduling is analogous to ATP (cf. section 7.8) but only load and unload time is considered, not plan and pick time and the rule strategy must be SCHEDL.

• *Time Window at the Customer*
Opening hours at the customer have to be modelled using the handling resource. This is a manual process that requires assigning the opening hours maintained in the customer master in SAP ERP™, to create a handling resource in SAP APO™ and to assign it to the customer. A BAPI exists for this.

8.6 Carrier Selection

The carrier selection is performed using after the planning of the shipments is finished and before the shipments are created in SAP ERP™. The carrier selection is performed according to the criteria
  • priority (from the transport service provider view of the carrier in the mode of the transportation lane),
  • costs or
  • allocation (business share per transportation line, not in total).
If one criterion is not maintained for all carriers, the next criterion is used. The carrier selection is performed either automatically using a carrier selection profile (transaction /SAPAPO/CSPRF) or manually. The carrier selection returns with all possible carriers listed in an order. This way it is easy to take the next carrier if one is not available resp. declines in the collaboration stage.

If a route contains several stages, only carriers are selected which are assigned to all lanes. For allocations the lane from start to end is relevant.

● *Allocations*
Note that the quota resp. allocation is per lane, i.e. it is not possible to model a business share with this functionality. The default planning area is 9ATPVS (the characteristic LOCNO represents the carrier). The MaxNr. in the transportation lane must not be initial

● *Continuous Move*
If one stage of a transport is already assigned to a carrier who has the flag for continuous move, the same carrier is selected for stages which are preceding or succeeding that stage. In this case there is no check for priority or costs, the only exception are allocation limits.

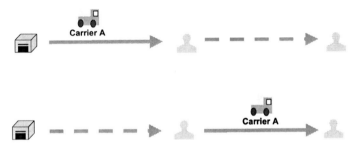

Fig. 8.18. Continuous move

The flag for continuous move is set in the location master of the carrier in the 'TSP'-view.

8.7 Collaboration

It is possible to communicate the result of the carrier selection in a collaboration scenario. In this case the requests for the shipments are sent to the transport service provider via internet or IDOC.

The prerequisite is the maintenance of collaboration partners with the transaction /SAPAPO/CLP_SETTINGS analogous to the configuration of collaborative forecasting (see chapter 4.7), the definition of EDI-partners (partner type CR, transaction WE20) and of the IDOC port (transaction WE21).

The requests for the shipments can be accepted or rejected by the transport service provider as shown in figure 8.19.

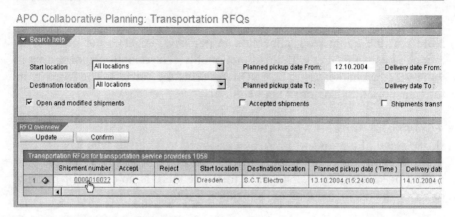

Fig. 7.19. Collaborative carrier selection

8.8 Release and Transfer to SAP ERP™

The shipments are transferred to SAP ERP™ either in an interactive mode from the planning board with the menu path *Goto → Transfer Planned Shipments to OLTP* or in background with the transaction /SAPAPO/VS551. The status of the transfer is checked with the transaction /SAPAPO/VS60.

8.9 Dynamic Route Determination

Dynamic route determination uses pre-defined routes in order to provide a more precise transportation and shipment scheduling during the ATP check of sales orders (or quotations). This way schedules and resource capacities are already considered during the ATP check, and a shipment is created. This shipment is likely to remain unchanged only in case of full truck loads, else TP/VS might change it during optimisation. It is possible to display different transportation alternatives during the ATP check, sort them by costs or punctuality and select one interactively. The routing guide is also used for interactive scheduling in the TP/VS planning board.

Part IV – Distribution

9 Distribution and Supply Chain Planning Overview

9.1 Distribution and Supply Chain Planning Scenarios

In many industries – e.g. consumer goods or chemicals – the same products are stored in different warehouses to reduce delivery times and optimise transports. In this case there is a supply network on finished product level and planning is required to supply the warehouses and to determine the appropriate safety stock levels.

Distribution planning between the plants and warehouses gains increasing importance since many companies change their processes from non coordinated local inventory management to a global inventory management in order to reduce their inventories. By concentrating safety stocks from multiple local DCs to a central DC and changing the responsibility for the inventories at the local warehouses combined with service level agreements and a VMI (vendor managed inventory) process for the local warehouses, significant stock reductions are achieved. The focus for distribution and supply chain planning is on make-to-stock production.

• Tasks Within Distribution and Supply Chain Planning
The supply planning for the warehouses has a more requirements planning oriented part, where the focus lies to propagate the demands of the supply network to the procuring plant in order to trigger the production or the external procurement, and a more execution oriented part, where the focus is to make the best of the given supply and demand situation – which often differs more or less from the initially planned situation. The latter is accordingly a short term task.

Regarding the requirements planning the main differentiator is whether a finite, i.e. a feasible – plan is required or whether a propagation of the demands to the procuring locations is sufficient. In a single sourcing network with a make to stock process where production capacity is usually not the bottleneck – that is production can be adjusted within reasonable time to meet the demands – a simple propagation of the demands to the

J. T. Dickersbach, *Supply Chain Management with APO*,
DOI: 10.1007/978-3-540-92942-0_9, © Springer-Verlag Berlin Heidelberg 2009

procuring location is sufficient. This includes a netting of the respective inventories and the consideration of transportation times and lot sizes.

If however the production capacity or the capacity of an external supplier is in many cases a bottleneck which affects the available quantities at the target locations and/or sourcing alternatives exists within the network, production capacities have to be considered at least at a rough cut level. In this case, where an integrated distribution and production planning is performed, we are talking about supply chain planning.

Figure 9.1 gives an overview about the planning tasks. The supply chain structure in this figure is not generic, because independent demand (i.e. forecast or sales orders) could be relevant for any DC or any plant, and suppliers might as well be a bottleneck.

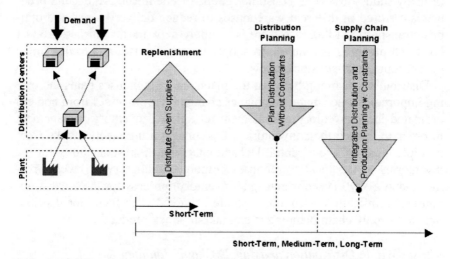

Fig. 9.1. Distribution planning, supply chain planning and replenishment

Note that only supply chain planning provides a feedback at the DCs about the feasible quantities. Else after the production planning process a feedback planning step has to be performed, e.g. using the supply distribution functionality of CTM.

There are two processes within replenishment: deployment and transport load building. Deployment is concerned with the fair share of quantities to the requesting parties in cases of shortage or surplus. The only constraints are the available quantities. Transport load building is one step closer to execution and focuses on the creation of truck loads, where the task is to adjust the planned stock transports to the available trucks/transport means and take their capacity restrictions into account. Though the quantities can be changed interactively during TLB, this is not the main focus of TLB.

The planning for the quantities – both as distribution resp. supply chain planning and deployment – usually lies within the responsibility of the supply chain planning organisation, whereas transport load building is often the responsibility of the warehouse resp. the delivery execution. Figure 9.2 shows a typical process chain.

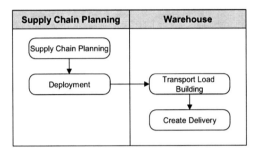

Fig. 9.2. Process chain for distribution and supply chain planning

9.2 Applications for Distribution and Supply Chain Planning

To leverage the advantages of a supply network, the transparency of the current stock and demand situation is a prerequisite. Regardless of ownership and responsibilities, distribution planning is performed based on demand and stock information with the result of planned stock transfers. The most important issues are usually the netting of the local stocks and the consideration of safety stocks, sourcing options, transportation times and lot sizes for the planned stock transfers. Nevertheless it is possible to create stock transfer orders manually as well. If required, it is possible to consider restrictions regarding storage capacity and even handling capacity for goods issue and goods receipt too. In SAP APO™ there are four applications that support distribution and production planning:

- the SNP heuristic,
- the SNP optimiser,
- CTM (with SNP or PP/DS master data),
- the PP/DS heuristics.

Though PP/DS is also capable to create stock transfer orders, the application designed for distribution planning is SNP. If not explicitly mentioned, the following descriptions relate to the SNP model.

Only the SNP optimiser and CTM are suited to realise the benefits of integrated distribution and production planning. The SNP heuristic is an infinite, level by level planning procedure, and though PP/DS is theoretically able to create feasible plans considering capacity restrictions in one step, it is strongly recommended not to use it in any complex environment – especially not across locations (see chapter 13.1). Figure 9.3 visualises the alternative applications in SAP APO™ and their respective scope.

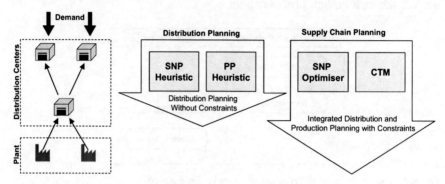

Fig. 9.3. Applications for distribution and supply chain planning

Note that the applications for distribution planning can not be used for supply chain planning and that the applications for supply chain planning are not suited for distribution planning without production planning (if they are applied in a straightforward way – there are workarounds though). The main features of these applications are listed in table 9.1:

Table 9.1. Features of the applications for distribution

	SNP Heuristic PP Heuristic	SNP Optimiser	CTM
Feasible Plan	No[1]	Yes	Yes
Dynamic Sourcing	No	Yes	Yes
Distribution Plng. w/o Prod. Plng.	Yes	No	No
Shortage Distribution	n.a.	by Cost	by Priority
Transport Capacity	No (Infinite)	Yes	No
Storage Capacity	No (Infinite)	Yes	No

[1] Requires a subsequent process - only via deployment in short term (or by raping deployment) or supply distribution

Distribution planning with the SNP Heuristic is often used in environments where multiple sourcing is not an issue and production is usually able to meet the demands, so that the main task is to calculate the demands for production planning taking local inventories, transportation times, safety stocks and lot sizes into account. SNP Optimisation allows a complete consideration of the supply chain determinants – e.g. multiple sourcing and costs and capacities for production, transport, storage and handling. The aim of the SNP optimiser is to find a global optimum for the supply chain based on costs and penalty costs. CTM pursues a priority based approach with (first come first serve).

• Feasible Distribution Plans
The SNP heuristic creates planned distributions without checking their feasibility. Since the distribution demand element at the source location is a distribution supply element at the target location, a possible shortage at the source location is not distributed to the target location. If a feedback is desired whether it is possible to meet the demand in the target location (e.g. for the availability check), there are basically three possibilities to create a feasible plan. These are CTM, the SNP optimiser or to use a multi-step approach (e.g. SNP heuristic and CTM supply distribution).

• Lot Sizes
For the calculation of the lot size usually the lot size of the product of the target location is used. A more specific way to define the lot size is in the transportation lane per product and means of transport. The distribution planning applications as SNP Heuristics, Optimiser, CTM etc. use lot sizes to increase the order quantities. Deployment on the other hand decreases the order quantities to the lot size.

Lot sizes are also used for the selection of a transportation lane, e.g. that a transportation lane is only valid for lot sizes between 20 and 500. If the order lot size is above 500, a different transportation lane has to be used. To use the lot size restriction for the transportation lane in the SNP optimisation, it is necessary to choose the options for discretisation in the optimiser profile. A discretisation method has to be selected, the discretisation for 'minimum transport lot size' has to be activated and the flag 'discretisation until end/detailed' resp. an end date for the discretisation has to be set. Differing from the SNP heuristic and the PP/DS heuristic, the lot size restrictions in the 'product'-view of the transportation lane are not used, but the lot size profile in the 'product specific means of transport'-view. The lot size profile is maintained with the transaction /SAPAPO/SNP112.

Like the SNP optimiser, CTM uses the lot size profile of the 'product specific means of transport'-view. These restrictions might incline CTM to cause excess coverage in the target location.

• Comparison of the Applications for Distribution Planning
The four applications in SAP APO™ that are able to plan stock transfers – the SNP heuristic, the SNP optimiser, CTM and the PP/DS heuristic – do

not have the same properties. Some of their main properties are listed in table 9.2.

Table 9.2. Application properties regarding stock transfer planning

Function	SNP Heuristic	SNP Optimiser	CTM	PP/DS Heuristic
Safety Stock	Yes	Yes[1]	Yes	Yes
Lot Size of Product Master	Yes	Rounding[2]	Yes	Yes
Lot Size of Transportation Lane (Selection)	Yes	Yes[1,3]	Yes[3]	Yes
Transport Duration	Yes	Yes	Yes	Yes
Stock Transfer Horizon	Yes	Yes[1]	Yes[4]	No
Selection of Transport Method (Dates vs. Costs)	Yes	Yes	Yes	Yes
Transport Resource	No[5]	Yes	No[5]	Yes
Handling Resource	No[5]	Yes	No[5]	Yes
Storage Resource	No[5]	Yes	No[5]	No[5]
Quota Arrangements	Yes	No[6]	Yes	Yes
Procurement Priority of Transportation Lane	Yes	No[6]	Yes	Yes

[1] setting in optimiser profile
[2] enhancements in note 483910 or using the lot size profile in 'product specific transp. method', note 511782
[3] lot size profile in 'product specific transport method'-view required.
[4] can be overruled in control setting
[5] capacity load is calculated, but no finite planning
[6] sourcing according to total supply chain costs

• *Replenishment with SAP APO™*
For the replenishment process the alternatives are mainly whether to use the deployment heuristic or the deployment optimiser to determine the quantities plus optionally the transport load builder to take transport resource restrictions into account and arrange the stock transfer orders accordingly. The four alternative ways are shown in figure 9.4.

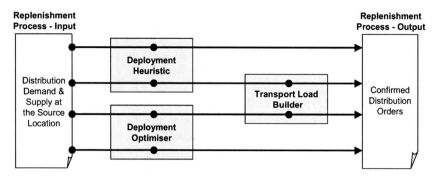

Fig. 9.4. Alternative ways for replenishment

Whereas deployment determines the quantities to be shipped, TLB is concerned with the assignment of those orders to one transport. All the three applications – deployment heuristic, deployment optimiser and TLB – are part of the SNP module.

9.3 Order Cycle for Stock Transfers

The stock transfer order has two aspects – as a demand in the source location and as a supply in the target location. The document types do differ in SAP ERP™ and SAP APO™, and there is more than one possibility for correspondence. The order life cycle in SAP APO™ and SAP ERP™ is shown in figure 9.5:

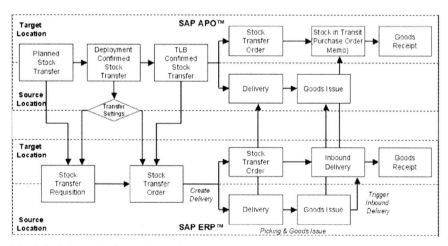

Fig. 9.5. Order cycle for stock transfer

The creation of the planned stock transfer, deployment and transport load building (TLB) is performed in SAP APO™. The execution part from the creation of the outbound delivery in the source location until the goods receipt in the target location is performed in SAP ERP™, and the information is displayed in SAP APO™.

The three order types in SAP APO™ for planned stock transfer, deployment confirmed stock transfer and TLB-confirmed stock transfer are matched to the stock transfer requisition and the stock transfer order in SAP ERP™ according to the settings for the SNP transfer (transaction /SAPAPO/SDP110). None of the SAP APO™ orders have to be transferred to SAP ERP™. If the orders are transferred, for the deployment confirmed stock transfer either a purchase requisition or a purchase order is created in SAP ERP™.

The outbound delivery for the stock transfer order is created with transaction VL10B in SAP ERP™, picking and the posting of the goods issue is done with transaction VL02N in the delivery. Using the message type LAVA the inbound delivery in the source location can be triggered with the posting of the goods issue (or manually with the transaction VL31N). For the use of deliveries it is necessary that a vendor is assigned to the source location and that an info record exists in SAP ERP™ - the configuration has to be the same as for stock transfers across company codes (as shown in figure 9.9). The inbound delivery is transferred to SAP APO™ as a purchase order memo with the significance of the stock in transit.

This order flow is independent whether it is within one company code or across company codes. If the source location and the target location are in different systems, the document flow on the SAP ERP™ side differs however as shown in figure 9.6. The communication between the systems is usually done via ALE.

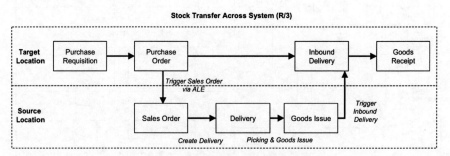

Fig. 9.6. Order flow for stock transfer on the SAP ERP™-side

The stock transfer across different systems affects the order flow in SAP APO™ as well – the planned stock transfer is substituted by sales order and purchase order.

● *ATP Categories for Stock Transfers*
The stock transfer orders that are created in SAP APO™ have different categories depending whether they are planned stock transfers created by SNP (heuristic, optimiser or CTM), deployment confirmed stock transfers created by deployment or TLB confirmed stock transfers created by TLB (transport load builder). Table 9.3 lists these categories.

Table 9.3. Stock transfer order categories

Order Category Source Location	Order Category Target Location	Order Created by Application (without Transfer)
BH	AG	PP/DS
BH	AG	CTM (PP/DS)[1]
BH	AG	SNP
BH	EA	CTM (SNP)[1]
EG	EF	Deployment
BI	BF	TLB

[1] default in CTM customising

For CTM in table 9.3 the default categories are listed. These can be changed with the transaction /SAPAPO/CTMCUST. A motivation to do so might be that deployment requires distribution orders with the categories BH/AG.

9.4 Integration of Stock Transfers to SAP ERP™

The transfer properties for stock transfer orders are defined with the transaction /SAPAPO/SDP110 as listed in table 9.4. Note that the settings for the application SNP affect in-house production as well.

Table 9.4. Order transfer options to SAP ERP™

Application	Transfer to SAP ERP™ as Order Type	Transfer Frequency
SNP	Purchase Requisition	Immediate, Periodic or No Transfer
Deployment	Purchase Requisition or Purchase Order	Immediate, Periodic or No Transfer
TLB	Purchase Order	Immediate, Periodic or No Transfer

The multiple categories and transfer options provide several possibilities to configure the process from a planned stock transfer in SAP APO™ to a purchase order in SAP ERP™. Figure 9.7 illustrates some of these.

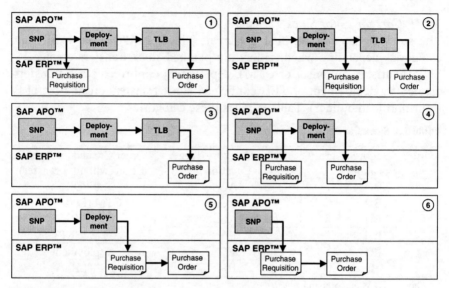

Fig. 9.7. Scenarios for distribution order life cycles

In this figure 'SNP' stands for any of the applications SNP heuristic, SNP optimiser or CTM.

• *Stock Transfer Order*

At this stage in distribution planning the corresponding SAP ERP™ object to the planned stock transfer is the purchase requisition – if any. The order dates of the stock transfer orders in SAP APO™ and the corresponding purchase requisitions resp. purchase orders in SAP ERP™ might differ, since the order is first scheduled in SAP APO™, then transferred to SAP ERP™ and scheduled again in SAP ERP™ according to the delivery date. The scheduled dates in SAP ERP™ are not transferred back to SAP APO™. Stock transfers are executed according to the order dates in SAP ERP™, therefore it is important to keep the scheduling in SAP APO™ and SAP ERP™ consistent. Figure 9.8 provides an overview of the different scheduling procedures in SAP APO™ and in SAP ERP™. In SAP ERP™ the two cases – with and without shipment and transportation scheduling in SAP ERP™ – have to be distinguished.

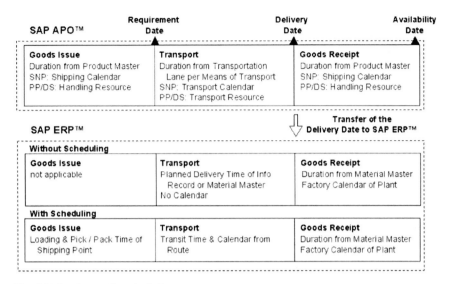

Fig. 9.8. Stock transfer scheduling

Note 420648 gives the recommendations listed in table 9.5 for a consistent modelling.

Table 9.5. Recommendations for consistent modelling of stock transfers

	Goods Issue	**Transport**	**Goods Receipt**
Without Shipment and Transportation Scheduling in SAP ERP™	Do not use goods issue duration	Use planned delivery time in SAP APO™ as well via BAdI[1] Do not use a transport calendar in SAP APO™	Transfer goods receipt duration from SAP ERP™ to SAP APO™ and use identical calendars[2]
With Shipment and Transportation Scheduling in SAP ERP™	Goods issue time = pick / pack time + load time, identical calendars, no rounding in SAP ERP™	Use identical transportation duration & calendars	Transfer goods receipt duration from SAP ERP™ to SAP APO™ and use identical calendars[2]; do not use goods receipt hours in SAP ERP™

[1] note 379006, [2] note 333386

The delivery dates in SAP APO™ are usually at 12:00 and in SAP ERP™ at 00:00. Other causes for inconsistencies between SAP APO™ and SAP

ERP™ as well as between the source and the target plant in SAP ERP™ are explained in the notes 333386 and 76301.

The transport duration in SAP APO™ is calculated using the entry from the transportation lane, whereas SAP ERP™ uses the planned delivery time of the material master of the target location. If a product is procured from different plants, SAP ERP™ is not able to handle the different transport duration. For this case note 441622 describes a correction to substitute the planned delivery time in the SAP ERP™ order with the transport duration from SAP APO™.

• *Stock Transfer Across Company Codes*
Since SAP APO™ does not know any company codes, there is no difference in SAP APO™ whether stock transfers are planned within one company code or across company codes. On the SAP ERP™ side however cross company stock transfers require the additional settings shown in figure 9.9. The assignment of the sales area and the customer to the plant is made with the maintenance view V_001W_IV.

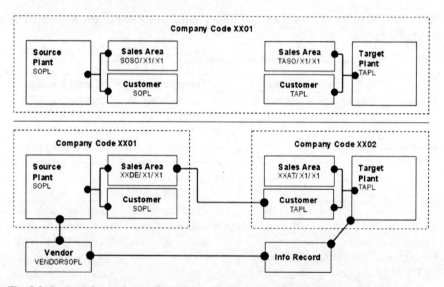

Fig. 9.9. Settings for stock transfer within and across company codes in SAP ERP™

The assignment of the vendor to the source plant is done in the 'partner functions'-view of the vendor in the menu path '*Extras → Additional Purchasing Data*'. The customer representing the target plant is assigned to the sales area of the source plant.

• *Cross System Stock Transfer*

Stock transfer across two SAP ERP™ systems is modelled by a purchase order in the target plant and a sales order in the source plant. The settings for stock transfer across systems is shown in figure 9.10.

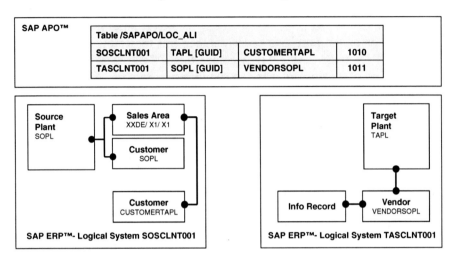

Fig. 9.10. Settings for stock transfer across systems

For the substitution of the source plant by the according vendor at the transfer of the sales order to SAP ERP™ and the substitution of the customer by the target plant at the transfer of the sales order from SAP ERP™ to SAP APO™, the mapping table /SAPAPO/LOC_ALI has to be maintained with the transaction /SAPAPO/LOCALI.

The order flow across the systems after creating a stock transfer order in SAP APO™ is shown in figure 9.11 for a transfer as purchase order after deployment run (see case 4 in figure 9.7).

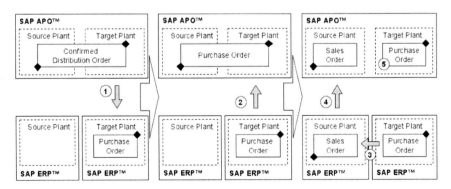

Fig. 9.11. Order flow for stock transfer across systems

The integration of a stock transfer order across two SAP ERP™ systems takes place in five steps:

1. Transfer of the confirmed distribution order as purchase order to the SAP ERP™ system of the receiving plant. The source is substituted from the supplying plant by the vendor according to table /SAPAPO/LOC_ALI.
2. Transfer of the purchase order number and the category type from SAP ERP™ to SAP APO™ (key completion).
3. Creation of a sales order in the supplying system via ALE or manually. The sales order contains the number of the purchase order in the field for the external purchase order number as a reference to the purchase order of the receiving system.
4. Transfer of the sales order from SAP ERP™ to SAP APO™.
5. Deletion of the requirement node of the purchase order in SAP APO™.

The sales order and the purchase order are not linked any more in SAP APO™. Changes in the sales order are transferred to SAP APO™ but do not have any impact on the purchase order (since purchase orders are not allowed to be changed in SAP APO™). The purchase order has to be adjusted in SAP ERP™ (e.g. triggered by an order confirmation). The adjusted purchase order is transferred to SAP APO™ as well.

Critical points in this scenario are the consistency of all three systems after order changes and the restoring of the stock transfers in case of an initial transfer to SAP APO™. These cases as well as the other scenarios for the distribution order life cycles should be checked in the feasibility study for the project.

9.5 SNP Planning Book

The SNP planning book is the preferred tool to display the planning situation in the supply network and to perform interactive distribution planning. The transaction for the SNP planning book (which is similar to DP) is /SAPAPO/SNP94. SNP uses the same planning book structure as DP, though there are some specific settings. In the planning object structure the flag for SNP planning has to be set, which selects the standard characteristics for location, product, PPM, activity, resource and transport lane.

In the planning area the key figure details contain entries regarding the key figure semantics, the key figure functions, the category group and the category, figure 9.12.

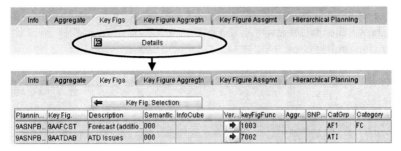

Fig. 9.12. Key figure settings for SNP

The key figure semantics define whether the key figure contains time series data or orders (the default is time series, semantics for orders start with LC in their description). The entry for the category group defines the ATP categories which are read from live cache into the key figure and the entry for the category defines the category of the order which is created in live cache from an entry in the key figure. The key figure functions contain additional coding and might overrule the previous settings.

In SNP all key figures contain live cache orders. For the key figures with an order semantic it is possible to define the according category or category group. Though generally key figures with a time series semantic can be used as well, an eye should be kept on these during testing.

The planning functionality for SNP (supply network planning, capacity planning, transport load builder and deployment) is selected in the planning book. The standard planning book for SNP is 9ASNP94, which contains the necessary macros for the calculation of the total demand, the projected stock etc. Figure 9.13 gives an overview of the standard settings for SNP.

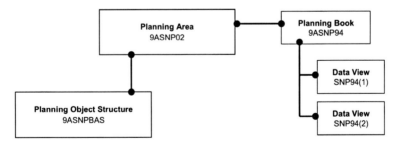

Fig. 9.13. Standard settings for the SNP planning book

Quantities and capacities are displayed in separate data views, because the selection criteria are different. In the quantity view always the locationproduct should be chosen for display, in the capacity view the resource is usually the relevant characteristic to select the capacity consumption. Figure 9.14 shows the data view for the planning of quantities SNP94(1) of the standard planning book 9ASNP94.

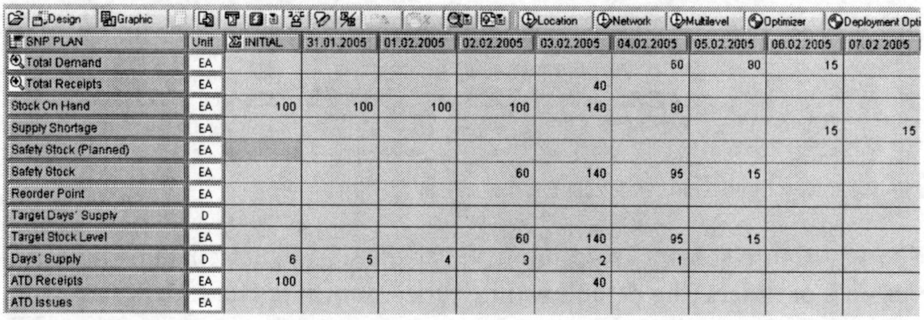

SNP PLAN	Unit	INITIAL	31.01.2005	01.02.2005	02.02.2005	03.02.2005	04.02.2005	05.02.2005	06.02.2005	07.02.2005
Total Demand	EA							60	80	15
Total Receipts	EA						40			
Stock On Hand	EA	100	100	100	100		140	90		
Supply Shortage	EA								15	15
Safety Stock (Planned)	EA									
Safety Stock	EA					60	140	95	15	
Reorder Point	EA									
Target Days' Supply	D									
Target Stock Level	EA					60	140	95	15	
Days' Supply	D	6	5		4	3	2	1		
ATD Receipts	EA	100					40			
ATD Issues	EA									

Fig. 9.14. SNP planning book

The different types of demand are subsumed in the key figure 'total demand' as the different types of receipts are in the key figure 'total receipts'. Both can be expanded to show more details.

The key figure 'stock on hand' represents the projected available stock and is calculated using the actual stock, the total demands and the total receipts. The categories which contribute to the actual stock are defined in the location master. If the supply does not cover the demand, the difference (including initial stock) is displayed as 'supply shortage'. The demand for the safety stock – in this example calculated using two safety days' supply – does not contribute to the supply shortage. The ATD-receipts and issues are used for deployment and are defined in the location master as well. Both the total demand and the total supply are calculated by macros using several other key figures. Details for the orders are displayed with right mouse click within the key figure for the demand resp. receipt as shown in figure 9.15.

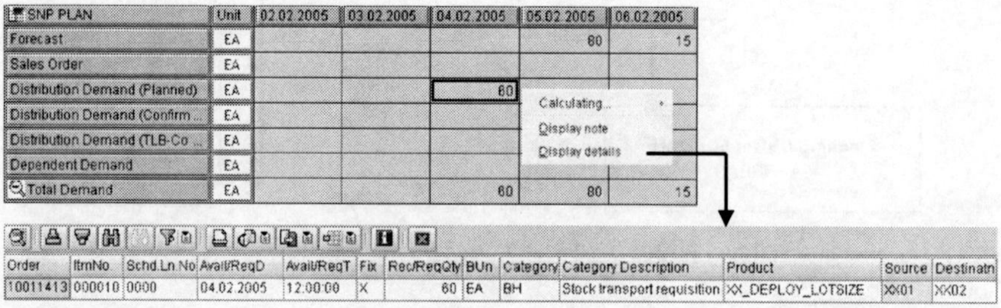

SNP PLAN	Unit	02.02.2005	03.02.2005	04.02.2005	05.02.2005	06.02.2005
Forecast	EA				80	15
Sales Order	EA					
Distribution Demand (Planned)	EA			60		
Distribution Demand (Confirm...	EA					
Distribution Demand (TLB-Co...	EA					
Dependent Demand	EA					
Total Demand	EA			60	80	15

Calculating...
Display note
Display details

Order	ItmNo	Schd.Ln.No	Avail/ReqD	Avail/ReqT	Fix	Rec/ReqQty	BUn	Category	Category Description	Product	Source	Destinatn
10011413	000010	0000	04.02.2005	12:00:00	X	60	EA	BH	Stock transport requisition	XX_DEPLOY_LOTSIZE	XX01	XX02

Fig. 9.15. Order details

In the standard planning book three key figures for stock transfer planning are provided which represent the order statuses after distribution planning, deployment and transport load building, as listed in table 9.6.

Table 9.6. Stock transfer categories in the SNP planning book

Source Location (Demands)		Target Location (Receipts)	
Key Figures for Distribution Demand	Categories	Key Figures for Distribution Receipt	Categories
Planned	BH, EB, ED	Planned	AG, EA
Confirmed	EG	Confirmed	EF
TLB-Confirmed	BI	TLB-Confirmed	BF

For SNP the planning book is more standardised than in DP, though changes are possible just the same. Some of the SNP functionality however depends on macros, and changes in the planning book might affect some vital macros for functions like the SNP heuristic or the deployment, therefore any change should be done with great care.

Another important difference to DP is that navigation attributes are not allowed for SNP (see also note 453644).

SNP assumes make-to-stock planning. However, note 443953 describes how to display make-to-order requirements.

10 Integrated Distribution and Production Planning

10.1 Cases for Integrated Planning

In traditional logistics concepts distribution planning and production planning are carried out completely independent of each other. Though the idea of SCM implies no separation of the planning according to functions, in many cases, especially when single sourcing is given, a hierarchical step-by-step approach – first distribution planning and afterwards production planning – is sufficient.

Depending on the supply chain this hierarchical approach might not be appropriate to exhaust the optimisation potential. This is mainly the case in multi-sourcing environments with multiple production sites where sourcing decisions are made according to the available capacity. The more the bottleneck is located at the beginning of the material flow, the more complex become the sourcing decisions and the decisions where stock is kept (probably even at semi-finished stage).

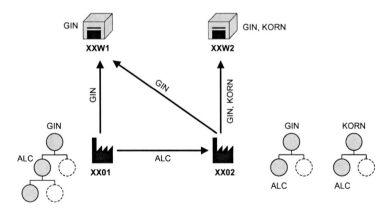

Fig. 10.1. Example for integrated distribution and production planning

The example in figure 10.1 contains a planning problem that might illustrate the potential for integrated distribution and production planning. The

J. T. Dickersbach, *Supply Chain Management with APO*,
DOI: 10.1007/978-3-540-92942-0_10, © Springer-Verlag Berlin Heidelberg 2009

supply chain contains two DCs and two plants, where DC XXW1 delivers GIN and DC XXW2 GIN and KORN. GIN is produced in both plants XX01 and XX02, KORN is only produced in plant XX02. The key component for both products is ALC, which is produced only in plant XX01 and is either processed there or sent to plant XX02. The DC XXW1 is supplied from both plants.

In this rather small example already some planning decisions exist where distribution and production must be considered simultaneously to achieve a good plan. Some of these decisions are:

- sourcing for location XXW1: procure GIN from XX01 or XX02 – depends on the load in XX01 and XX02,
- location prioritisation for location XX02: produce GIN for XXW1 or XXW2 – depends whether XX01 has still capacity to supply additional quantities to XYW1 (and on the market priorities),
- product prioritisation in location XX02: produce GIN or KORN – depends on the product priorities,
- ship ALC to XX02 or process it in XX01 – depends on the production costs, the market and the product priorities and the demand structure in XY02,
- load balancing for GIN between the plants XX01 and XX02 – depends on the production costs, the lot size dependent production costs and the capacity extension costs and possibilities,
- trade off between big lot sizes to reduce set-up costs and small lot sizes to reduce storage costs.

The main constraints which have to be considered are demand fulfilment according to priorities and finite capacities.

10.2 SNP Optimiser

10.2.1 Basics of the Supply Network Optimiser

The basic idea of the SNP optimiser is to plan the entire supply chain – distribution, production and procurement – at optimal costs by modelling the complete supply chain as linear equations and solve them by linear programming (LP) or mixed integer linear programming (MILP). The difference between the two lies in the consideration of discrete decisions, as lot size intervals.

The equations have the structure displayed in the following lines. As a simplification the time dependency is neglected. The objective function is

$$\text{Min } \{ \Sigma_{\text{Product}} [(D - S) * \text{Penalty} + S * \text{SupplyChainCost}] \} \qquad (10.1)$$

where D represents the total demand quantity, S the total supply quantity and all entries are product and partially location dependent. The supply chain costs contain costs for production, procurement, transport, storage and handling, the penalties are costs for lateness, non delivery and safety stock violation. The constraints for the plan are e.g. (related to the example in figure 10.1) the limited quantities that can be produced in the plants:

$$\text{Stock}_{\text{GIN-XY01}} + P_{\text{GIN-XY01}} \leq 15000 \qquad (10.2)$$

$$\text{Stock}_{\text{GIN-XY02}} + \text{Stock}_{\text{KORN-XY02}} + P_{\text{GIN-XY02}} + P_{\text{KORN-XY02}} \leq 15000 \qquad (10.3)$$

$$\text{Stock}_{\text{ALC-XY01}} + P_{\text{ALC-XY01}} \leq 20000 \qquad (10.4)$$

$$P_{\text{GIN-XY01}} + P_{\text{GIN-XY02}} + P_{\text{KORN-XY02}} \leq \text{BOM Ratio} * P_{\text{ALC-XY01}} \qquad (10.5)$$

where P stands for the produced quantity.

The penalties for non-delivery and lateness are maintained in the product master either globally or location specifically. The supply chain costs are maintained in the master data for product, PDS resp. PPM, resource capacity variant and transportation lane. The big advantage of the SNP optimiser is that it takes the costs for production, transport and storage (and – if required – even for handling) into account. By setting these costs it is possible to model decisions as

- extend production capacity in a plant or procure from a different plant considering increased production and transport costs,
- extend production capacity in a plant or procure externally,
- switch to a more expensive transport method to reduce the transportation duration (e.g. from truck to plane),
- if a transport capacity is already consumed, switch to another source,
- produce and ship just in time in order to minimise storage costs.

The more sourcing alternatives exist – probably even for different BOM levels – the more complicated the planning problem becomes. Linear optimisation is very well suited for this kind of problems.

10.2.2 Optimiser Set-up and Scope

SNP optimisation is either accessed from the SNP planning book (see chapters 9 and 15) or – which is the more usual way – defined as a background planning task in transaction /SAPAPO/SNPOP. Figure 10.2 shows the according settings for defining the scope and the rules for the optimisation.

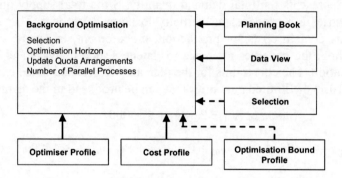

Fig. 10.2. Optimisation in the background

The optimiser profile is maintained with the customising path *APO → Supply Chain Planning → SNP → Profiles → Define SNP Optimiser Profiles. A*nd defines the constraints for the plan, whether and to what extent discretisation is performed and technical settings regarding the optimisation algorithm and controls to tune the performance.

• *Scope and Horizon*
The scope for optimisation is selected by products and locations. If a location product is missing – e.g. the input component of a PPM – the optimiser assumes its availability. Neither constraints nor costs result from excluded locationproducts. Orders are only created within the optimisation horizon.

• *Master Data*
Additionally to the maintenance of many costs in multiple master data objects (see next section), the usage of the SNP optimiser requires some other master data settings as listed in table 10.1.

Table 10.1. Master data requirements for SNP optimisation

Master Data	Entry
Location Master	Time zone
Product Master	Penalty costs for non-delivery & lateness ('SNP1'-view)
	Maximum delay ('SNP1'-view)
Resource	Activity overlaps period – if activities take more than 24 h
Capacity Variant	Capacity variant 1, 2 and 3 optionally for capacity extension and minimum capacity utilisation

10.2.3 Costs and Constraints

Costs are used in the SNP optimisation to define the objective function. These costs are without any currency, and they are not uploaded from SAP ERP™ but have to be maintained in various master data. Table 10.2 lists the costs, their semantic and where they are maintained.

Table 10.2. Costs for optimisation

Cost Semantic	Where Maintained	Relates to Cost Profile
Production Cost	PDS/PPM – Single Level Cost	Production
	Definition for 1st & 2nd Variant of Resource	Production Capacity
Procurement Cost	Product	Procurement
Transport Cost	Transport Lane – 'Means of Transp.'-View	Transport
	Transport Lane – 'Prod. Spec. Means'-View	Transport
	Definition for 1st Variant of Resource	Transport Resource
Storage Cost	Product – 'Procurement'-View	Storage
	Definition for 1st Variant of Resource	Storage Cap.
Handling Cost	Definition for 1st Variant of Resource	Handling Cap.
Safety Stock	Product – 'Procurement'-View	Safety Stock
Lateness Cost	Product - Penalty for Non-Delivery	Non-Delivery
	Product - Penalty for Delay	Delay

The costs for procurement, safety stock, storage and handling, which are maintained in the product master, relate to the base unit of the product, the cost for transportation is defined for any common unit of measure of the products. The penalty costs for the safety stock are multiplied with the number of days which the safety stock is not available. Whether storage costs are calculated as stock quantity multiplied by the number of buckets or additionally multiplied by the number of days per bucket depends on the settings described in note 544877. The procurement costs in the 'product'-view of the transport lane are not used by the optimiser. Note that the (single level) costs in the PDS resp PPM relate to the base quantity of the PDS resp. PPM. In the product master the costs for delay and non-delivery are maintained for three demand types. These correspond to the demand priorities 1 (customer demand), 5 (corrected forecast demand) and 6 (forecast demand). The priorities are set in the optimiser profile per demand type. With the transaction /SAPAPO/SNP106 the supply chain costs for a plan which has been calculated by an optimisation run are displayed.

Probably the most critical issue in the application of the SNP optimiser is the appropriate maintenance of the costs. If the relations of the costs are not maintained right, some rather unexpected results are received, e.g. no production at all because supply chain costs exceed the penalties for non delivery or permanent transport because storage costs are higher. The optimiser is very sensitive to inappropriate settings of the costs.

The costs are multiplied according to the weighting per cost category in the cost profile (transaction /SAPAPO/SNP107). The advantage of the cost profile is the possibility to influence the planning result without having to change lots of master data. The danger however is that – as described above – small changes of the weights might confuse the cost ratios and lead to undesired results (see also note 420650). Therefore it is recommended to use only the factors zero or one.

● *Constraints*

The constraints for the optimisation are the demands, the capacities, the material availability and the production and stock transfer horizons. Most of these constraints are controlled in the optimiser profile as listed in table 10.3.

Table 10.3. Constraints for the SNP optimiser

Constraint Type	Control in Optimiser Profile
Production Resource Capacity	Finite or Infinite Planning
Transport Resource Capacity	Finite or Infinite Planning
Storage Resource Capacity	Finite or Infinite Planning
Handling Resource Capacity	Finite or Infinite Planning
Customer Demand	No Control – Always Priority 1
Forecast Demand	Priority
Dependent Demand	Priority
Safety Stock	Priority
Production Horizon	Consider or Do Not Consider
Stock Transfer Horizon	Consider or Do Not Consider
Component Availability	No Control – Always Considered if Included

It is possible to define the priority per demand type. Another setting in the optimiser profile regarding the prioritisation is the checkbox 'hard prioritisation', which causes customer requirements to be covered first in a separate optimisation run.

Whether finite or infinite planning is performed depends entirely on the selection of the constraints in the optimiser profile. The flag in the resource master is ignored – the optimiser plans either all resources or no resources finite.

The quota for sourcing is a result of the SNP optimisation. However, in some cases it is desired that the contractual agreements that are modelled in the quota arrangement are considered. Therefore an option exists in the profile to control whether quota arrangements shall be considered during the optimisation run.

10.2.4 Discretisation

The optimisation problem becomes significantly more complex if it is not linear any more. Some cases in which a discretisation is required are

- lot size dependent cost, e.g. decreasing costs due to less set-up or increasing costs if procurement contracts are exceeded,
- technical restrictions require different master data, e.g. if the production process or the resource changes for huge lot sizes,
- fixed resource consumption, e.g. for set-up,
- technical restrictions require a fix lot size, a lot size rounding or a minimum lot size and
- extension of the standard capacity.

Each discretisation parameter and each bucket for which discretisation is used complicates the planning problem – discrete optimisation is in general a NP-complete problem. Therefore as less discretisation parameters and as less 'discrete' buckets should be applied as possible. Note 454433 describes how to limit the number of buckets which are planned with discretisation using the 'end bucket date' setting. Since the SNP optimiser creates a medium-term plan, each discretisation step should be questioned whether it is really necessary. Lot sizes for example will be considered in the further order processing – e.g. at the conversion to PP/DS orders – anyhow. If the capacity extension is used, the costs increase with the quantity. The costs for additional capacity relate to the consumed capacity and are calculated only for the additional capacity, as shown in figure 10.3.

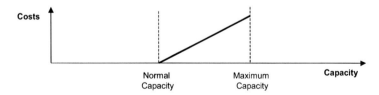

Fig. 10.3. Cost function for additional capacity

The SNP optimiser uses three different capacities: the normal capacity, the minimum capacity (which the optimiser tries to load independent whether there is demand or not) and the maximum capacity, which is used in case the normal capacity is not sufficient (and requires additional costs). These capacities are modelled as capacity variants. The maximum capacity is always the capacity variant 1, the normal and the minimum capacity are assigned to the variant number in transaction /SAPAPO/RESC01. For the consideration of the maximum and minimum capacity the assignment as 'active variant' is not required. Figure 10.4 shows the definition of the variant semantic and its display in the capacity view of the SNP planning book (see chapter 14.1).

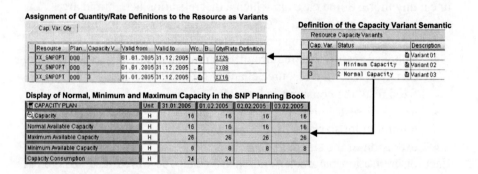

Fig. 10.4. Capacities for the optimiser

Decreasing production costs can be modelled in two ways: one is assigning a cost function to the PPM, the other one is to use two PPMs – one with a higher cost, the other one with lower costs but a minimum lot size. Depending on the modelling the curve for the cost per quantity is different, figure 10.5.

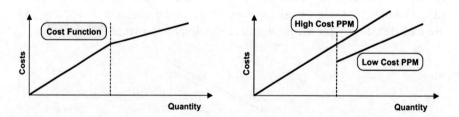

Fig. 10.5. Cost function for decreasing costs

Cost functions are defined with the transaction /SAPAPO/SNPCOSF. The steepness of the cost function depends on the lot size interval. Cost functions are created for procurement, production or transport with the according type and are accordingly assigned to the product master (procurement), the PDS resp. PPM (production) and the means of transport view of the transportation lane (transport). However it is not recommended to use cost functions.

Depending on the desired discretisation effect – e.g. minimum lot sizes – the according settings in the optimiser profile and the master data are required. The respective discretisation effects are described in the chapters for distribution planning and for production planning. In each case it is necessary to select the discrete optimisation in the optimiser profile. The discretisation steps are activated on daily, weekly or monthly basis. If mixed buckets are used, this way the horizon for the discretisation is limited to improve the performance, figure 10.6.

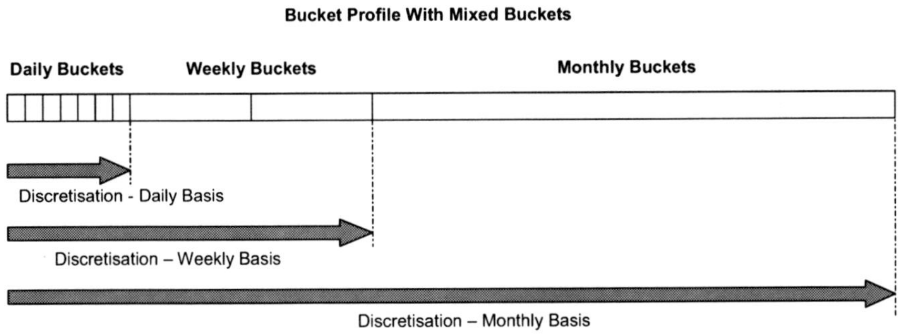

Fig. 10.6. Activation basis for discretisation

If discretisation is used, usually time or product decomposition has to be applied. The maximum lot size for orders is not a real discretisation step, because it is applied only after the optimiser has calculated the solution. Anyhow in many cases it is not necessary to model maximum lot sizes in SNP. Note 503294 provides additional information on lot sizes in SNP optimisation.

10.2.5 Technical Settings

The available optimisation methods are primal or dual simplex method. As usual with optimisation algorithms, it depends on the problem which one to use. The optimisation performance is tuned using time and product decomposition, which divide the optimisation problem in a couple of subsets. The risk of optimising these subsets instead of the complete scope is that only sub-optimal solutions are reached. Time decomposition is defined in bucket numbers, product decomposition in percentage of the products. Percentages between 1 and 30 are recommended.

For optimisation the planning situation is read from the live cache and transformed into an appropriate model for the optimiser. This model is saved in the input log. The solution of the planning problem is calculated by the optimisation solver, taking the control parameters from the optimiser profile into account. The result of the optimisation run is logged as well. Finally the according orders are created. The structure of the optimiser is shown in figure 10.7.

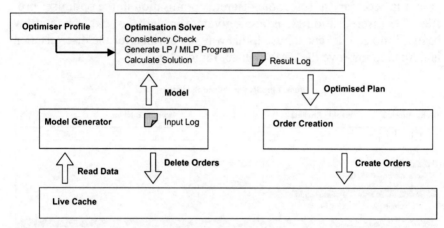

Fig. 10.7. Structure of the SNP optimiser

To prevent the order deletion, it is possible to set up a kind of change planning with the parameters described in note 578352.

• *Logging*
Input log, result log and message log are displayed with the transaction /SAPAPO/SNPOPLOG. As shown in figure 10.7, the input data and the optimiser result write different logs, which are selected per optimisation run. Note 509732 contains the link for the detailed description of the log file entries. For even more detailed information it is possible to set an optimiser trace. The trace file is customised with the transaction /SAPAPO/OPT10 and displayed with the transaction /SAPAPO/OPT11.

The performance for the optimisation depends on the number of locations, products and buckets. A way to reduce the load for optimisation is therefore to apply aggregated buckets – e.g. by decreasing the granularity with the time horizon, i.e. using mixed buckets. Since setting up an appropriate optimisation scenario is a rather complicated task, note 579373 refers to a special consulting service for optimisation.

The main problems for the use of the SNP optimiser are the consistent maintenance of the relevant costs, the difficulties to understand the results and the missing possibilities to integrate interactive planning steps into the scenario.

10.3 Capable-to-Match

10.3.1 CTM Planning Approach

The basic idea of capable-to-match (CTM) is to perform an iterative approach where demand elements are prioritised, supply elements are categorised and demands are matched with the supplies – according to the search strategy with existing stocks and planned receipts and/or with production. This implies that CTM planning leads to a first come (or better: highest priority) – first serve approach and shortages are not distributed evenly. The CTM approach is therefore best suited when demand priorities exist – whether demand type (e.g. special orders), customer priorities (e.g. for key customers), product priorities (highest profit) or others. Another feature of CTM which has to be kept in mind is that CTM is not an optimisation but a heuristic. This means that there will be no re-planning of orders which are already created for other demands and that the solution will not necessarily be optimal. These functional parts within the CTM – demand prioritisation, supply categorisation and the matching resp. new planning – are shown in figure 10.8.

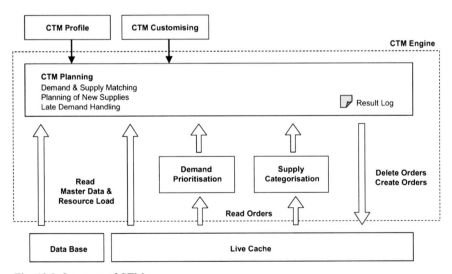

Fig. 10.8. Structure of CTM

The CTM planning takes place in a separate planning engine (which has to be set up in customising like an optimiser). In the first step the master data and the existing orders according to the planning scope – defined by the master data selection – is loaded into the CTM engine, where demand prioritisation, supply categorisation and order creation are performed according to the planning mode and the search strategy. Note that the CTM planning is a location-by-location procedure, so there is no global demand prioritisation and no global supply categorisation across the supply chain, but all supply categories and the production are taken into account for each location according to the search strategy.

Nearly all the settings which are relevant for the CTM planning are made within the CTM profile (transaction /SAPAPO/CTM). The CTM profile defines the planning parameters and triggers the CTM planning as well. The scope of the planning is defined by the master data selection with transaction /SAPAPO/CTMMSEL. Figure 10.9 provides an overview of these settings.

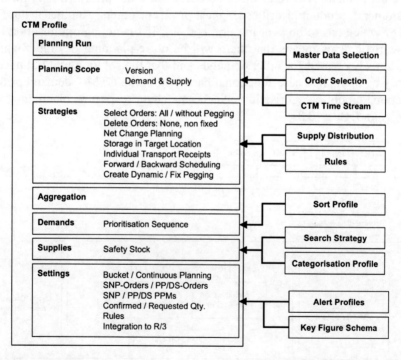

Fig. 10.9. CTM profile

Differing from the normal use of the icons, a new CTM profile is created with the button 'other planning profile'.

• *Master Data Selection*

Inappropriate master data selection is one of the most frequent causes for undesired results. The master data objects locationproduct, transportation lane and PPM define the planning scope and are selected in the master data selection profile. Stock transfer is only carried out if the locationproducts of both the target and the source location are included in the selection. For production all input products of the PDS resp. PPM have to be included as well. As a consequence, all related products to a planned product have to be included up to the externally procured component. A very useful tool to avoid errors due to inappropriate master data selection is the master data checker within the CTM profile. If a product is procured via stock transfers from another location, it is necessary to include the according transportation lane and the product in the source location into the master data selection, or else purchase requisitions are created. The master data selection is maintained with the transaction /SAPAPO/CTMMSEL and assigned to the CTM profile.

• *Explanation*

In order to help understanding the result of the CTM planning run an explanation log is offered. The detail of the explanation depends on the settings in the explanation profile. The explanation profile is created with the transaction /SAPAPO/CTMEXPL.

10.3.2 Prioritisation, Categorisation and Search Strategy

The demand prioritisation is performed according to one or more characteristics. These characteristics and their sort sequence are either selected within the 'demand'-view of the CTM profile or defined in the sort profile (as in backorder processing, cf. section 7.9). There are 255 priorities available, therefore several demands might have the same priority.

• *Supply Categorisation*

There are two alternative ways to perform the supply categorisation, either by assigning ATP categories and/or category groups to the supply categories or by defining limits per locationproduct and assigning the limits to the supply categories as shown in figure 10.11. The choice between the alternatives as well as the assignment of the categories is done within the categorisation profile (transaction /SAPAPO/CTMSCPR). The prerequisite for both alternatives is the definition of the supply categories with the transaction /SAPAPO/SUPCAT. Figure 10.10 shows the maintenance of the categorisation profile.

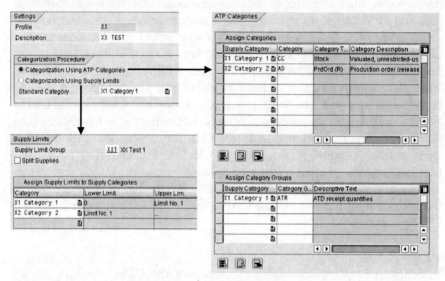

Fig. 10.10. CTM categorisation profile

For the supply categorisation via supply limits up to five inventory limits are specified per locationproduct with the transaction /SAPAPO/CTM02. In the second step the supply categories are assigned to the supply limits in the categorisation profile as shown in figure 10.11.

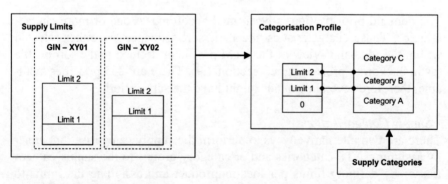

Fig. 10.11. Supply categorisation

The main purpose of the supply categorisation is to keep the inventory level above a certain limit, i.e. to start production before using stock. For this the definition of one supply limit is sufficient.

• *Search Strategy*

The search strategy (transaction /SAPAPO/CTMSSTRAT) defines the sequence in which the supplies are matched with the demands. Production is always included as a strategy step. If no production is desired, the PPMs resp. PDSs have to be excluded from the master data selection.

All the planning steps defined in the search strategy – the matching of the supplies and the production – are carried out location-by-location. The sequence of the locations is determined by the priorities. Figure 10.12 illustrates the impact of the search strategy on the planning result. Since CTM usually performs finite planning and the horizon for planning is restricted by the assigned time stream, the production step does not necessarily create sufficient supply for all the requirements, so that inventory levels (i.e. supply categories) might be preserved for the case that the demand exceeds the production capacity. This way it is possible to model a kind of safety stock.

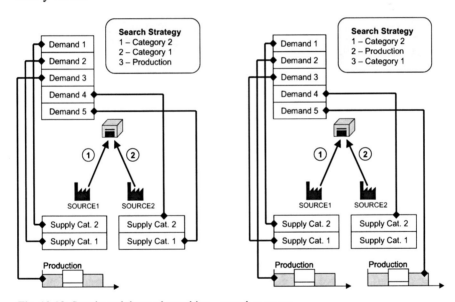

Fig. 10.12. Supply and demand matching – search strategy

In this example location SOURCE1 has a higher priority than SOURCE2, therefore first all the planning steps are performed for location SOURCE1. Production however is only able to cover one demand.

If the indicator for 'split supplies' in the categorisation profile for the supply limits is not set, huge supply quantities – e.g. the stock – are not classified according to the defined supply categories but completely assigned to the first category.

10.3.3 CTM Planning

CTM planning is performed demand by demand. First the path for the sources of supply with the highest priority (resp. quota arrangement) are searched and depending on the setting 'shortage allowed' in the CTM customising (transaction /SAPAPO/CTMCUST) partial solutions – i.e. supplies which do not meet the required quantity – are not discarded. This might lead to multiple orders as figure 10.13 shows.

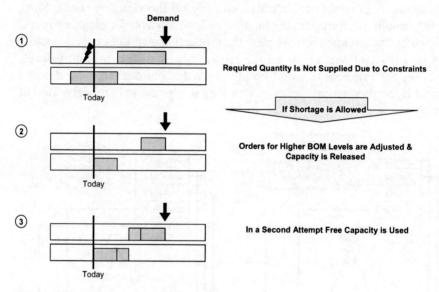

Fig. 10.13. Order split due to partial solutions

Note that this behaviour only happens if there is a partial solution and the system tries in a second attempt to load the remaining capacity and not if there are alternative sources of supply.

In the CTM profile it is possible to choose between backward and forward scheduling. Note that in the case of backward scheduling the supply for the demand with the highest priority is scheduled the closest towards the demand date and has therefore the least time buffer.

● *Distribution*
Presumed that there are no capacity constraints on the supply side (only production resources are considered in CTM), distribution is calculated according to quota arrangements or priorities. It is not possible to mix these – if quota arrangements exist, priorities are not taken into account anymore. The quota arrangements relate to the sum of the demands, not to

a single demand – there is no supply split for single demands. In the control parameters (menu path '*Control Data* → *Control Parameters*') it is possible to overrule the quota arrangements.

CTM takes the priority of the transportation lane (the procurement priority), the location priority and the product priority (from the 'SNP2'-view) into account. These priorities are used for the decisions as shown in figure 10.14.

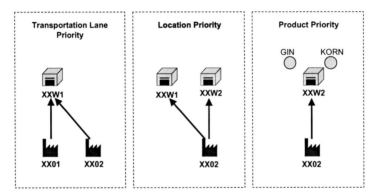

Fig. 10.14. Priorities for CTM

The transportation lane priority is used for classical sourcing decisions. Assuming that there is sufficient supply, the transportation lane determines where to procure from. The location priority and the product priority gain significance in the case of shortage, when the demands with the highest priorities are covered first. For the transportation lane the highest priority is zero, for the product one is the highest priority (and zero the lowest). To use location and product prioritisation, the characteristic LOCPRIO for the location priority and the characteristic MATPRIO for the product priority have to be included as sort criteria for the demand.

• *Production*

Production is usually the last step in the search strategy. The PDS resp. the PPM is selected according to its ability to produce in time and according to the quota arrangements or priorities. If quota arrangements are defined, no priorities are used anymore. The priorities are calculated using the fixed and the variable multi-level costs (the PDS resp. PPM priorities are not used).

If the procurement type of the product is 'X' (both in-house production and external procurement allowed), external procurement is only chosen if no valid PDS resp. PPM exists. Within the 'settings'-view of the CTM profile it is possible to substitute procurement types for the planning run.

In the control parameters it is possible to overrule the production horizon. The lot size methods lot-for-lot, fix, minimum and maximum lot sizes are supported. Periodic lot sizes can be created using the aggregation functionality for the demands.

CTM is able to use both SNP and PP/DS master data. In the 'settings'-view of the CTM profile it is possible to choose which kind of master data should be used and whether bucket-oriented or time-continuous planning is performed. Though it is possible to perform bucket-oriented planning with PP/DS master data, there are quite a few restrictions in this – fix durations, variable resource consumption and mixed resources are required, no transfer to SAP ERP™ is possible and subsequent processing (e.g. interactive planning) is apt to cause problems.

CTM planning is usually finite, but within the 'strategy'-view of the CTM profile it is possible to plan according to the resource settings or generally infinite. This is however only possible for multi resources. Production resources are the only resource types which are taken into account by CTM, neither transport nor storage resources are considered. Differing from the SNP optimiser, CTM does not use capacity variants of the resource as capacity extension.

Set-up is in general not supported by CTM. If no products are assigned to the set-up activity, the set-up activities are ignored. To consider set-up nevertheless, the options described in note 506700 have to be used. Sequence dependent set-up is not supported.

Requirements in the past are only considered if they are included in the planning horizon. In this case however receipt elements are created in the past. This is prevented by the use of the production resp. stock transfer horizon.

• *Scheduling with SNP Master Data*
If the order size exceeds the bucket capacity (SNP master data), the orders are split according to the daily free buckets. If set-up is modelled as a fix resource consumption, this is calculated per bucket, i.e. for each bucket the set-up is considered.

Fix or minimum lot sizes have to be used with great care because in this case the orders are not split anymore but still planned finitely. This might lead to a very low utilisation and to shortages due to unplanned orders. By increasing the bucket size (cf. section 20.5) the effect might be decreased, but the danger of a too low utilisation still remains.

• Limitations of CTM with PP/DS Master Data
CTM is a one step finite planning and might therefore not lead to a sufficiently good schedule, e.g. regarding sequence dependent set-up, resource
utilisation etc. The resource utilisation from the resource master is not used
for PP/DS (for SNP resources it is considered). With release SCM 2008 it
is also possible to consider shelf life (as a constraint) and characteristics in
CTM.

• Late Demand Fulfilment
If backward scheduling is used and the required quantity can not be met
to the required date, the late demand fulfilment starts. Late demand is
allowed in the CTM customising. Within the 'strategy'-view of the CTM
profile three late demand fulfilment strategies are available. The logic of
these strategies is not exactly intuitive. In order not to get into too much
detail only the standard procedure is therefore explained and visualised in
figure 10.15.

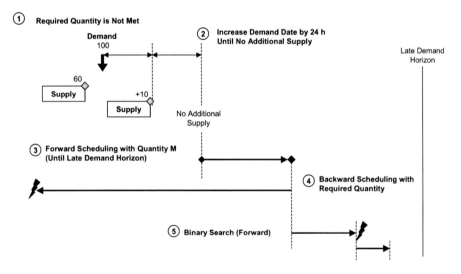

Fig. 10.15. Late demand fulfilment (standard procedure)

If the required quantity is not met, the demand date is internally increased
in 24 h-steps until another increase does not lead to any improvement
regarding the supply quantity. At this point forward scheduling starts
(step 3), but not for the full quantity but only for the quantity M which is
composed of several minimums as described in formula 10.6:

$$M = \text{Min} \{RQty, \text{Min} \{100, \text{Max} \{MinLS_{Prod}, MinLS_{PDS/PPM}\}\}\} \quad (10.6)$$

where RQty is the requested resp. remaining demand quantity after the previous steps, $MinLS_{Prod}$ is the minimum lot size of the product master and $MinLS_{PDS/PPM}$ the minimum lot size of the PPM resp. the PDS. If this succeeds, from the found date a backward scheduling is performed with the full quantity (step 4). As the last resort – if the previous steps have not been successful – a binary search is performed forwards up to the late demand horizon. In this case any existing surplus supply element before the late demand horizon is used and therefore an earlier production is inhibited. This might lead to late demand supply.

If several sourcing alternatives exist, all alternatives are checked to create a receipt in time. When the system changes to forward scheduling only the first sourcing alternative is considered.

For late demand handling within the CTM customising it is possible to choose between domino and airline strategy. The airline strategy disregards the priorities partially. This might be desired if the objective is to increase the number of deliveries in time, whereas it is not so important if a demand is fulfilled a little bit or quite a lot too late.

• *Aggregation*
By activating the aggregation option in the according view of the CTM profile orders are not matched individually anymore but aggregated according to the CTM time stream. It is possible to select whether demands resp. supplies will be planned at the beginning , at the middle or at the end of the CTM time stream bucket. Figure 10.16 visualises the impact of the aggregation.

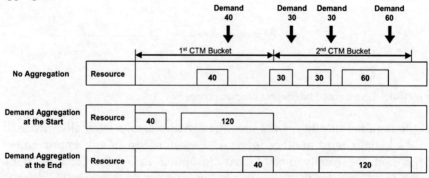

Fig. 10.16. Aggregation

The demands of the buckets are cumulated and scheduled at the beginning or at the end of the bucket. Scheduling the demand at the beginning of the bucket provides a buffer for planning (as periodic lot sizes do), while scheduling the demand at the end of the bucket accepts lateness already in planning.

10.3.4 CTM Planning Strategies

The selection of orders for planning depends on the settings for order selection and order deletion. Fixed orders are generally not deleted in an automated planning run. Orders are fixed in CTM, if they are manually fixed or have the status of a production or a purchase order. Other conditions to regard orders as fixed are if they are

- outside the planning horizon or
- not included into the master data or
- pegged to orders outside the planning horizon or
- contain components which are not included into the master data selection.

PP/DS orders are fixed for CTM (if SNP master data is used), but SNP orders are not fixed for CTM (if PP/DS master data is used).

Figure 10.17 illustrates the effect of the possible combination of the settings for the order selection and the order deletion on the planning mode for an example where three demands (D1, D2 and D3) are pegged to the supplies S1, S2 and S3; S4 is a surplus supply. In the initial situation demand D2 and supply S2 are connected by fixed pegging, and supply S3 is fixed. The options for order selection are 'all orders', 'only orders without fix pegging' or 'only orders without pegging' at all. For order deletion the possible settings are 'no deletion', all 'non fixed orders' or to 'delete the order tree', that is all non fixed orders which are pegged to the selected demands.

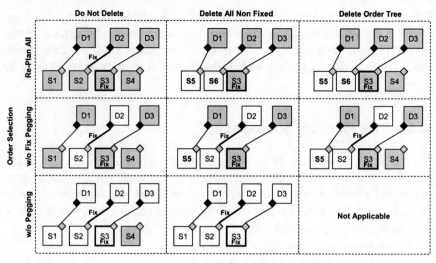

Fig. 10.17. Planning mode

The selected orders are marked grey. The selection is the same in each row, differences exist because orders are deleted or new ones are created. The left column – where no orders are deleted – provides the best information about which orders are selected. In figure 10.17 a case with surplus supply is shown, because in this case there are more differences in the planning result. Uncovered demand would be covered by any of these settings. The main difference in these combinations is whether the surplus is deleted and whether regenerative planning is performed. Another parameter which influences the regenerative planning is the flag for 'net change', which selects only changed orders for planning. It should only be used in combination with 'replan all orders' and 'delete order tree', else it might not produce the desired results – e.g. in combination with 'delete all orders' all supplies are deleted, but only supplies for changed demands are planned again. To delete all non fixed orders for a selection, the flag for 'end planning run after order selection' has to be set.

• *Integration to SAP ERP™*
In the 'settings'-view of the CTM profile three transfer options exist: no transfer, transfer only new orders and transfer new and deleted orders. For a reflection of the SAP APO™ planning in SAP ERP™ the option 'transfer new and deleted orders' has to be chosen. If 'transfer only new orders' is selected, transferred orders will not be deleted in CTM – independent of the settings for order deletion. For the duration of the CTM run it is recommended to stop the CIF inbound queue.

• *Store Receipts at Target Location and Transport Receipts Individually*
These two parameters relate to stock transfers. The parameter 'store receipt
at target location' controls whether the products are stored rather at the
source location or at the target location, figure 10.18. This parameter is
only relevant if backward scheduling is used.

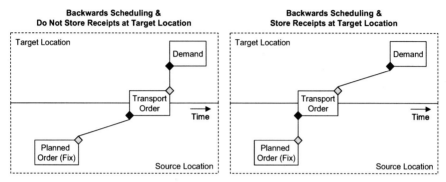

Fig. 10.18. Store receipt at target location

Another parameter regarding the distribution controls whether the number
of stock transfer orders relate to the number of demands (i.e. one stock
transfer per demand) or to the number of supplies (i.e. immediate transfer
of each supply), figure 10.19.

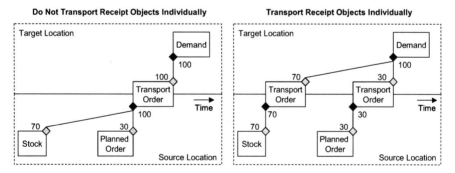

Fig. 10.19. Transport receipt objects individually

• *CTM Customising*
In the customising for CTM (transaction /SAPAPO/CTMCUST) the settings
regarding the created categories, the late demand handling, logging and
other technical settings are specified. Using the setting for maximum early
delivery the storing for inventories is restricted.

10.3.5 Supply Distribution

Supply distribution offers the option to push supplies to the target locations even when there is no demand for this. This might be required if the production plant does not keep the stock itself but passes it immediately on to the DC. Supply distribution is either included into the CTM planning as a subsequent planning step by assigning a variant to the CTM profile or executed as a stand alone functionality with transaction /SAPAPO/CTM10. The prerequisite for this functionality is that outbound quota arrangements are maintained for the source locations.

All supplies without pegging are pushed to the target locations. If the supply distribution is used in combination with a SNP heuristic, backward pegging should be allowed to avoid shortages in the source location and double receipts in the target location.

11 Distribution Planning

11.1 Master Data for Distribution Planning

The supply chain network is defined by locations and transportation lanes. The master data in this chapter focuses on locations, transportation lanes, calendars and transport resources. Table 11.1 provides an overview about the relevant master data for distribution planning.

Table 11.1. Master data for distribution planning

Master Data	Mandatory for Distribution Planning	Integration to SAP ERP™
Location	Yes	Yes
Product	Yes	Yes
Transportation Lane	Yes	if Info Record exists (Cross Company Stock Transfers)
Transport Method	Yes	No
Transport Resource	No	No
Time Stream	No	No
Quota Arrangement	No	No

Within the some of the master data some entries are mandatory for distribution planning. These entries listed in table 11.2.

Table 11.2. Overview of the master data maintenance

Master Data	Entry	Required for
Location	Stock Categories	Netting in SNP
Product	Weights & Measures	Stock Transfer with Transport Resource
Transportation Lane	Transport Method	Stock Transfer

- *Transportation Lanes*

The transportation lane defines the material flow between locations and is the prerequisite for any stock transfer. It is maintained with the transaction /SAPAPO/SCC_TL1 and contains the three views 'product', 'means of transport' and 'product specific means of transport' as shown in figure 11.1. Make sure to press enter before saving to prevent a loss of data.

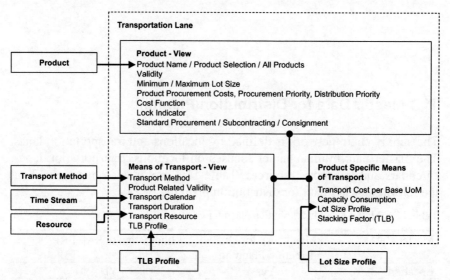

Fig. 11.1. Settings in the transportation lane

In the 'product'-view the transportation lane is restricted regarding the products and the lot sizes. It is even possible to lock a transportation lane, if it should temporarily not be used for distribution planning.

In the 'means of transport'-view one or more transport methods (e.g. truck or plane) are assigned to the transportation lane. On transport method level the transport duration, the transport calendar, the cost for the transport method and – if required – the transport resource are assigned. The TLB profile for transport load building (cf. chapter 12) is maintained on this level as well.

Though it is not a mandatory field, without a transport method it is not possible to create a stock transfer order. Transport methods are selected by the planning applications according to their ability to procure in time, their costs and – if the SNP optimiser is used – the available capacity of the transport resource. Only if a distribution receipt is created manually, it is possible to select the transport method. There is no possibility to change the transport method manually nor even to display it in the distribution order. The only way is to choose the transportation lane as display characteristic and select orders according to the transportation lane and the transport method.

The 'product specific means of transport'-view contains optional entries, for example for the transport resource capacity consumption or the product specific costs and the lot sizes for the optimiser.

The 'product specific means of transport'-view can not be maintained for the selection 'all products', but either for individual products or for a mass selection.

• *Transport Resource*

The prerequisite for using a resource as transport resource in the transportation lane is that it has the type 'T'. Either bucket, single mixed, multi mixed or scheduling resources can be assigned. If a transport resource is assigned to the transportation lane, the weights and measures in the product master ('attributes'-view) have to be maintained. Figure 11.2 shows the capacity consumption for a transport resource. Capacity is only consumed if the according entries exist in the 'product specific means of transport'-view. If PP/DS is used for distribution planning, a transport resource is necessary for correct scheduling.

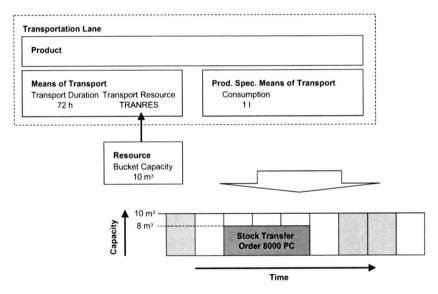

Fig. 11.2. Transport resource capacity requirements

If stock transfers have to be scheduled by the PP/DS optimiser, a multi-mixed resource has to be used as transport resource.

11.2 SNP Heuristic

The SNP heuristic is the preferred tool for distribution planning without production planning. Its advantage is its simplicity: The SNP heuristic calculates planned stock transfers to cover the demand quantity taking the safety stocks, the lot sizes and the transport duration into account – but not the capacity. Therefore the result is easy to understand, but the distribution plan might not be feasible. As the result of the SNP heuristic in the worst case a lateness due to transport durations is possible, but no shortage. Section 14.2 provides additional information about the SNP heuristic.

11.3 Planned Stock Transfers

For the scheduling of stock transfer orders the duration for goods issue, transport and goods receipt is taken into account. Goods issue and goods receipt are maintained in the product master in days (used by both SNP and PP/DS), while the transport duration is maintained in hours. Each of these durations is represented by an own activity in the stock transfer order. For their scheduling the transport calendar and the shipping calendars of the target and the source location are taken into account as shown in figure 11.3.

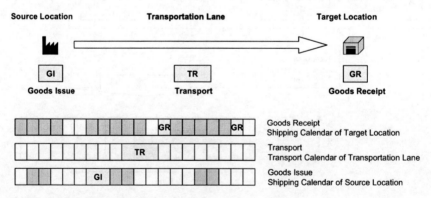

Fig. 11.3. Scheduling of stock transfer orders

The stock transfer horizon defines a frozen horizon in which no stock transfers are planned, figure 11.4.

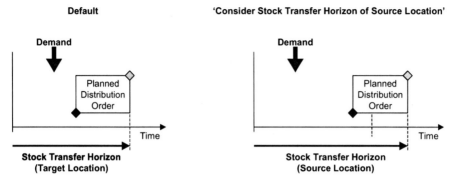

Fig. 11.4. Stock transfer horizon

The stock transfer horizon is maintained in the 'SNP2'-view of the product master. In the version master it is possible to determine that the stock transfer horizon from the source location is used instead from the target location.

• Calendar and Time Stream
The relevant calendars for SNP are the time stream objects which are created with the transaction /SAPAPO/CALENDAR. Usually the time stream is created for all calendar days, and the working days are defined in the factory calendar with the transaction SCAL. For this case weekly periods with days one to seven and start at 00:00:00 and end at 24:00:00 are appropriate. The 'periods' button defines which periods are generated in the live cache for the correspondence of time continuous order dates and SNP time buckets.

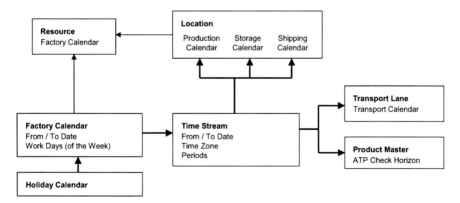

Fig. 11.5. Calendars in SAP APO™

Figure 11.5 shows the assignment of the factory calendar to the time stream and the assignment of factory calendar and time stream to the master data.

● *Time Zones*

Time zones are maintained in the location master and in the time stream. If there is a mismatch in the time zones between the location master and the time stream of the location, this might lead to a violation of the non-working days in scheduling. The time zone used for stock transfer resp. planned order creation in SNP is the one maintained in the time stream (shipping and transport resp. production), while the resources use the time zone of the location. For SNP the time zone had to be UTC (see also note 420648), but meanwhile other time zones are supported as well. The time zone of the location is the time zone of the SAP ERP™ system (transaction STZAC) if the location is transferred from SAP ERP™. A mismatch in the time zones leads also to an offset between the order dates displayed in the SNP and in the PP/DS transactions.

11.4 Stock in Transit

After the goods issue has been posted in the source location, the goods are on the way to the target location. To reflect this change in the status of the stock transfer, the stock transfer order can be replaced by a purchase order memo. The prerequisite for this is that the configuration on SAP ERP™ side contains a vendor that is assigned to the source location and an info record. The procedure is driven by the stock transfer execution in SAP ERP™ starting with the creation of an outbound delivery in the source location. Differing from the creation of deliveries for sales orders, the deliveries for stock transfer orders are created with the transaction VL10B. After picking, the goods issue is posted for the delivery with the transaction VL02N. It is possible to configure the SAP ERP™ system to create an inbound delivery in the target location with the goods issue using the message type LAVA. Alternatively an inbound delivery can be created manually with the transaction VL31N (the vendor that is assigned to the source location and the stock transfer order number are used as keys). For the goods receipt at the target location (transaction MIGO_GR) the inbound delivery must be referenced. The number of the inbound delivery can be found in the 'confirmation'-view of the stock transfer order. Figure 11.6 shows the order flow.

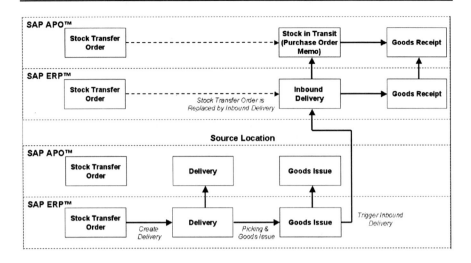

Fig. 11.6. Order flow for stock in transit

The usage of the inbound delivery in SAP ERP™ is the only way to model the stock in transit in SAP APO™. The usage of the 'stock in transit' in SAP ERP™ is also integrated with SAP APO™, but it is not helpful because it is not planning relevant, it does not reduce the stock transfer order and it does not contain any information about the planned receipt date.

11.5 Storage and Handling Restrictions

Especially in consumer goods industries where a high volume of goods (both literally and in numbers) is dealt with, storage capacity might become a constraint. The planned capacity consumption by actual and planned stock is evaluated in the capacity view of the SNP planning book (see chapter 14.1). Figure 11.7 shows this principle.

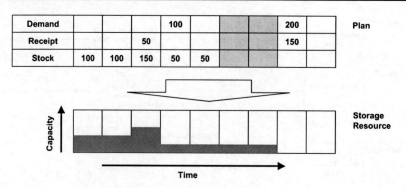

Demand				100				200	
Receipt			50					150	
Stock	100	100	150	50	50				

Plan

Storage Resource

Fig. 11.7. Storage resource

Planned receipts and issues within one period do not affect the storage capacity consumption. The assignment of the storage resource is done in the location master. Storage capacity is consumed according to the storage consumption entry in the GR/GI view of the product master.

The restriction in the model is that only one storage resource per location is possible. This might be a problem if different qualities of warehouses are used in one location, e.g. for frozen and for non-frozen products. Since physical stock does not disappear during non working days, the time stream for storage should not contain any non working days.

The storage resource is in no way suited to model storage restrictions due to container resources in the production.

• *Handling Capacity*
Capacity consumption of handling resources for goods issue and goods receipt is calculated using the handling capacity consumption entries in the product master. The capacity is represented by handling resources (resource type 'H'), which are assigned to the location (one for goods receipt and one for goods issue). If the handling capacity is checked in SNP, the handling resource has to be either a bucket or a mixed resource. Capacity consumption is calculated as order quantity times capacity consumption (both in the base unit of measure) for each day of the goods receipt resp. goods issue time (as specified in the product master).

The handling resource is also needed for the scheduling of the goods receipt and the goods issue times. For PP/DS a bucket resource is sufficient for scheduling, though it is recommended to use a calendar resource.

11.6 Sourcing

In a multi-sourcing environment distribution planning decides from which source the demands are covered. Except for SNP optimisation, which decides the sources based on total supply chain costs, either quota arrangements or priorities are used. If none of them are maintained, simply the first sourcing alternative is chosen.

• *Priorities*
Sourcing according to priorities represents a clear preference from where to procure, and other sources are only used as an exception. Accordingly SNP and PP/DS heuristics will always choose the source with the higher priority. Different sources are only chosen manually. CTM however switches to alternative sources with a lower priority in case of capacity or material constraints. The priorities for sourcing are maintained in the transportation lane (transaction /SAPAPO/SCC_TL1) as procurement priorities, where zero is the highest priority.

• *Quota Arrangements*
If it is intended to source from multiple locations on a regular basis, quota arrangements have to be used. Quota arrangements relate to the location and are created with the transaction /SAPAPO/SCC_TQ1. For distribution planning the inbound quota arrangements are the relevant ones (outbound quota arrangements are used for deployment). The concerned products are assigned to the quota arrangement, and the ratio of the sources is defined per product by double click. If more than one transport method is assigned to the transportation lane, the SNP heuristic interprets the ratio of the location as for each transport method. As an example, if there is a one to one quota arrangement defined between two sources and one of the transportation lanes contains two transport methods, the demands are divided by three, so that real ratio between the locations is two to one.

The quota arrangements are used differently in the applications SNP heuristic on one hand and the PP/DS heuristic and CTM on the other hand. Figure 11.8 illustrates this difference for three demands in different time buckets and a quota arrangement of one to two.

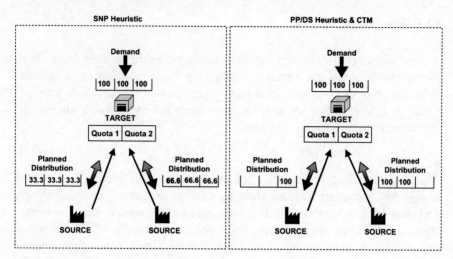

Fig. 11.8. Quota arrangements

Quota arrangements are maintained not only per location but per means of transport as well. If quota arrangements are maintained, priorities are ignored.

12 Replenishment

12.1 Deployment

12.1.1 Deployment Overview

If the supply chain behaves exactly as planned – i.e. neither changes in the demand nor unpredicted deviations of the supply happen – deployment is not necessary. Since we are living in an imperfect world, both demand and supply will usually differ from the planned quantities when it comes to execution. If the demand exceeds the supply, it has to be decided which demands – in case of the supply chain network: which locations – will be covered and to what extent. This is exactly the scope of deployment.

Depending on the supply chain structure these decisions are more or less complex. In a hierarchical, single sourcing structure a simple fair share rule is sufficient. In a multi-sourcing structure however the decisions become more complex. This case is dealt with later on in this chapter concerning deployment optimisation.

The basic idea of deployment is to convert planned stock transfers into confirmed stock transfers (confirmed distribution requirements resp. receipts) according to the available supplies, the demands, the deployment strategy and the fair share rule. The available supplies are defined by the difference between the ATD (available-to-deploy) relevant receipts and issues. The ATD-receipts and the ATD-issues are category groups which are assigned to the location master and/or to the locationproduct master. If an entry exists for the locationproduct it is used, else the entry from the location is taken. Stock usually contributes to the ATD-receipts, whether other receipt elements as production orders or purchase orders are included in the ATD-receipts as well depends on the business scenario. Regarding the ATD-issues, deliveries and confirmed distribution requirements should be included in any case. If a location serves for the supply of warehouses as well as for the direct shipment to customers, an important issue is the question whether (third party) sales orders are included into the ATP-issues

J. T. Dickersbach, *Supply Chain Management with APO*,
DOI: 10.1007/978-3-540-92942-0_12, © Springer-Verlag Berlin Heidelberg 2009

category group or not. Including sales orders means that they are prioritised over distribution requirements, since the available quantity for distribution is reduced by the confirmed quantity of the sales orders. On the other hand, an exclusion of the sales orders means that the total available quantity is used for distribution, i.e. sales orders have the lowest priority. Depending on the categories in the check mode for deliveries for the ATP check (cf. chapter 7), this leads either to a lower prioritisation of the sales orders or to conflicts because quantities are planned one way in deployment but handled 'first come first serve' for delivery. A workaround to handle both distribution requirements and sales orders with the same priority is described later on. Having the ATD-quantity defined, it is distributed according to the deployment settings defined in the locationproduct master. Safety stock is ignored by deployment (at least by the deployment heuristic). Since safety stock is modelled in SAP APO™ as a demand and not a supply element, this means that safety stock settings do not have any impact on the available quantity.

12.1.2 Deployment Heuristic

The deployment heuristic is a source location by source location approach to distribute the ATD quantities. Deployment is either carried out online in the interactive planning book or in the background with the transaction /SAPAPO/SNP02. For each source location a separate background deployment planning run is required. The structure of the deployment heuristic settings for background planning is shown in figure 12.1.

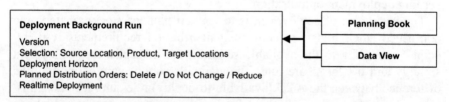

Fig. 12.1. Settings for the deployment heuristic

Deployment is already a step towards execution and therefore based on short term data. What short term means is defined by the three horizons 'deployment horizon', which is entered in the set up for the deployment run, and the 'deployment pull horizon' and the 'deployment push horizon' in the 'SNP2'-view of the product master of the source location. The deployment horizon defines the maximum horizon for which orders are read. The deployment pull horizon defines the horizon for the relevant requirements (ATD-issues) and the deployment push horizon defines the horizon for relevant ATD-receipts (e.g. production orders). Figure 11.2 visualises the significance of the deployment pull- and the deployment push horizon.

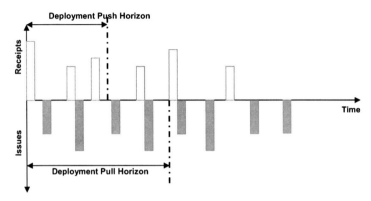

Fig. 12.2. Deployment horizons

The focus of the deployment is the short term, therefore a distribution requirement that is close to today might 'steal' the ATD-quantities from a distribution requirement further in the future. This behaviour might not be desired if both requirements are quite close to today. For this purpose the SNP checking horizon can be applied. The logic of the SNP checking horizon is to take all issues (e.g. deployment confirmed distribution requirements) into consideration before using the ATD-receipts for the deployment confirmation of new requirements, figure 12.3.

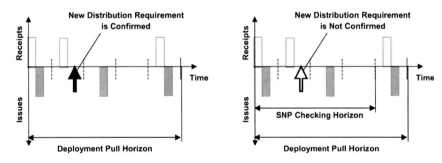

Fig. 12.3. SNP Checking horizon

• *Deployment Strategy*

The point in time of the stock transfer – i.e. whether stock is rather kept at the source or at the target location – is defined by the deployment strategy.

Available strategies are pull (blank), pull/push (P), push by demands (X), push by quota arrangement (Q) and push taking the safety stock horizon into account (S). The deployment strategy is maintained in the 'SNP2'-view of the product master of the source location.

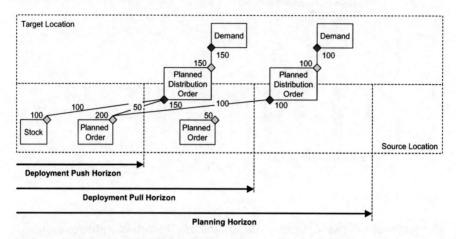

Fig. 12.4. Example for deployment strategies

Given the example as shown in figure 12.4 above with planned distribution orders, the deployment run creates the results as shown in the figures 12.5 to 12.9 depending on the deployment strategy settings. With the 'pull' strategy the distribution orders are confirmed according to the requirement date of the planned distribution orders at the source location – i.e. no re-scheduling is performed, as figure 12.5 shows.

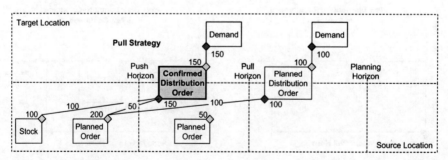

Fig. 12.5. Pull deployment

Using the 'pull/push' strategy, the confirmed distribution orders are scheduled as early as possible. Figure 12.6 visualises this for the same example.

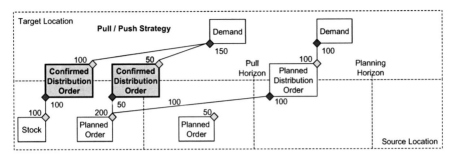

Fig. 12.6. Pull/push deployment

The difference between the strategies 'pull/push' and 'push to demand' is, that using the strategy 'push to demand' the deployment pull horizon is overruled by the planning horizon, as shown in figure 12.7.

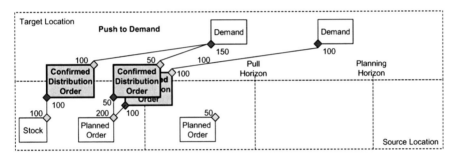

Fig. 12.7. Push to demand

If the deployment strategy 'push by quota' is used, all ATD-receipts within the deployment push horizon are shipped to the target locations according to the outbound quota of the source location (which is maintained with the transaction /SAPAPO/SCC_TQ1) regardless of the requirements in the target locations, see figure 12.8.

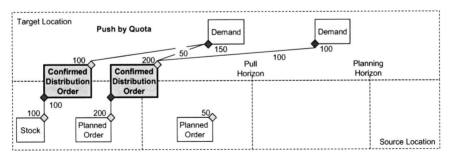

Fig. 12.8. Push by quota arrangement

The deployment strategy 'safety stock push horizon' behaves basically like the deployment strategy 'pull/push' with the only difference that ATD-quantities which 'cover' the safety stock are not deployed immediately but with a delay. This delay is calculated as the requirement date of the planned stock transfer minus the safety stock horizon (a setting in the 'SNP2'-view of the product master), figure 12.9.

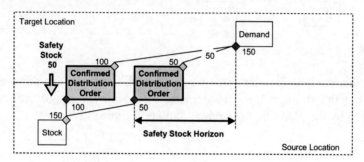

Fig. 12.9. Safety stock push horizon

If no push is allowed for deployment, because certain locations have to be delivered just in time, this is defined with the according setting in the location master of the target location.

• *Fair Share*
Until now it has been only described which elements are taken into account for deployment. In a supply network in most cases a source location supplies to more than one target location. During deployment the requirements are processed in the order of their requirement date (in the granularity of the assigned data view), so that shortages affect always the requirements farther in the future. For requirements which are due in the same bucket, the fair share rule defines which requirements are fulfilled and to what extent. The provided rules are percentage distribution by demands (A), percentage fulfilment of target (B), percentage division by quota arrangement (C) and division by priorities (D). Figure 12.10 shows the difference between fair share rules A and B. While fair share rule A divides all ATD-receipts proportionally according to the demands of the target locations, fair share rule B tries to keep the absolute quantity of the shortages the same. This implies that in the example in figure 12.10 up to an ATD-receipt of 200 all supplies are deployed to TARGET1.

Fair share rule C does not use the demands of the target locations but the outbound quota arrangements of the source location as a proportional factor. Distribution according to priorities using fair share rule D covers the demands of the target locations according to their priorities. The distribution priorities of the transportation lane are used for this.

If the deployment heuristic is used immediately after a finite distribution planning (i.e. SNP optimisation or CTM), naturally no fair share situation exists. In this case the deployment heuristic only confirms the planned distribution orders.

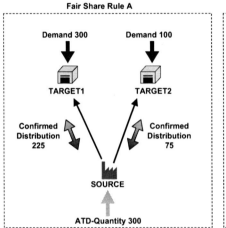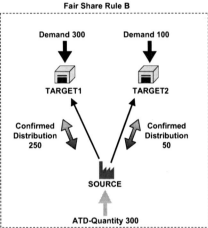

Fig. 12.10. Fair share rules A and B

The deployment relevant settings in the 'SNP2'-view of the product master are grouped into profiles which provide an alternative option for master data maintenance. Table 12.1 lists the fields and the according profiles.

Table 12.1. Deployment relevant settings in the product master

Field in Product Master	Profile	Transaction
Deployment Pull Horizon	Demand	/SAPAPO/SNP101
Deployment Push Horizon, Safety Stock Horizon	Supply	/SAPAPO/SNP102
Fair Share Rule, Deployment Strategy	Deployment	/SAPAPO/SNP111

• *Confirmed Distribution Orders*

Confirmed distribution orders are not changed any more by subsequent deployment runs, even in the case that the ATD-quantity is decreased in the meantime. In this case the planning situation has to be adjusted manually, e.g. supported by alerts.

If deployment it is not able to confirm the full planned quantity, the planned distribution order is reduced by the confirmed quantity – if the according option is selected in the deployment run. Other options are not to change orders – in this case the deployment run is merely a simulation – or to delete orders, which results in the deletion of all planned distribution orders within the horizon, independent whether they are confirmed, partially confirmed or not confirmed at all.

● *Distribution Order Categories*
The deployment functionality requires distribution orders of the categories AG for receipt and BH for requirement – else no orders are selected. If deployment is performed after a CTM run, make sure that the categories are appropriate (transaction /SAPAPO/CTMCUST).

● *Lot Sizes for Distribution*
Lot sizes are used in deployment to round down the quantities. In the global settings for SNP with the customising path *APO → Supply Chain Planning → SNP → Basic Settings → Maintain Global SNP Settings* it can be chosen whether the lot sizes of the target location product master, the transportation lane (lot size profile in the 'product specific means of transport'-view) or none is used. Since deployment may change the means of transport and the originally required quantity, the rounding of the distribution planning is not sufficient for all cases.

● *Real-time Deployment*
Real-time deployment does not use the planned distribution orders of the last distribution planning run, but performs a SNP heuristic to determine the current requirements.

● *Means of Transport Selection*
Deployment optimisation creates always new orders with a new selection of the means of transport; deployment heuristic tries to use the same means of transport as the planned stock transfer if the bucket is the same, else it selects the means of transport anew.

● *Fair Share Between Distribution and Sales Orders*
If a plant or a central DC is not exclusively dedicated to procure other DCs but keeps inventory to deliver customers directly as well, the question of the prioritisation of sales orders versus distribution requirements arises. If sales orders do have a higher priority than distribution requirements, this is

modelled by including the sales order categories into the ATD-issues. In many cases however it is desired to treat direct customers and inter-company customers (represented by the distribution requirements) equally. From SAP APO™ 5.0 on the fair share between distribution orders on one hand and sales orders or forecasts is controlled in the SNP-view of the product master.

12.1.3 Deployment Optimisation

The main difference between the deployment optimisation and the deployment heuristic is that the existing stock transfer requisitions are not taken into account but the distribution requirements and the ATD-quantities of the defined scope are recalculated as a basis to create confirmed distribution orders according to the deployment settings. Since the sourcing decisions of the distribution planning are discarded by the deployment optimiser, the decision whether the heuristic or the optimisation is used has an impact on the significance and the requirements for the distribution planning.

The deployment heuristic uses the distribution requirements calculated by the SNP heuristic, CTM or the SNP optimiser as a basis and does not take changes in the demand or the supply since the last planning run into account. The only exception is the use of the real-time deployment, which carries out a SNP heuristic for the relevant locationproducts anew. For single-sourcing this is a suitable solution to deploy according to the actual demands, but for multi sourcing structures there are clear disadvantages compared to the deployment optimisation because of the fixed ratio of the requirements according to the quota arrangements.

The deployment optimiser is called with transaction /SAPAPO/SNP03. The structure of the deployment optimisation is similar to the SNP optimisation. Both use the same objects for the optimiser profile, the cost profile and the cost settings. Consequently the log file is displayed with the same transaction /SAPAPO/SNP106 as well.

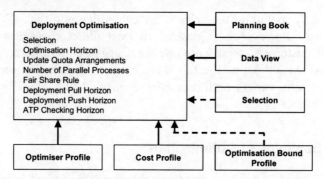

Fig. 12.11. Structure of the deployment optimisation

The optimiser is able to delete confirmed stock transfers within the planning horizon. Another difference to the deployment heuristic is that all locations – source and target – have to be included into the scope. The discretisation options and the prerequisites are analogous to distribution planning with the SNP optimiser.

The deployment optimiser provides the fair share strategies 'distribution based on costs' (blank), 'percentage distribution by demand' (A) and 'percentage fulfilment of target' (B). The strategy setting in the product master is overruled. Though the strategies A and B have the same name as for the heuristic, they behave differently with the optimiser. Fair share rule B cumulates the requirements per target location within the deployment pull horizon and distributes the quantities according to the proportions of the cumulated demand, figure 12.12.

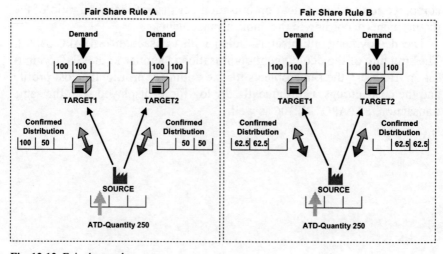

Fig. 12.12. Fair share rules

Depending on the storage costs in the target location the deployment orders are created just in time or earlier.

Deployment based only on costs tends to corner solutions with zero distributions. Figure 12.13 illustrates this behaviour for equal costs compared with the fair share rules A or B.

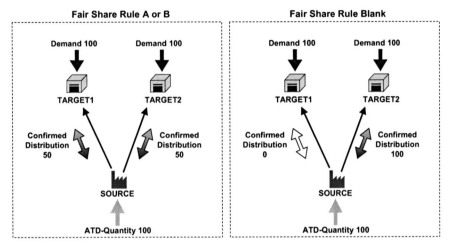

Fig. 12.13. Fair share according to costs

For multi-sourcing environments the deployment optimiser is suited to take deviations between supply and demand into account and balance or distribute shortages.

12.2 Transport Load Builder

The position of the transport load builder (TLB) in the replenishment process is a short term planning tool to combine confirmed distribution orders to truckloads or other transport units according to the capacity restrictions. The use of the TLB is an optional step in the distribution and replenishment planning. If TLB is used, the integration to SAP ERP™ should be customised the way that only TLB confirmed orders are transferred as purchase orders (transaction /SAPAPO/SDP110, see chapter 9).

The TLB planning follows the deployment run and uses confirmed distribution orders as input. The orders are selected by transportation lane and transport method and are combined to TLB confirmed distribution orders according to the settings in the TLB-profile. To skip deployment and use planned distribution orders as input for TLB, changes in the category assignments within the planning area according to note 514947 are required.

With SAP APO™ 4.1 the logic for the TLB was changed and enhanced. A new algorithm is used with more options for upsizing or downsizing and to include and bring forward deployment orders. Additional features of the new TLB engine are parameter to control the product arrangement across shipments (straight loading vs. load balancing), the use of further parameters for capacity restrictions and the consideration of loading groups (as sort criteria). For upgrade installations either the new or the old functionality is available.

• *Procedure for the Transport Load Building*
The procedure for TLB is to load all selected deployment orders according to the restrictions in the TLB-profile. If 'straight loading' is used, the orders are sorted according to the loading group, while 'load balancing' tries to distribute the products evenly onto different truckloads. Figure 12.14 shows an example for a transport capacity of 100 units:

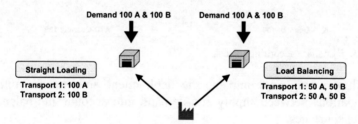

Fig. 12.14. Straight loading vs. load balancing

The advantage of straight loading is that the same products resp. the products with the same loading group are loaded onto the same truck, which increases the efficiency in loading and unloading. With load balancing the products are evenly distributed, which minimises the risk of missing a product if a shipment fails.

If straight loading is used and the minimal load quantities are not met, a re-distribution takes place where some shipments will load the trucks only to the lower limit. If there is still no valid result, upsizing or downsizing is used as shown in figure 12.15.

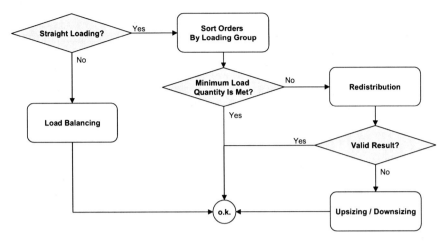

Fig. 12.15. Procedure for straight loading

Whether upsizing or downsizing is used is determined by a threshold value and two alternative algorithms. If upsizing is used, additional deployment orders are included which fulfil the two criteria:

- deployment order lies within the pull-in horizon
- the settings for 'maximum coverage period' and 'shipment upsizing' in the product master allow the preponement.

The settings to control the procedure for transport load building are maintained in the transportation lane and in the product master as shown in figure 12.16.

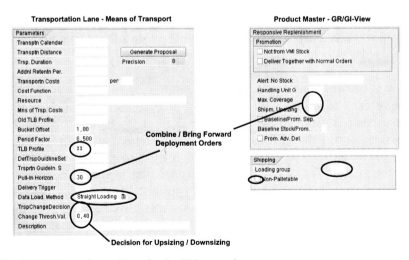

Fig. 12.16. Master data settings for the TLB procedure

• Horizons for TLB

The most important horizons for TLB are the planning horizon and the pull-in horizon. The TLB planning horizon defines which distribution orders are taken into account for the TLB run, and the TLB pull-in horizon defines which orders might be scheduled forward and is maintained in the transportation lane itself. Looking from the earliest order, other distribution orders within the TLB pull horizon are combined as shown in figure 12.17.

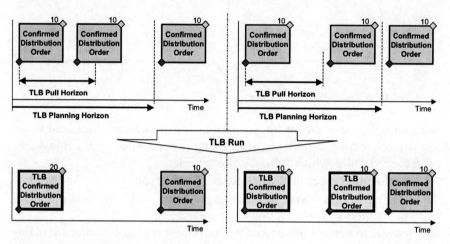

Fig. 12.17. TLB horizons

With the new TLB algorithm the days of coverage at the target location (a setting in the locationproduct master) can be used as well to influence the decision whether a deployment order is moved forward and combined with another order or not.

• Capacity Restrictions for TLB

Capacity constraints are only considered as defined in the TLB-profile. Other capacity restrictions – e.g. the capacity of the transport resource – are ignored during the TLB run. If transport resources are a capacity constraint, these have to be considered in the distribution planning resp. the deployment run. The relevant capacity restriction in the TLB profile is the minimum of the three following constraints: the maximum volume, the maximum weight and the maximum number of pallets. A lower limit exists as well to inhibit uneconomical transport orders. The volume and the weight limits are checked using the volume and the weight properties in the 'attributes'-view of the product master. The standard parameters are weight, volume and pallets. Additional parameters can be defined with transaction /SAPAPO/TLBPARAM (take note 710198 into account).

To consider the pallet restriction it is necessary to maintain pallets as an alternative unit of measure in the product master. The pallet restrictions are maintained in the TLB profile as pallet floor spots. The capacity consumption of the pallet floor spots takes the stacking factor of the product master into account, figure 12.18.

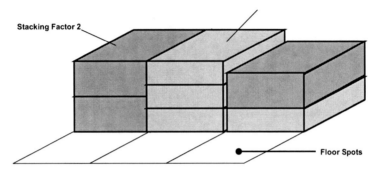

Fig. 12.18. Pallet restriction

Taking the stacking factor into account, the pallet constraint is described by the formula

$$\text{MaxPallet}_{\text{TLB Profile}} \geq \Sigma \left(\text{QtyPallets}_{\text{Product}} / \text{Stacking Factor}_{\text{Product}} \right) \quad (12.1)$$

where $\text{QtyPallets}_{\text{Product}}$ is the order quantity per product in pallets. This capacity calculation assumes that pallets with different products are stacked together, which is not the case in all businesses – in the pharmaceutical industry e.g. restrictions exist regarding the combination of products for transport purposes. Incompatibilities of these kind can not be modelled with the TLB functionality. The loading group is considered as a soft constraint.

The TLB profile is defined with the transaction /SAPAPO/TLBPRF and contains the information regarding the capacity restrictions for combining distribution orders to TLB confirmed distribution orders. The TLB profile is assigned to the 'means of transport'-view of the transportation lane.

● *TLB Planning Run*
If no volume, no weight or no unit of measure for pallets are maintained in the product master, the respective dimension is not regarded as a constraint for the according distribution orders. If a distribution order exceeds a capacity limit of the TLB profile, the order is split into one or more orders which suit the maximum capacity and an order for the remaining quantity.

The TLB run is carried out in the interactive mode either from the SNP planning book or directly via transaction /SAPAPO/SNPTLB. Interactive planning allows creating TLB confirmed orders manually. In the 'detailed item'-view it is possible to change orders. Figure 12.19 shows the planning book for TLB.

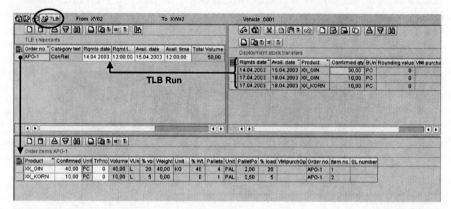

Fig. 12.19. Interactive TLB

The TLB planning in the background is carried out with the transaction /SAPAPO/SNP04. The structure of the TLB planning in the background is shown in figure 12.20.

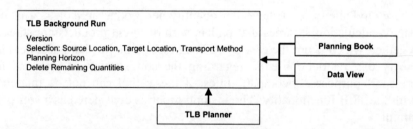

Fig. 12.20. Structure of the TLB planning in the background

Part V – Production

13 Production Overview

13.1 Production Process Overview

From the process point of view the production planning scenario consists mainly of the production planning itself, i.e. the creation of planned orders resp. purchase requisitions for the demand, the detailed scheduling to create a sequence for the orders which is sufficiently feasible (usually in the short-term) and the execution of the production order. In many cases it is desired to have at least a rough feasibility check of the production plan in the mid-term as well. Figure 13.1 shows the process chain for a make-to-stock production without any distribution planning, starting with a forecast from Demand Planning.

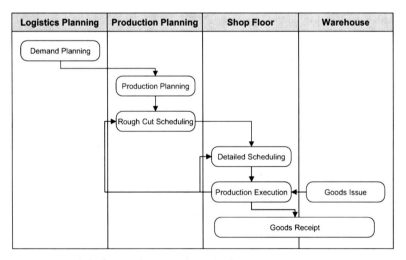

Fig. 13.1. Process chain for a make-to-stock production

Sometimes there is an organisational separation between production planning – i.e. the determination of the quantities which have to be produced (and externally procured) – and the detailed scheduling, where the sequence of the orders and operations is determined. In other cases – especially if the scheduling tasks are more complicated – both processes are performed by the same organisation. The detailed scheduling is usually only performed for the short term horizon. Traditionally both roles – production planning and scheduling – are limited to the plant level. If a supply chain contains alternative production possibilities the process becomes more complex and usually an integrated distribution and production planning is required – at least for the rough-cut plan, cf. chapters 10 and 14.

Depending on the business requirements it might be favourable to create a rough-cut production plan for the mid-term or even long-term before the more detailed production plan. This approach represents the idea of a hierarchical planning, i.e. to plan with less detail for the farther future and has the benefits of avoiding unnecessary information and unnecessary system load and to have a tighter integration with the distribution planning. On the other side there is the downside of integrating two different planning granularities which should not be underestimated.

● *Requirements for the Production Plan*
The requirements for production planning vary widely. Some criteria for a good production plan are
- to meet the demands,
- to consider the resource capacities and the material availabilities,
- high utilisation of the resources
- low set up efforts
- to minimise the stocks (produce just in time) and
- to minimise the work in progress.

The technical constraints, the number of orders and operations and the complexity of the material flow might lead to a very constraint plan. Especially in these cases the stability of the plan becomes another important aspect: Small changes in the circumstances (e.g. a delay in the production or an unexpected resource downtime) should not cause a complete replanning and scheduling.

● *Dimensions for the Complexity of Production Planning*
One dimension for the complexity of the production planning lies within the processes and process steps which are used – a single production planning step and a simple scheduling heuristic for one BOM-level is less complicated than a global and a local production planning in different granularities and more elaborate scheduling heuristics and/or a sequence optimisation for several BOM-levels performed by different planners. The different responsibilities for the interaction between rough cut production

planning, detailed production planning and detailed scheduling are usually separated by disjunctive horizons and a kind of exception handling is used to deal with urgent matters.

The speciality of production planning is that there is a second dimension for the complexity which is not or only to a much lower degree found in the other processes. This dimension is the complexity of the physical production process and its modelling for planning. Examples for the complexity are

- Time constraints between production steps (within an order or between BOM-levels, e.g. a maximum duration between activities in steel mills because metal has to be processed before solidification)
- Overlapping production (e.g. a continuous production of a commodity with a huge lot size and an overlapping packaging)
- Processes which involve containers (e.g. in chemical, pharma or consumer industries)
- Alternative resources, secondary resources (e.g. for labour) and resource compatibility for alternative production sequences (e.g. etching and coating can be performed on four alternative resources each, but if etching is done on resource 1, coating has to be done on resource 2 or 4)
- Oven processes and synchronisation of the orders (e.g. a furnace is opened only at the end of a heating cycle)
- PRTs (e.g. dies for moulding)
- Sequence dependent set-up (e.g. small set-up from white to black, huge set-up from black to white)
- Batch pureness for production (e.g. for GMP compliance in pharmaceutical industries).

The main benefit using PP/DS compared to PP in SAP ERP™ is the enhanced possibilities to create feasible plans – including sequence optimisation. Its main difficulties are due to the reflection of complex production structures and the manifold master data settings. In SAP APO™ there are more possibilities to model the technical constraints of the production process than in SAP APO™, and each technical constraint increases the complexity of the planning problem. The temptation to model each constraint is the downside of these possibilities, since the complexity of the model might stretch the system possibilities to their limits, both by provoking errors and decreasing the performance – especially combined with a generous use of 'finite planning' of the resources and 'automatic planning' of the product.

● *Appropriate Modelling*
One of the challenges is to find the appropriate modelling for planning processes, because the definition of too many constraints will tend to provide results which are unsatisfactory in terms of utilisation, stability of the plan and maintenance of the solution – if the model becomes too complex it is difficult to identify mistakes. Another aspect of complex modelling is that the requirements for master data maintenance become more severe, and in most cases the impact of negligent master data maintenance is underestimated. And the approach to overkill a planning problem by a complex modelling still has the risk of being incomplete because manual planning often contains optimisation steps and plausibility checks which are not modelled.

The lessons learned in many implementation projects were to simplify and to resist the temptation of complex modelling. Therefore it is strongly recommended to keep the constraints as less as possible in modelling and to use a two step approach to create a feasible plan, i.e. infinite production planning first and finite scheduling on the key resources afterwards.

Generally the complexity of a production planning task increases with the number of BOM-levels, number of operations and finite resources, the use of fixed resp. minimum and periodic lot sizes and sequence dependent set-up.

● *Separate Steps for Production Planning and Detailed Scheduling*
An important recommendation for the modelling of the production planning process is to use an infinite production planning first and a subsequent finite scheduling in a separate step. Though APS systems allow theoretically a one step-approach, experience shows that this tends to cause multiple problems both from the business point of view (loser products, low resource utilisation) and the system point of view (e.g. performance issues).

● *Order Cycle*
The result of the production planning step are planned orders which contain the information about the dependent demand and the capacity requirement (in SAP APO™). Depending on whether the planned order was created by a SNP planning run or by a PP/DS planning run, the planned order has different categories. Both are transferred to SAP ERP™ as planned orders just the same, but only the PP/DS planned orders can be converted into production orders from SAP APO™ (in SAP ERP™ the SNP planned orders can be converted to production orders as well, cf. chapter 18).

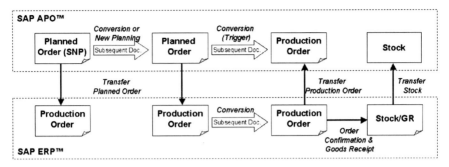

Fig. 13.2. Order cycle for production

If PP/DS is used then the SNP planned order is either converted into a PP/DS planned order or deleted by the PP/DS planning run. These options are explained in more detail in section 14.7. Figure 13.2 provides an overview about the order cycle for production.

The production order is reduced by the order confirmation and remains in SAP APO™ until it is technically completed, cf. chapter 18.

13.2 Applications for Production Planning

13.2.1 Scenario and Property Overview

Production planning has probably the most scenario variants. Major distinguishers in the production process are

- whether some kind of master production schedule resp. a rough-cut planning is used and – if so – whether there is an organisational separation between the planning processes,
- whether a make-to-stock (with or without forecast consumption) or a make-to-order production is used,
- what requirements exist regarding the feasibility of the plan for the medium-term and the short-term horizon and
- how much manual effort resp. automation in the creation of the plan is desired.

Regarding the level of automation experience shows that the expectations are often set too high – a complete automation is not possible in most cases, and the planning result of SAP APO™ should be considered as a proposal that must be checked by the planners. This proposal is usually nevertheless a big help.

The information for the feasibility of a plan can be derived in different ways with different level of detail:

- by infinite planning and checking the capacity requirements. This is mainly interesting if the key driver for the production is the (planned) demand and production or material capacity is not the constraint in the medium-term
- by bucket oriented finite planning, where capacity bottlenecks cause a delay of the product availability. This provides a good overview about the feasible production plan, but does not take sequence dependent constraints into account.
- by detailed scheduling, which provides a feasible plan considering all the modelled constraints but requires the most effort as well. In most cases it is not necessary to perform a detailed scheduling to get the information whether a production plan is feasible or not. This option should only be used if there are significant technical sequence dependent constraints which do not allow an approximation via the bucket approach and/or if no bucket related planning is used.

• Production Planning Applications

SAP APO™ offers different applications for production planning and detailed scheduling. Production planning is supported by the modules SNP and PP/DS with two different levels of detail. These are

- rough-cut and bucket-oriented planning (SNP) and
- time-continuous planning (PP/DS).

These two levels of detail require a different set of production master data. Different PPM resp. PDS and resources are necessary – if multi resources are used, the properties for both PP/DS and SNP are included in one object. Detailed scheduling with the purpose of creating a sequence for order execution is only supported by PP/DS since SNP is limited to bucket-oriented planning and there is no sequence within a bucket. Figure 13.3 shows the different applications for production planning and detailed scheduling in SAP APO™.

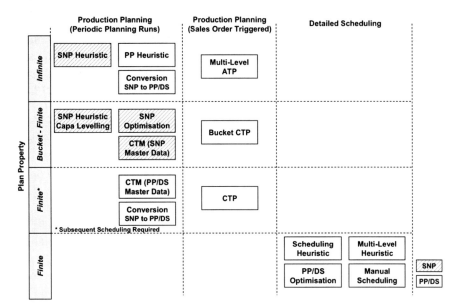

Fig. 13.3. Applications for production planning and scheduling

Even if a finite plan is created by a PP application, it is in any case recommended to perform a subsequent process step for detailed scheduling at
least for the short term horizon. Otherwise the quality of the plan will not
meet the requirements. The process of creating a feasible plan is described
in more detail in section 13.3.

Based on the variety of the applications there are many alternatives how
to create a feasible plan in SAP APO™. Some of the scenarios are listed
below. Each one has its advantages and disadvantages.

1. Distribution and medium-term production planning with the SNP
   heuristic, the SNP optimiser or CTM with SNP master data, short-
   term production planning by conversion of the SNP orders to PP/DS
   orders and detailed scheduling with the DS heuristics or the PP/DS
   optimiser.

2. Distribution planning with the SNP heuristic, medium-term production planning with the SNP optimiser, short-term production planning by conversion of the SNP orders to PP/DS orders and detailed
   scheduling with the DS heuristic or the PP/DS optimiser.

3. Production planning for the medium-term as well as for the short-
   term horizon with the higher level of detail in PP/DS with the PP
   heuristics or CTM with PP/DS master data and detailed scheduling
   with the DS heuristic or the PP/DS optimiser. Sourcing decisions are
   either already done in DP or with rules-based ATP or are rather simple

and calculated according to quota arrangements. An advantage of this scenario is that only one set of master data has to be maintained (for PP/DS, not for SNP). Especially with rather complex production scenarios (frequent master data changes, alternative resources, sequence dependent set-up times, variant configuration, shelf life) this scenario has advantages.

4. Demand planning in SAP ERP™ – especially when there is already a sufficient solution using SOP or long-term planning this is an alternative to the use of DP. Production planning is performed with the PP heuristics, scheduling with the DS heuristics or the PP/DS optimiser.

5. Demand planning and production planning in SAP ERP™ – SAP APO™ is used only for scheduling. The scheduling is restricted to production orders unless the functionality for 'MRP Based Detailed Scheduling' with SAP ERP™ 5.0 Enhancement Pack 2 is used (transaction CFDS in SAP ERP™). The advantage of this scenario is that only products and resources are required as master data.

6. Only medium-term production planning on rough-cut level is performed in SAP APO™ – either by the SNP heuristic, the SNP optimiser or CTM with SNP master data. The benefit is having a capacity check and purchase requisitions for components with long lead time. Planned orders are transferred to SAP ERP™, where production planning and scheduling on detailed level for the short-term horizon takes place.

7. Production planning is triggered by the sales order entry either via CTP or multi-level ATP (cf. chapter 16). Scheduling is performed with the scheduling heuristics or the PP/DS optimiser.

Figure 13.4 visualises the process flow across the applications for these:

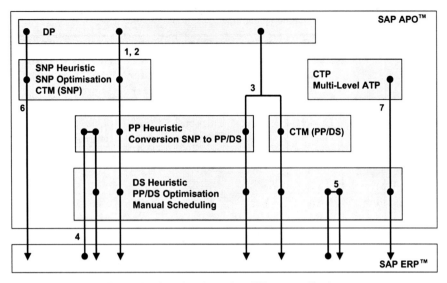

Fig. 13.4. Scenarios for production planning using different applications

The applications for production planning and their level of detail are listed in table 13.1. The applications SNP optimiser and CTM have been described before in chapter 10.

Table 13.1. Applications for production planning and their level of detail

Application	SNP Master Data	PP/DS Master Data
SNP Heuristic	X	
SNP Optimiser	X	
CTM	X	X
PP/DS Heuristic		X

The PP/DS optimiser – another application in this area – is used only for scheduling, not for the creation of orders. Each of these applications has its strengths and its weaknesses to solve given planning problems. In table 13.2 some of their properties are listed.

Table 13.2. Properties of production planning applications

Function	SNP Heuristic	SNP Optimiser	CTM (SNP Master Data)	CTM (PP/DS Master Data)	PP/DS Heuristic	PP/DS Optimiser
Order Creation	Yes	Yes	Yes	Yes	Yes	No
Scheduling	Bucket	Bucket	Bucket	Continuous	Continuous	Continuous
Set-Up	Yes	Yes	Yes[1]	Yes[1]	Yes	Yes
Sequence Dependent Set-Up	No	No	No	No	Yes	Yes
Finite Planning	Manually[2]	Yes	Yes	Yes	Restricted	Yes
Infinite Planning	Yes	Yes[11]	Yes[12]	Yes[12]	Yes	Yes
Alternative Resources	Yes[3,5]	Yes[5,4]	Yes[5]	Yes[6]	Yes[6,7]	Yes[6]
Altern. PDS/PPM	Yes[3]	Yes[4]	Yes[3]	Yes[3]	Yes[3 or 7]	No
Extend. Capacity	No	Yes	No	No	No	No
Parallel Operation	No	No	No	No	Yes	Yes
Prod. Horizon	Yes[8]	Yes[9]	Yes[9]	Yes[9]	Yes[10]	No

[1] note
[2] using capacity levelling functionality
[3] according to quota arrangements
[4] according to total cost minimum
[5] modelling as alternative PPMs
[6] modelling as alternative modes within one PPM
[7] alternatives are only taken into account in combination with finite planning
[8] production planning only outside the production horizon
[9] production horizon can be disregarded, else production planning only outside
[10] production planning per default only within the production horizon
[11] setting in the optimiser profile, relates to the entire scope
[12] with the use of planning parameters

The complexity of the applications, their tolerance to imperfect master data and configuration and the difficulty to interpret the solution varies widely. Table 13.3 gives a rough categorisation about the difficulties for the implementation.

Table 13.3. Difficulties and risks of production planning applications

Property	SNP Heuristic	SNP Opti- miser	CTM	PP/DS Heuristic	PP/DS Opti- miser
Interpretation of Results	+ +	- -	-	+	-
Interactive Planning	o	-	o	+	o
Robustness of Design & Configuration	+	-	o	o	-
Robustness to Faulty Master Data	+	-	o	o	o

In the design of the PP/DS heuristic some traps and temptations exist – mainly in the use of one step finite planning – which may cause many difficulties. Therefore from APO 4.0 on finite planning with a PP heuristic is only possible in the case of an upgrade or via BAdI. In any case it is not recommended to use it as described in more detail in chapter 15.2.

For the SNP optimiser the values of the costs which have to be maintained in many objects (e.g. product master, PPM, resource variant) increase the sensitivity to faulty master data. Though we consider the implementation and the use of the SNP optimiser as rather difficult, it is the only application which is able to create optimised production plans regarding the entire supply chain costs and capacity constraints.

13.2.2 Lot Size

Lot sizes have a significant impact on the production planning result and the challenges for scheduling. Generally speaking, fixed, minimum and periodical lot sizes increase the complexity of the planning problem.

In many cases the lot sizes are defined by the production process. Some machines require fixed lot sizes, e.g. in chemical and pharmaceutical production processes due to batch production and validation. Still there are sometimes alternatives in the modelling of the lot size. Generally small lot sizes increase the number of orders, which affect both the effort for planning and execution. Big lot sizes on the other hand cause an increase of the lead time. As a synthesis there is the possibility of continuous consumption (see section 19.3), which can be applied with some restriction for production processes within one location. Figure 13.5 visualises the effect of the different lot size methods. The lot sizes are set per locationproduct in the product master.

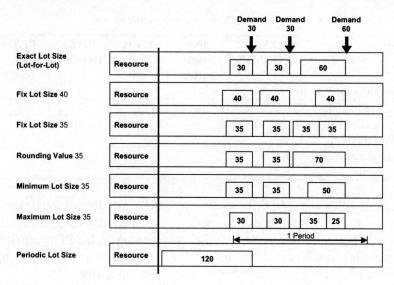

Fig. 13.5. Lot size methods

Periodic lot sizes group the demands of a period and create one order to cover the demands. If the demands change, the order is adjusted in date and quantity.

Not all lot sizes are supported by the SNP production planning applications. Depending whether discretisation is used for SNP optimisation, more or less possibilities exist within SNP, table 13.4:

Table 13.4. Lot size methods in SNP

Lot Size	SNP Heuristic	SNP Optimisation	SNP Optimisation (Discrete)	CTM (SNP Master Data)
Exact/Lot-for-lot	Yes	Yes	Yes	Yes
Minimum	Yes	No	Yes	Yes
Maximum	Yes	Yes	Yes	Yes
Fix	Yes	No	Yes	Yes
Rounding Value	Yes	No	Yes	Yes
Periodic	Yes	No	No	Yes[1]

[1] workaround

Though CTM supports minimum and fix lot sizes, we recommend to use them only very carefully in combination with SNP master data, since CTM performs a finite planning and does not split orders across buckets which might lead to a very low utilisation.

13.2.3 Scrap

Depending on the production process more or less scrap is produced, which might affect all or only some components. In SAP ERP™ several possibilities exist to maintain scrap in the material master, in the BOM and in the routing. Table 13.5 lists these scrap types and their behaviour.

Table 13.5. Scrap types in SAP ERP™

Scrap Type in SAP ERP™	Maintenance	Function
Assembly Scrap	Material Master of Output Material	Increases the total order quantity (Order Quantity * [1+Scrap]) and accordingly the dependent demands.
Component Scrap	Material Master of Component	Increases the demand for the component (Order Quantity * BOM-Ratio * [1+Scrap]). Assembly scrap adds to this.
Component Scrap	BOM on Component Level	Overwrites the component scrap of the material master. Same function as above.
Component Scrap & Net Flag	BOM on Component Level	Net flag annuls the assembly scrap, so that only the component scrap is used.
Operation Scrap & Net Flag	BOM on Component Level	Operation scrap has to be used together with the net flag and increases the dependent demand (Order Quantity * [1+Scrap]).

The impact of these different scrap types on the dependent demands for an order in SAP ERP™ is shown in figure 13.6. The ratio between output and input material is one to one in this example. The settings in the material master are displayed within the box for the material, settings in the BOM are displayed at the according branches.

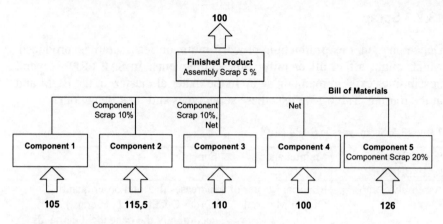

Fig. 13.6. Assembly and component scrap in SAP ERP™

In SAP APO™ there are only two possibilities to maintain scrap, one is the assembly scrap in the product master, the other one is the operation scrap in the PPM (which is maintained in the production activities).

The concept of scrap in SAP APO™ differs fundamentally from the concept in SAP ERP™. While in SAP ERP™ the dependent demand is increased by the assembly scrap,

$$\text{Input} = \text{Output} * (1 + \text{Scrap}_{R/3}) \qquad (13.1)$$

in SAP APO™ the output quantity is decreased by the assembly scrap:

$$\text{Output} = \text{Input} * (1 - \text{Scrap}_{APO}) \qquad (13.2)$$

$$\text{Input} = \text{Output} / (1 - \text{Scrap}_{APO}) \qquad (13.3)$$

Accordingly a fixed lot size in SAP APO™ represents the total order quantity, that is the sum of the output quantity plus the scrap (differing from SAP ERP™, where the fixed lot size equals the output quantity). Note 390850 mentions the motivation for having these different ways of handling scrap.

The practical implications are that the scrap factor has to be adjusted during the master data transfer:

$$\text{Scrap}_{APO} = 1 - 1 / (1 + \text{Scrap}_{R/3}) \qquad (13.4)$$

The adjustment of the assembly scrap value is performed during master data transfer in the CIF (see also note 334222), but problems might occur caused by the rounding inaccuracy. If a fixed lot size is used, the value of the fixed lot size has to be adjusted as well to achieve the same result as in SAP ERP™:

$$\text{Fix Lot Size }_{APO} = \text{Fix lot Size }_{R/3} / (1 - \text{Scrap}_{APO}) \qquad (13.5)$$

The fixed lot size is transferred from SAP ERP™ without any conversion and the scrap is adjusted during transfer. Again, due to rounding problems it might be a bit difficult to get the same quantities as in SAP ERP™.

Component scrap (both from the BOM and from the material master) and operation scrap from the BOM are exploded during CIF transfer and are modelled by a modified BOM ratio. A restriction regarding the modelling of scrap in SAP ERP™ is that the net flag is not considered in SAP APO™. With a modification as described in note 350583 however this can be achieved for the PPM (not for the PDS) as well.

Operation scrap from the routing is transferred into the operation scrap of the production activity without any adjustment. Its impact is explained using the order structure in figure 13.7.

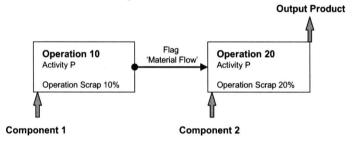

Fig. 13.7. Example for operation scrap

Depending on the setting for the material flow in the activity relations in the PDS or PPM, operation scrap is calculated as shown in table 13.6.

Table 13.6. Impact of the material flow

| | Scrap for Component | |
	Material Flow	No Material Flow
Component 1	$1 / \{(1\text{-Scrap}_{Op.20}) (1\text{-Scrap}_{Op.10})\} - 1$	$1 / (1\text{-Scrap}_{Op.10}) - 1$
Component 2	$1 / (1\text{-Scrap}_{Op.20}) - 1$	$1 / (1\text{-Scrap}_{Op.20}) - 1$

By default the material flow flag is set, which causes a cumulation of the scrap and represents the case that the first component is processed in the second operation as well.

13.3 Feasible Plans

Though it is theoretically possible to create feasible plans using the production planning heuristics, this is not recommended both from the business point of view and from the system performance view. Finite planning represents a simultaneous creation of planned orders and their scheduling. Production planning is in no means suited for any kind of scheduling optimisation, therefore a scheduling process step will be required anyway. The business implication is that using finite scheduling the utilisation of the resource will usually decrease due to a scattered loading and the ability to produce in time is likely to decrease as shown in figure 13.8.

Note that using the strategy 'squeeze in' is not suited for mass processing. Another issue is that production planning is performed product-by-product, which implies that if finite planning is used, the last product to be planned will get the worst schedule. If finite planning is used nevertheless, it is strongly recommended not to have more than one finite resource per order.

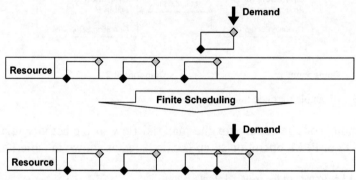

Fig. 13.8. Resource utilisation and lateness with finite scheduling

As an alternative to infinite planning it is possible to create orders in a de-allocated mode with the use of a BAdI as described in note 362208. For releases lower than SAP APO™ 4.0 creating orders in de-allocated mode has the advantage that if sequence dependent set-up is used scheduling problems due to an uncertain sequence are excluded.

Another aspect of the feasibility of a plan is whether the sequence of the material flow is regarded, that is whether the order for the component ends before it is required for the order for the finished product. To prevent the scheduling into the past, the result of the production planning run might violate the material flow sequence (this is even more likely if finite planning is applied). Though it is possible to enforce production planning to respect the material flow sequence using the multi-level MRP heuristic, it is strongly recommended to use only the single-level MRP heuristic and resolve any material flow violations in the subsequent scheduling step. The difference between the single-level MRP heuristic SAP_MRP_001 and the multi-level MRP heuristic SAP_MRP_002 is that SAP_MRP_001 is executed only for one level according to the flag 'single-level'. SAP_MRP_002 plans all levels downwards and adjusts the planning for the top levels immediately as shown for an example in figure 13.9.

In this comparatively small example already six planning runs are necessary for multi-level MRP instead of three for single-level MRP. If more than one scheduling attempt is necessary for several components, especially in crossing or diverging material flows, this ratio gets even worse.

For the reasons described above, production planning should preferably be performed infinite or in de-allocated mode and only with single-level MRP. Creating a feasible plan therefore requires a subsequent scheduling step to restore the material flow sequence and for the finite scheduling of the operations.

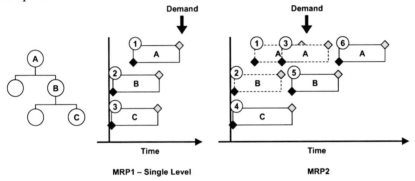

Fig. 13.9. Single- and multi-level MRP

SAP APO™ offers different possibilities to create feasible plans on detailed level. Among these are the scheduling heuristics, the PP/DS optimiser, CTM and manual scheduling. The appropriate scenario to create a feasible plan depends – as always – on the requirements and the circumstances, e.g. which resources, which BOM levels and which components have to be planned finite and for which horizon. Another question is whether it is necessary to have the information immediately available (e.g. to confirm sales orders to the customer on the telephone). If this is the case, the sales order confirmation process during the ATP check has to be regarded as well (cf. chapter 16) and the check of allocations, multi-level ATP or CTP have to be examined.

• *Feasible Plans on Detailed Level*
Feasible plans on detailed level are created either by manual scheduling, by scheduling heuristics or by the PP/DS optimiser – or by a combination of these. Though the approach to use the optimiser for the complete planning problem is very tempting, it does have the downside that the result is difficult to understand and to control.

Usually the approach to analyse the planning problems on the resources and to use the appropriate planning methods per resource group (e.g. per product group and/or BOM-level) is more successful. Generally the PP/DS optimiser is suited for complex scheduling problems on key resources, the scheduling heuristics for medium to simple problems on key resources and the infinite service heuristics (bottom-up and top-down) to adjust the orders on the non-bottleneck resources.

• *Feasibility vs. Utilisation*
The downside of an automatically created feasible plan is that the result in terms of resource utilisation, output or lateness might be insufficient – this is the more likely the more constraints are modelled. One possibility to adjust the plan is manually, but it might be quite difficult to find the cause for the delay resp. low utilisation, especially because any resource conflict is propagated to the finished product level. As an alternative it is possible to encounter for some problems by de-allocating operations – both with scheduling heuristics and the PP/DS optimiser. The required settings are described in the respective chapters.

13.4 Master Data for Production

13.4.1 Production Master Data Overview

The relevant master data for production planning are mainly the location, the product, the resource and the PDS resp. the PPM. The PDS and the PPM describe both the operations and their capacity and component requirements for in-house production. The PDS and the PPM are alternatives and will be referred to as 'plans' for the sake of simplicity.

• *Resources*
SNP and PP/DS require a different view of the capacity – SNP in buckets, and PP/DS as a time continuous capacity. Therefore the resources for SNP and PP/DS have different properties and are either different objects (bucket resources for SNP, single- or multi-resources for PP/DS) or combine

both properties in one object (single-mixed or multi-mixed). In the capacity of the work center resp. the resource in SAP ERP™ it is possible to control which resource type (single/multi, single/multi-mixed or bucket) shall be created in SAP APO™ during CIF-transfer, figure 13.10.

The prerequisite for this is that the transfer is set active on SAP ERP™ side with transaction CFC9. Until SAP APO™ 4.0 it is still necessary to use the include ZXAPOBAPIUSERU02 of the user exit EXIT_SAPL10004_001 within the enhancement APOBP002 to create mixed resources.

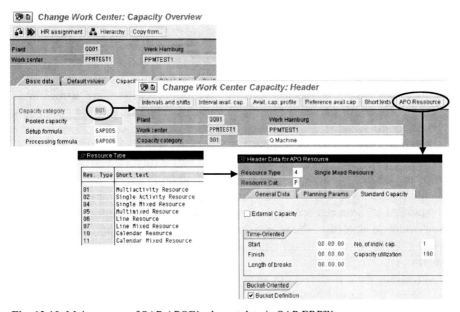

Fig. 13.10. Maintenance of SAP APO™ relevant data in SAP ERP™

• *Plans (PDS resp. PPM)*

The PDS and the PPM contain the information about the BOM and the routing and correspond to the production version in SAP ERP™. The applications PP/DS and SNP require separate master data objects.

Figure 13.11 provides an overview about the structure for the PDS and the PPM. The structure of the PDS and the PPM are similar.

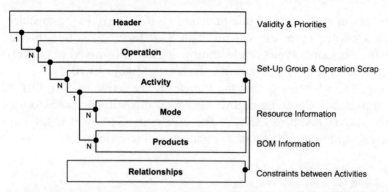

Fig. 13.11. Structure of the PDS and the PPM

Since SAP APO™ 4.1 the PDS (in SAP APO™ 4.0: RTO) exists as an alternative master data object to the PPM. Nevertheless there are some functional differences between these objects – mainly that the PDS supports Engineering Change Management. Another one is that the PDS is only created during transfer from SAP ERP™ – within SAP APO™ it is only possible to display it with the consequence that it might be already for prototyping necessary to implement BAdIs to maintain the SAP APO™ -specific fields. Notes 357178 and 507025 provide examples for enhancements. The functional differences are explained in more detail in section 13.4.5. It is important to note that there will be no more developments for the PPM. Note 705018 provides information for the migration from PPM to PDS as well.

For both the PPM and the PDS a production version in SAP ERP™ is required. Whilst the PPM transfer creates always a PP/DS-PPM, it is possible to transfer a production version as PP/DS-PDS and/or as SNP-PDS. The SNP-PPM is created either manually or by a generation report in SAP APO™ using the PP/DS-PPM as input (in this case mixed resources are required). For DP both the PDS and the PPM are created by respective generation reports. Figure 13.12 visualises the different options.

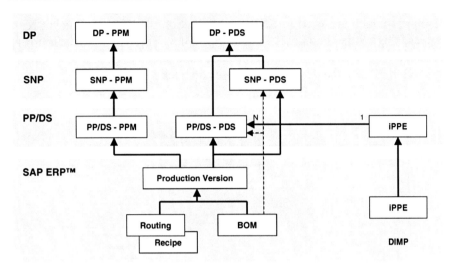

Fig. 13.12. Transfer of PPM and PDS

For special cases it is sufficient for the PDS to transfer only a BOM:
- phantoms (these do not have to be transferred for PPMs)
- subcontracting BOMs if the production is modelled in the plant
- any BOM if no scheduling takes place in SAP APO™, e.g. for multi-level ATP.

An alternative way to create the PDS is from the iPPE, if an iPPE is used.

Table 13.7. Priorities and costs for PPM and PDS

Field	SNP Heuristic	SNP Optimiser	CTM	PP/DS Heuristic
Priority	1st	---	---	1st
Multi-Level – Variable	---	---	1st	---
Multi-Level – Fix	2nd	---	1st	---
Single-Level – Var.	---	1st	---	2nd
Single-Level – Fix	---	1st	---	2nd

In case of alternative PPM or PDS the selection is performed either according to the priority or according to the cost. These fields are used differently in the planning applications, table 13.7.

13.4.2 Resource for SNP

Since SNP is an application for bucket-oriented planning, the resources for SNP offer a certain capacity per time bucket – usually per day. The resources which are used in SNP for production planning are bucket resources, single mixed and multi mixed resources. Because mixed resources combine the properties of bucket resources and single resp. multi resources as used in PP/DS, the description focuses on bucket resources.

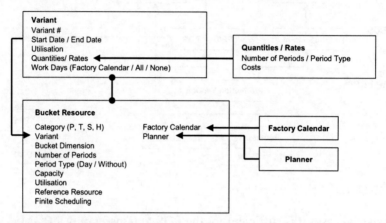

Fig. 13.13. Settings for a bucket resource

The bucket resources provide the capacity – e.g. the amount of working hours per day – of the resource which is used for the capacity consumption of the orders. For the scheduling of the orders the factory calendar is assigned to the resource. The capacity is defined in any unit of measure, and it is usually defined per day. The standard capacity is defined in the resource master.

If the capacity changes regularly – e.g. by reducing the number of shifts during vacations – and the time dependency of the capacities has to be considered, the use of capacity variants is helpful. These variants assign different capacities (defined in the entity 'quantities/rates') for time intervals to the resource. Figure 13.13 shows these settings for the bucket resource. Another way to define the capacity is the use of a reference resource, which helps to reduce the effort for the master data maintenance. The use of reference resources is described in section 13.4.5 for the PP/DS resources.

Though 'finite planning' can be flagged in the resource master, the SNP heuristic uses always infinite planning and the optimiser plans finitely resp. infinitely according to the optimiser profile setting. In the 'strategies'-view of the CTM profile it can be selected whether CTM regards all resources as finite, as infinite or as defined in the resource master.

The entry for the utilisation in the resource master diminishes the defined capacity (except for CTM) and can be used e.g. to model unpredicted downtimes.

The resources are maintained with the transaction /SAPAPO/RES01. For production the resource type has to be 'P'. The resources are maintained per version or model independently. Only the version dependent settings are used for scheduling and capacity calculation. If resources are created in SAP APO™ interactively (i.e. not via CIF), they are created model independently and have to be assigned to the relevant model interactively.

• *Resource Aggregation and Integration with SAP ERP™ and PP/DS*
The idea of SNP is to perform an aggregated planning. An aggregation regarding time is inevitable with the time bucket approach, but there is additionally the possibility to aggregate the resources resp. the work centers as well. There are two aspects to this aggregation, the business view and the system integration view.

The condition sine qua non for an aggregation of resources is that it makes sense for rough-cut planning from a business point of view. An example for this might be a set of similar production machines, e.g. a set of punches. If these machines are more or less identical and sequence dependent set-up does not apply, these machines might be aggregated throughout all SAP applications. If there are differences which have to be considered in detailed planning and/or sequence dependent set up is used, the production machines have to be modelled individually in PP/DS for scheduling purposes.

If the prerequisites for an aggregation in SNP are fulfilled the next question is the integration to PP/DS and SAP ERP™. If the SNP master data is maintained manually in SAP APO™, any kind of aggregation is possible – but both the SNP resources and the SNP-PPM have to maintained manually and the horizons for planning with SNP and planning with PP/DS have to be disjunctive (because in this case there is no integration of the capacity load to PP/DS).

If mixed resources are used, the work centers are transferred from SAP ERP™ and are created as mixed resources. In this case the modelling is rather a one to one relationship.

13.4.3 PDS and PPM for SNP

In this chapter we explain the basic structure of the PDS and the PPM for SNP (which is similar), how to generate SNP-PPMs from PP/DS-PPMs and what to consider for integration with SAP ERP™.

• *PDS and PPM for SNP*

The production data structure (PDS) and the production process model (PPM) are by far the most complicated master data objects in SAP APO™. The objects for SNP can be regarded as a soft introduction to the more complex PDS resp. PPM for PP/DS. The basic idea of the PDS and PPM is to group all the master data information that is needed to create an order into one object – corresponding to the production version that combines the BOM and the routing resp. the recipe in SAP ERP™.

The information is mainly stored on the levels 'header', 'operation', 'activity', 'mode', 'product' and 'relationship' as shown in figure 13.11. Related to the activity, input and output products are maintained in the 'component'-view and scheduling and capacity relevant data in the 'mode'-view. The relationship between the activities is the third information on this level, which is in the case of SNP always a chain with sequential predecessor and successor relations, figure 13.14.

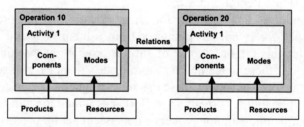

Fig. 13.14. PDS and PPM structure for SNP

The parameters for the input resp. output components are mainly their quantity and whether they are required resp. available at the start or at the end of the activity. The capacity consumption and the component requirements can be maintained as variable (i.e. depending on the order quantity) or as fixed (independent of the order quantity). All variable parameters – variable (component) consumption and variable capacity consumption – relate to the quantity of the output product. For the scheduling of the activity only a fixed duration is allowed, which should be a multiple of a day. Figure 13.15 gives an overview about the information on activity level.

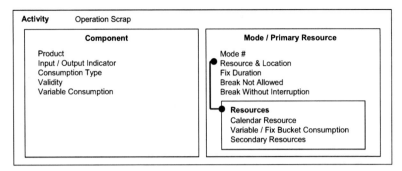

Fig. 13.15. PDS and PPM information on activity level

- *PDS for SNP*

Differing from the PPM for SNP, the PDS for SNP is transferred directly from SAP ERP™ and is not generated from the PP/DS-PDS. The structure of the SNP-PDS is simpler than the structure of the PP/DS-PDS. The structure of the SNP-PDS is used by the applications SNP, CTM and DP.

In a production version it is possible to maintain routings for detailed and rough-cut planning. For the PDS transfer it is possible to select from which routing the PDS will be generated (differing from the PPM transfer where always the routing for detailed planning was selected). This might be a modelling option for the transfer of SNP-PDS. The PDS is displayed with the transaction /SAPAPO/CURTO_SIMU.

Since the correspondence between the settings maintained in the routing on SAP ERP™ side and the settings required for the SNP-PDS is less obvious than for the PP/DS-PDS, the transformation logic has to be maintained in the method CALC_BUCKET_CONSUMPTION of the BAdI /SAPAPO/CURTO_SNP. /SAPAPO/CURTO_SNPDEF is a default implementation to calculate the bucket consumption from the standard values for the duration.

The PDS for SNP supports ECM only for components of the BOM. ECM for operations is not supported in SNP. In the case of extensive use of ECM with relevance for rough-cut planning it is recommended to maintain separate routings for rough-cut planning and transfer these as previously mentioned.

- *PPM for SNP*

To be precise and to confuse things a little bit more, there are actually two entities, the 'plan' and the 'PPM', where the plan contains the information regarding the production process itself and the PPM is used to access the plan. To create an order, first a PPM is selected according to the product, the location, the validity and the lot size. This PPM is linked unequivocally to one plan. In the case of by-production one PPM for each output product is linked to the same plan. Note that in most transactions the plan is used for selection, but there are also cases in which only the PPM is used. Unfortunately the distinction between plan and PPM is not always consistent. According to the usage in speech in the following the whole complex is referred as PPM.

PPMs are created with the transaction /SAPAPO/SCC05. The usage for SNP is 'S' and for DP 'D'. Because the structure of the DP-PPM is a simplified SNP-PPM without any scheduling relevant data (i.e. without modes), only the SNP-PPM is described. It is recommended to use as less detail as possible for the modelling of the production process in SNP. This means to include only the most critical operations and only the most critical components into the PPM.

• Generation of PPMs
Since SNP-PPMs are not transferred from SAP ERP™, they have to be created in SAP APO™. This can be done either manually or by generating SNP-PPMs from PP/DS-PPMs. In both cases it has to be considered for the process design that there is no integration of PPM changes at all, so that an enhanced PPM master data process has to be defined.

The planning in SNP uses a different data model than PP/DS (time buckets versus continuous time), therefore the PP/DS-PPM can not be converted 'one to one' but has to be interpreted – and for interpretation there is always some liberty. SAP offers two reports to generate SNP-PPMs, with lot size margins (transaction /SAPAPO/PPM_CONV_310) and without lot size margins (transaction /SAPAPO/PPM_CONV). Table 13.8 describes the properties of these reports. The generation of the PPMs is performed via creating a PP/DS planned order. Some other common properties of these reports are

- the input components are always linked to the start of the first activity, the output product is linked to the end of the last activity
- secondary resources in the PP/DS-PPM are considered as additional secondary resources
- order internal relationships and parallel sequences are modelled as additional secondary resources
- for components with limited validity note 513868 describes a way of modelling.

Note that the use of mixed resources is the prerequisite for both reports.

Table 13.8. Properties of the SNP-PPM generation reports

Feature	Generation Without Lot Size Margins	Generation With Lot Size Margins
Lot Size Interval	Manual maintenance	As in PP/DS-PPM
Validity (PPM & Comp.)	As in PP/DS-PPM	
Costs	Single (Var) * Lot Size; Multi (Var)* Lot Size; Multi (fix)	
Interpretation of Fix Component Requirement in PP/DS	Variable component requirement	
Modelling of Operations and Activities	New operation with change of the primary resource. New activity with change of the secondary resource or the material flow	One activity – the last resource is the calendar resource, others are secondary resources
Fix Duration	Calculated from lot size and resource availability (maintained manually)	1 day*
Resource Consumption	Bucket consumption of PP/DS-PPM	Durations in PP/DS-PPM
Interpretation of Fix Durations in PP/DS-PPM	Variable capacity consumption	Fix capacity consumption
Set-Up Durations	Variable capacity consumption	Added to the production activity
Sequence Dep. Set-Up	Set-up duration from 'blank' to set-up group/key	
Interpretation of Alternative Modes	Only one is considered	For each combination a separate PPM is created

* BAdI /SAPAPO/PPM_CALC allows to change the calculation of the fix duration

To reduce the level of detail for SNP planning, the products and resources which are marked as not SNP relevant in the respective master data are not considered for the PPM generation.

The usability of these reports for automated batch jobs is limited – the generation report without lot size margins is not batch enabled at all and for the generation with lot size margins the relevant PPMs have to be marked manually. Some helpful notes on this topic are 323884, 514842, 513868 and 516260.

● *Assignment of the Master Data to the Model*
Resources and PPMs which have not been transferred from SAP ERP™ but created manually or per report in SAP APO™, have to be assigned explic-

itly to the model. This is possible from both the resource and the PPM maintenance or within the supply chain engineer (transaction /SAPAPO/SCC07).

13.4.4 Resources for PP/DS

In PP/DS scheduling and capacity consumption are not separate steps, but the capacity is consumed by the scheduled operation. Therefore the basic property of a PP/DS resource is the working time, which depends on the standard working hours, the breaks and the factory calendar. If a resource is available without any breaks, the working time should be maintained as from 00:00:00 to 24:00:00 (instead of 23:59:59).

Another key property of the resource is the control flag for finite scheduling. If this flag is set, still a finite scheduling mode in the strategy profile is required to really perform finite scheduling. If the flag for finite scheduling is not set in the resource, the resource is always considered as infinite by the PP/DS heuristics (not necessarily by the PP/DS optimiser and CTM) and no alerts for overload are created. Whenever possible, the resource should be marked as infinite.

In PP/DS two types of production resources are used, the single resources and the multi resources (and the corresponding mixed resources). Resources are maintained with the transaction /SAPAPO/RES01. Single resources represent the most common type of a machine, where only one operation is executed at a time and the criterion for scheduling is whether a free slot for at least about the duration of the operation is available.

• *Resource Capacity*
Usually the available working time is modelled per shift, and the shifts are assigned to a shift sequence to model circumstances like less working hours on a Friday. The shift sequence is assigned to the resource as a variant. This way it is possible to use shift sequences time dependent – for example two shifts in summer instead of three. Figure 13.16 shows the settings for single activity resources and the capacity variants.

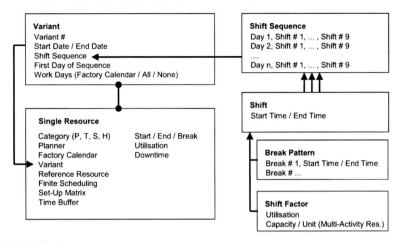

Fig. 13.16. Single resource

• *Finite Resources*

Two parameters determine whether finite scheduling is performed – one is the planning mode in the strategy profile, the other the 'finite scheduling' flag in the resource. Only the bottleneck resources should be marked as finite to avoid performance and scheduling problems due to too many constraints. If a resource is not marked as finite, it is neither in production planning nor in interactive scheduling planned finitely, but there are nevertheless possibilities to create a finite plan on infinite resources if desired (e.g. using the optimiser). Another implication of the flag 'finite scheduling' is that no overload alerts are created for infinite resources, table 13.9.

Table 13.9. Finite strategy and finite resource

	Planning Strategy Finite	Planning Strategy Infinite
Resource Finite	No Overload, Else Alert	Overload, Alert
Resource Infinite	Overload, No Alert	Overload, No Alert

In the capacity variant the shift sequences are assigned to the resource for specified time intervals. The setting 'first day' defines which day of the shift sequence matches the start day of the interval. Figure 13.17 visualises this matching for a three-day sequence, where the first day of the validity interval starts with the capacity of the second day of the sequence.

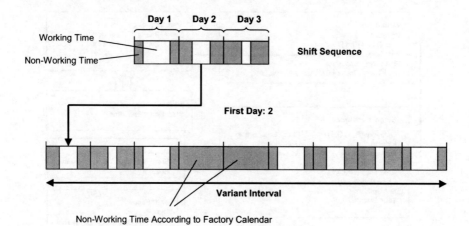

Fig. 13.17. Assignment of shift sequence to validity interval

Another way to define the resource capacity is using a reference resource. This is helpful in cases where the capacity is frequently adjusted for a set of resources in the same way. If a reference resource is used, only the capacity of this resource has to be adjusted. Figure 13.18 shows the logic to determine the valid resource capacity.

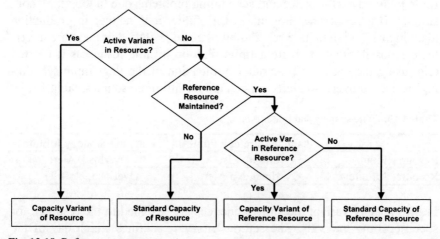

Fig. 13.18. Reference resources

• *Buffers*

There are mainly two possibilities to model time buffers in SAP APO™, these are reducing the utilisation of the resource with the effect that each operation is proportionally prolonged and using a time buffer for the resource.

The time buffer causes the operation to be scheduled the specified buffer time in advance. SAP APO™ does not support a time buffer reduction like SAP ERP™.

● *Multi Resources*
The capacity of a single resource is completely defined by the working time, and when an activity is scheduled on it, its capacity is completely blocked with this activity. Differing from this multi resources have a second capacity dimension, which allows more than one activity to be scheduled finitely at the same time on the resource – depending on the capacity consumption for the second capacity dimension.

An example for the use of a multi resource is a furnace, which is available for certain working times and has a restricted volume. The required heating duration in the furnace is usually independent of the lot size, but the volume is a hard constraint for the lot size. The volume of the furnace limits the number of the products (depending on their volume) and not the number of the orders. Figure 13.19 illustrates this example.

The consumption of the second resource capacity dimension (in this example the volume) is defined in the PPM.

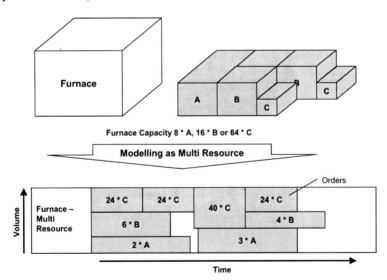

Fig. 13.19. Modelling of a furnace as a multi resource

Another parameter for the scheduling on multi resources is the flag for synchronisation. The synchronisation determines whether the operations have to start and to finish at the same time. Figure 13.20 shows the same load as in figure 13.19 above, only scheduled using the option for synchronisation.

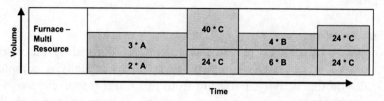

Fig. 13.20. Synchronisation

Because operations are scheduled in parallel, no defined sequence between predecessor and successor operations exists. Therefore sequence dependent set-up is not possible for multi resources and accordingly no set-up matrix can be assigned.

• *Calendar Resources*
Calendar resources are resources without capacity especially designed for the usage as handling resources, where only the calendar properties are required for the scheduling of the goods receipt in PP/DS. Using any other kind of resource for a handling resource leads to performance and object selection problems, because each order with a goods receipt time has an activity on that resource. Calendar resources are a separate resource kind and have their own tab in the transaction /SAPAPO/RES01.

• *Resource Types*
For better performance the scheduling is performed in the live cache. Therefore the resources are loaded into the live cache as well. If the order creation is aborted, one cause might be that the resources are not properly loaded into live cache. With the reports /SAPAPO/CRES_CREATE_LC_RES and /SAPAPO/OM_RESOURCE_GET_ALL it is possible to check the resource status in the live cache. The resource types in the live cache are listed in table 13.10.

Table 13.10. Resource types in live cache

Resource Type	Resource Type in Live Cache
Single Resource	2
Multi Resource	1
Single-Mixed Resource	4
Multi-Mixed Resource	5
Bucket Resource	3
Shipment Resource	8

• Resource Integration
Unless specified differently in the settings for the SAP APO™ resource within the work center capacity, the settings for the number of individual capacities or the flag 'can be used by several operations' in the work center capacity determine that the work center is transferred as a multi resource. The number of individual capacities is transferred as capacity.

It is not possible to change a multi resource to a single resource or vice versa. If the resource has the wrong type after transfer, it has to be deleted in SAP APO™, which requires the steps of deleting the resource from all PPMs, deleting the resource from all models and finally deleting the resource itself. The name of the resource in SAP APO™ is the concatenation of W[work center]_[plant]_[capacity category] – e.g. the corresponding of a work center 4711 in plant 1000 of the capacity category 001 is W4711_1000_001. The entries for the working times, the factory calendar, the utilisation and the flag for finite scheduling are transferred from the SAP ERP™ capacity to SAP APO™. If more than one capacity category is used – e.g. both machine and labour – for each capacity a resource is created.

13.4.5 PDS and PPM for PP/DS

As for SNP, the PDS and PPM for PP/DS consist of operations and activities, components (products) and modes (resources) which are assigned to the activities and relationships between the activities. The object for PP/DS is however more complex than for SNP – there are more settings and set-up is a separate activity.

• Operation and Activity
The operation is used rather as a grouping entity, the relevant data for planning is maintained on activity level, as
- activity type (e.g. production ['P'] and set up ['S']),
- whether the set-up duration is calculated sequence dependent and
- operation scrap (for activities of type 'production').

The entities for the component requirements and the operation duration and resource assignment (the mode) are defined on activity level, too.

• Components
On component level the information is maintained whether the product is an input or output, the quantity of the product and the consumption type (at the start, at the end or continuously).

The BOM ratio is maintained as variable consumption (related to the output quantity, i.e. the base quantity of the PDS resp. PPM) and/or as fix consumption (independent of the order quantity, e.g. catalyst products). If the input/output indicator is set to 'O', the 'consumption' has the significance of a production.

For the PP/DS-PPM the component view is structured into the 'logical components'-view and – for each logical component – the 'alternative component'-view. The name of the latter is a little bit misguiding, because alternative components can not be modelled within one PPM, but different products with different validities are grouped to a logical component.

• *Mode*

The mode determines - analogous to the routing in SAP ERP™ - the duration of the activity and on which resource the activity is scheduled. Alternative resources are modelled as alternative modes. Entries on this level are

- the mode priority,
- the resource and the location,
- the variable and the fix duration (related to the base quantity) and
- scheduling constraints (e.g. break not allowed, production within a shift).

The setting to inhibit breaks might cause errors if the required time slot for an operation is larger than the available time slot without breaks. In the 'resource'-view to each primary resource it is additionally defined

- whether it is a calendar resource (this setting is independent of the resource type 'calendar resource'): only calendar resources are relevant for scheduling. Per mode only one calendar resource is possible.
- for multi resources: the variable and the fix capacity consumption: in the example of a furnace in figure 13.19 the variable capacity consumption for the production of A is 1/8 of the furnace capacity, 1/16 for B and 1/64 for C. If variable capacity consumption is used, the duration has to be fix – the increase of both capacity consumption and duration with the order size is not supported. A proposition for the integration of the capacity consumption from the SAP ERP™ routing is to use the resource capacity as base quantity of the operation and apply it to calculate the variable consumption in a user exit. If the multi resource is used to model the labour, the capacity consumption is fixed, but the duration is variable. A proposition for the integration from SAP ERP™ is to transfer the ratio of labour time to work time as fix consumption via user exit.

- For bucket resources: the variable and the fixed bucket consumption, which is used in the SNP capacity evaluation and should therefore correspond to the values maintained above, and optionally
- secondary resources, if it is desired to load additional resources with the operation, e.g. for the modelling of the labour requirements.

Figure 13.21 provides an overview of all these settings on activity level.

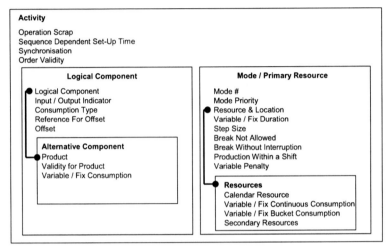

Fig. 13.21. Settings on activity level for PP/DS-PPMs

• *Relationships*

In the 'relationship'-view of the PPM the dependencies between the activities are defined by

- the reference type, which defines the sequence of the activities (start-start [0], start-end [1], end-start [2] and end-end [3]),
- whether minimum and/or maximum time constraint exist and the value of these constraints,
- whether the time buffer of the resource is used ('reference subtype'),
- whether the mode of the next activity is chosen freely or restricted regarding identical primary resource or identical mode number ('mode linkage').

The flag for 'material flow' defines whether material is passed on between the activities and therefore the operational scrap has to be cumulated. The influence of this setting is described in the paragraph about 'scrap'.

• *Header*

On header level the same settings as in SNP are possible. The PPM is selected according to the settings product, location, lot size, validity date and procurement priority. The lot size interval of the PPM corresponds to the lot size interval of the production version. Note that the lot size in the PPM is used to select the PPM and has no influence at all on the order lot size. The order lot size is only influenced by the lot size settings in the product master.

• *PDS vs. PPM*

The main advantages of the PDS compared to the PPM is that PDS supports the engineering change management and the integration of object dependencies (see Dickersbach 2005 for the latter). With SAP APO™ 4.1 the PDS does not yet support all the functions the PPM does, but the PPM will not be developed any further and the PDS will. Some of the key differences in the supported functionality are listed in table 13.11. Note 517264 contains the link to a detailed functionality matrix where PDS and PPM are compared for different releases.

Table 13.11. Supported functionality – PPM vs. PDS in SAP APO™

Supported by PDS and Not Supported by PPM	Not Supported by PDS But Supported by PPM
ECM (Date Effectivity)	Product Flow (for Container Resources)
Order BOM & Project BOM	Maintenance in SAP APO™
Integration of Object Dependencies	
Rapid Planning Matrix (with DIMP)	

• *PDS for CTM with PP/DS Master Data*

If CTM is to be carried out with PP/DS, a PDS of the type 'CTM' is generated from the PP/DS-PDS at the point in time of the transfer of the PP/DS-PDS. The prerequisite for this is that the method CREATE_CTM_PDS of the BAdI /SAPAPO/CURTO_CREATE is be activated.

• *Navigation in the Production Process Model*

The navigation through the PPM might be a little difficult. Sometimes using the icons can be helpful. Figure 13.22 gives a legend to these:

Fig. 13.22. Icons for navigation within a PPM

● *Integration of the Production Version*
To transfer the information of the BOM and the routing resp. recipe from
SAP ERP™ to the SAP APO™ PPM, first production versions have to be
created in SAP ERP™. Other prerequisites for the integration are that the
materials and the work centers are transferred (and that the according integration models are active, cf. chapter 25) and that the control key in the
routing is relevant for scheduling. Some typical integration problems are
described in note 448085. If routings are transferred, the relationships between the activities are by default activity chains with a minimum time
constraint of zero. For recipes the relationships are created as defined in
the recipe. If no relationships have been defined, the activity network is
not complete and – if the PPM is used – the PPM is not activated.

The name for the PDS is the concatenation of the material, plant, production version, supplier (for subcontracting) and usage. The name of the
PPM is a concatenation of the routing type ('N' for routing, '2' for recipe),
the routing group, the group counter and the production version.

● *Base Units*
The variable durations of the activities, which are transferred from the
standard values of the operation of the routing, are adjusted to the base
quantity of the BOM. Figure 13.23 visualises the correspondence of the
base units.

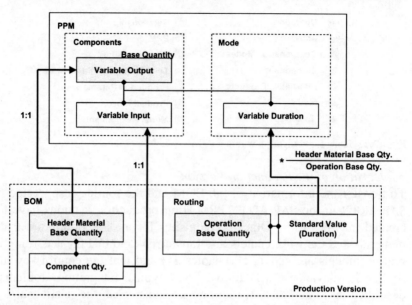

Fig. 13.23. Base quantities

If the BOM base quantity is much smaller than the base quantity of the operation, rounding problems occur or the operation duration is even zero. To calculate the fix and the variable duration for the activities, the standard value key and the scheduling formulas of the work center and the standard values of the operations are used.

• *Modelling of the Production Process*
As described above, there are many possibilities to model details and constraints. Since production planning is rather sensitive to performance problems and has to deal with the most complex calculations anyway, it is strongly recommended to keep the modelling as simple as possible. Each additional constraint makes planning more complicated.

• *Phantoms*
For the PDS the BOM for the phantom is transferred explicitly (without production version). For the PPM however the material master for the phantom itself is not transferred, but the components of the phantom are exploded and added to the PPM, so that the BOM level of the phantom is skipped. The restrictions in the handling of phantoms are that a phantom may only be used once within a BOM and that changes in the BOM for the phantom do not create any change pointers for the change transfer of the PPM (see also note 436395).

● *Secondary Resources*

If a work center contains more than one capacity, for each capacity a resource is created. The capacity category that is selected as basis for scheduling is transferred as primary resource and the other ones are transferred as secondary resources.

● *ERP Sub-operations*

Sub-operations in SAP ERP™ are modelled in SAP APO™ as operations of type 3, but there are some limitations. In SAP ERP™ sub-operations are not scheduled individually, but they inherit their scheduled dates from the operation to which they are assigned taking into account the relation type and offset. SAP APO™ does not apply this type of dependent scheduling but schedules the operations of type 3 individually which may lead to different results to SAP ERP™.

13.4.6 Integration to PP-PI

The structure of the recipe differs from the routing. Whereas in the routing most scheduling relevant information is assigned at operation level – work centre, standard values and standard values for the secondary resources – the recipe contains operations and phases. The resources are assigned to the operation but the standard values are maintained at phase level. Each operation contains one or more phases. Secondary resources are defined in the detail view of the operation (here secondary resources are separate resources and not an additional capacity of the work center) and the standard values are assigned in the detail of the secondary resource assignment. Another difference to the routing is that the phases have to be connected with order internal relationships explicitly.

● *PDS resp. PPM and Recipe*

Both operation and phase are transferred to SAP APO™ as operations, but with different operation types (operation type 1 corresponds to the operation of the recipe, operation type 2 to the phase). The corresponding structures of recipe and PDS resp. PPM are shown in figure 13.24.

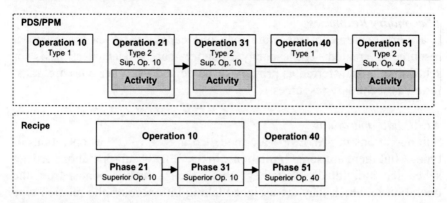

Fig. 13.24. Integration of the recipe to the PPM

The introduction of the dummy operation in the PDS and the PPM corresponding to the operation in the recipe is necessary because the confirmation in SAP ERP™ is done on operation level. The relationships within the PDS and the PPM are the same as maintained in the recipe.

• *Secondary Resources*
The modelling of secondary resources in the recipes is very different from the modelling in routings, and so is their integration to SAP APO™. The capacity requirement of the secondary resource is transferred as an additional operation (in SAP APO™ of the operation type 3) as shown in figure 13.25.

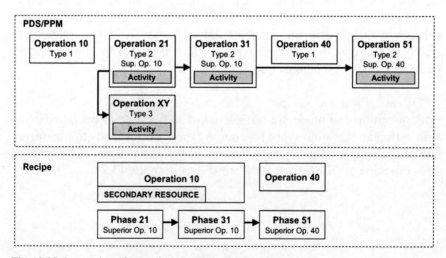

Fig. 13.25. Integration of secondary resources from recipes

The operation for the secondary resource is linked with a start-to-start relationship to the first phase of the operation. In the PDS this relationship is not displayed.

13.5 Dependencies to the SAP ERP™ Configuration

If production planning in SAP APO™ is going to replace production planning SAP ERP™, some points of attention should be checked to estimate the effort of rework that might be required in SAP ERP™. On one hand does SAP APO™ require some settings – mainly the use of production versions – on the other hand does SAP APO™ not support everything that is supported in SAP ERP™. Some of the limitations are

- production versions are mandatory,
- alternative sequences in the routing must have the same number of operations,
- resource hierarchies are not supported,
- dynamic buffers are not supported,
- only linear variable and fixed durations and capacity requirements are supported as formulas in the work center,
- distribution keys in the BOM (e.g. GLEI) are not supported for external procurement,
- serial numbers are not supported and
- dependent discontinuation is not supported.

The MRP planner is not transferred to SAP APO™ nor does it have a correspondence in SAP APO™. The PP/DS planner in SAP APO™ which is often used as a correspondence is not location specific (differing from the MRP planner).

● *Special Procurement Keys*
Special procurement keys are partially supported – table 13.12 provides an overview.

Table 13.12. Special procurement keys

Special Procurement Key	Modelling in APO
Consignment (10)	No impact for APO
Subcontracting (30, 31)	Described in Chapter 21
Stock Transfer Plant to MRP Area (45)	Described in Chapter 15
Direct Production/Collective Order (52)	Not Supported
Phantom in Planning (60)	Not Supported
Reservation from Alternate Plant (70)	Not Supported
Production in Different Plant (80)	Described in Chapter 15

14 Rough-Cut Production Planning

14.1 Basics of Rough-Cut Production Planning

The application for rough-cut production planning in SAP APO™ is SNP. The basic properties of SNP – e.g. the bucket-oriented planning and the planning book – are described in chapter 9. The objectives of rough-cut production planning in general are

- a rough-cut capacity check – either on finished product level or on component level,
- make-or-buy decisions based on capacity check resp. optimisation,
- calculation and check of the dependent demand for key components and
- the procurement of key components with long planned delivery time.

Within SNP the three applications SNP heuristic, SNP optimisation and CTM (with SNP master data) exist. The key question for deciding which one to apply is whether the information about the capacity load and some basic manual capacity levelling possibilities is sufficient or whether finite planning is really required. In the first case the SNP heuristic should be the appropriate application, else either SNP optimisation or CTM have to be considered. Each of these have a different approach to production planning including the scheduling of the orders. The common properties are described in the following.

• SNP Planning Book
The preferred tool to visualise the SNP results and perform interactive planning is the SNP planning book as described in section 9.5. Figure 14.1 shows the production planning specific part of the SNP planning book. Note that the SNP planning horizon which is blocked for SNP planning is highlighted for the key figure 'production (planned)'.

J. T. Dickersbach, *Supply Chain Management with APO*,
DOI: 10.1007/978-3-540-92942-0_14, © Springer-Verlag Berlin Heidelberg 2009

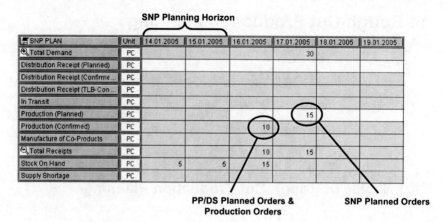

Fig. 14.1. SNP planning book

The respective key figures and the according categories are listed in table 14.1.

Table 14.1. Key figures and categories in the SNP standard planning book

Key Figure	Category Group
Dependent Demand	EL (SNP: Dependent Demand), AY (Dep. Demand)
Production (Planned)	EE (SNP: Planned Order)
Production (Confirmed)	AC (Production Order (Created)), AD (Prod. Order (Released)), AI (Planned Order), AJ (Planned Order (Fixed))

The categories for order reservation have to be added explicitly to the category group for the dependent demand. Note that planned orders created by PP/DS are regarded as fixed for SNP.

Though all settings can be changed like in DP, especially the changes in the planning book might affect some vital macros for functions like the SNP heuristic or the deployment.

● *Capacity View*
The standard view for capacity evaluations is SNP94(2) of the planning book 9ASNP94. In this case the capacity load is displayed for the resources as selected in the shuffler. Figure 14.2 visualises this view with the key figures for the absolute resource capacity consumption and the relative resource capacity level.

CAPACITY PLAN	Unit	17.01.2005	18.01.2005	19.01.2005	20.01.2005	21.01.2005	22.01.2005
Capacity	H	24	24	24	24	24	
Capacity Consumption	H	30	35	10			
Resource Capacity Level in %	%	125	146	42			

Fig. 14.2. Capacity view

As details to the 'capacity'-key figure a normal, a maximum and a minimum capacity are displayed. These are used for optimisation as described in section 10.2.4. In the second grid it is possible to display the resource load per product and per PPM resp. PDS and perform an interactive planning for these as shown in figure 14.3.

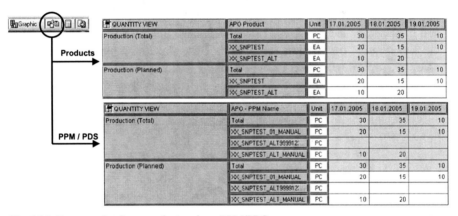

Fig. 14.3. Resource load per product and per PPM/PDS

The impact on the resource utilisation is shown online in the upper grid.

• *Demands in SNP*
It is possible to control whether the pegging relevant quantity, the requested quantity, the original quantity or the confirmed quantity is used in SNP (for the SNP heuristics). The selection of the relevant quantity is performed in the definition of the order category groups with transaction /SAPAPO/SNPCG. Chapter 5 describes how requirements strategies are supported in SNP and their impact on the relevant demand.

• *SNP Planned Order*
In SNP scheduling and capacity consumption are independent steps and use a different set of data (like in SAP ERP™). While the duration of an order is always a multiple of a day and independent of the order quantity,

the capacity consumption is usually variable (i.e. proportional to the order quantity) – but fixed capacity consumption is supported as well.

The order scheduling in SNP has two implications. One is that the orders take at least one day. Since SNP is a medium-term planning tool and is not designed for detailed scheduling, this is no real restriction. Having a fixed order duration, it is assumed that the order quantities do not vary widely enough to cause considerable prolongations of their production lead time. If this is not the case and the prolonged durations are relevant for planning, a possible solution is to use different sources of supplies (PDS resp. PPM) with different durations for the different lot size intervals.

The planned orders for SNP have the category EE and are created according to the lot size settings in the product master, so that more than one SNP planned order can be created within one bucket. The single orders and their details are displayed by right mouse click on the according cell of the key figure 'production (planned)'. Here the order number, the order quantity, the category, the PDS resp. the PPM and the availability date are displayed as shown in figure 14.4.

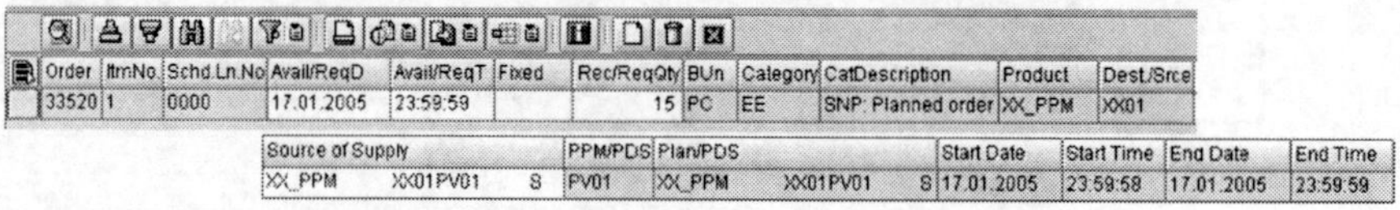

Order	ItmNo	Schd Ln.No	Avail/ReqD	Avail/ReqT	Fixed	Rec/ReqQty	BUn	Category	CatDescription	Product	Dest./Srce
33520	1	0000	17.01.2005	23:59:59		15	PC	EE	SNP: Planned order	XX_PPM	XX01

Source of Supply			PPM/PDS	Plan/PDS		Start Date	Start Time	End Date	End Time
XX_PPM	XX01PV01	S	PV01	XX_PPM	XX01PV01	S 17.01.2005	23:59:58	17.01.2005	23:59:59

Fig. 14.4. SNP order details

For more details, e.g. for the component demand or the operations, it is necessary to display the order with a PP/DS transaction.

SNP orders take part in pegging as any other orders. However pegging is neither taken into account by any SNP function (except CTM) nor can it be displayed by any SNP transaction.

14.2 SNP Heuristic

The SNP heuristic is a very simple way of infinite production planning on aggregated level per time period. The SNP heuristic tries to supply the quantities which are calculated using the key figures 'total demand' and 'total receipt'. Since both total demand and total receipt are calculated using the standard macro 'stock balance', there are possibilities to influence the behaviour of the heuristic.

The PDS resp. PPM for the planned order is selected according to the procurement priorities, and in case of equal priorities according to the fixed multi-level costs.

For the interactive mode – that is in the planning book – there are three ways to execute the SNP heuristic: either as 'location heuristic', 'multi-level heuristic' or 'network heuristic'. Figure 14.5 shows the difference in the planning scope regarding the locationproducts for the case that the heuristic is executed on the level plant and finished product. The dark locationproducts represent the ones that were planned.

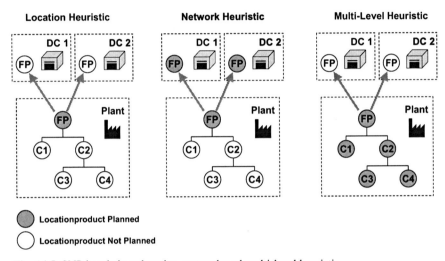

Fig. 14.5. SNP heuristic as location, network and multi-level heuristic

• *SNP Planning in the Background*
For the execution of the SNP heuristic in the background only the options for location heuristic and network heuristic are available. For the network heuristic it is possible to select the planning scope by product, for the location heuristic additionally by location and low level code. The calculation of the low level code is done with transaction /SAPAPO/SNPLLC or with the stage numbering algorithm (heuristic SAP_PP_020) as for PP/DS.

• *Net Change Planning*
From SAP APO™ 4.1 on the net change planning is applicable for the SNP heuristic as well. A separate entry for the SNP change planning relevance is used within the planning file (transaction /SAPAPO/RRP_NETCH).

• *Interactive Planning*
It is possible to create planned orders by entering the requested quantity into the according key figure in the planning book. These orders are fixed. The only place where this is displayed is in the SNP order details. Fixed orders are not deleted any more by the SNP heuristic. If an alternative PPM resp. PDS exist, a pop-up window requests a selection.

14.3 Capacity Levelling

The capacity requirements of an order are calculated during the explosion of the PPM or PDS and displayed in the capacity view of the planning book. Capacity levelling is performed either as a heuristic, as an optimisation or using a BAdI. The settings for the capacity levelling (including the maximum capacity in percent) are maintained either interactively per run or are stored in the capacity levelling profile with the customising path *APO* → *Supply Chain Planning* → *SNP* → *Define SNP Capacity Levelling Profiles*. In each case the levelling is performed only for the selected resource and does not take the dependency to the demand nor the availability of the components into account.

Capacity levelling is either performed in an interactive mode from the capacity view of the planning board or is executed in the background with transaction /SAPAPO/SNP05.

• *Capacity Levelling Heuristic*
The principle of capacity levelling is rather easy to comprehend as figure 14.6 shows. Overloads are distributed to other buckets backwards, forwards or both with options to prioritise levelling according to the order size or the product priority.

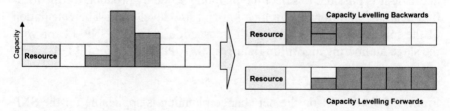

Fig. 14.6. Capacity levelling

The order size is adjusted during capacity levelling.

• *Capacity Levelling Optimisation*
Within the capacity levelling profile it is possible to choose the optimisation as capacity levelling method. The SNP optimiser as described in section 10.2 is used for this, but with the difference that the costs are created in the background according to the settings in the capacity levelling profile. The advantage is that no costs have to be maintained. If costs are already maintained, they are overruled.

14.4 SNP Optimisation for Production Planning

The properties of the SNP optimisation are already described in section 10.2 about the integrated distribution and production planning. To reduce the complexity of the optimisation problem and improve the performance, it might be a suitable approach to perform distribution planning using the SNP heuristic as described in section 11.2 and to concentrate on the optimisation of the production, especially if there are not too many sourcing alternatives in the supply network.

In this case the optimiser tries to meet the requirements for distribution demand (and – if customers are delivered directly from the plant – sales orders and forecast as well). The priority for the distribution demand is set in the optimiser profile. The main decisions or degrees of freedom for production planning with the SNP optimiser are

- the choice of the PPM resp. PDS,
- whether to extend the production capacity and
- in case of shortage: the prioritisation of the production according to penalties (per product and demand type).

An example for production planning using the SNP optimiser is the planning of dilution steps in chemical productions as in crop protection. In this case many products exist which contain the same ingredient, but with a different concentration. A product with a low concentration is produced either by a big dilution step or by several small dilution steps as shown in figure 14.7.

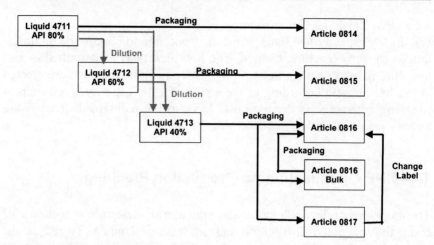

Fig. 14.7. Example for optimisation of production planning

The main issues here are the determination of the dilution steps and the prioritisation of the production lines. Liquid 4713 for example is produced in one or in two dilution steps, where two steps are more expensive than one step. On the other hand, if a small amount of liquid 4712 has to be produced anyhow, it might be favourable to choose two steps. These decisions get the more complicated the more alternative production possibilities (with different costs) exist.

As already mentioned in section 10.2, the SNP optimiser works best with completely linear problems. Depending on the production process a linear approach is not suitable any more and discretisation is necessary. Examples for the discretisation steps are the use of discrete lot sizes (e.g. to avoid a planned production of half pieces), fix and minimal lot sizes, lot size dependent costs and fix resource consumption. Table 14.2 lists the necessary settings for the respective discretisation. In each case additionally a discretisation method must be selected and the flag 'discretisation until end/detailed' resp. an end date has to be set in the optimiser profile.

Table 14.2. Settings for discretisation for the optimisation of production planning

Discretisation Step	Settings in the Optimiser Profile & the Master Data
Fix Lot Size (Product)	Integral PPM/PDS **PPM/PDS**: Discrete Output
Minimum Lot Size (Product)	Integral PPM/PDS **PPM/PDS**: Discrete Output
Rounding Value	Integral PPM/PDS **PPM/PDS**: Discrete Output
Maximum Lot Size (Product)[1]	-
Discrete Lot Sizes (Multiples of the PPM Base Qty.)	Integral PPM/PDS **PPM**: Discrete Output
Minimum Lot Size (PPM)	Minimum PPM/PDS Lot Size **PPM**: Discrete Output
Maximum Lot Size (PPM) [1]	Maximum PPM/PDS Lot Size
Fix Resource Consumption	Fixed Resource Consumption Production Capacity
Increase of Production Capacity	Discrete Production Capacity Increase Production Capacity **Resource**: Active Variant '2'

[1] no discretisation required

If the demand is below the minimum PPM resp. PDS lot size, the optimiser might increase the order quantity if the costs are less than the alternative. The maximum lot size of the PPM is interpreted by the optimiser as the maximum amount that can be produced per day (see note 503294). Procurement is always planned lot for lot, and reorder points are not taken into account (note 448986).

Fix resource consumption is manually calculated and displayed in the capacity view in any case. If the fix resource consumption should be taken into account for finite production planning, the discretisation step for fixed resource consumption has to be activated.

Like in the SNP heuristic, fixed orders are not deleted by the optimisation run.

14.5 Capable-to-Match (with SNP Master Data)

Like the SNP optimisation, CTM can be used for production planning at plant level for the same reasons at similar circumstances. The principles of CTM are described in section 10.3.

The maximum order lot size is calculated according to the provided resource capacity multiplied with the number of days for the duration in the PPM resp. PDS. If the required quantity exceeds this lot size, the order is split. The available capacity per bucket is defined in the resource master. Per default the capacity is related to one day, but analogous to the definition of the supplier capacity (see section 20.5) it is possible to create weekly (or larger) buckets for the resource. If the lot size of the product exceeds the maximum order lot size that is supported by the bucket, the order is not created. Therefore we recommend to use exact lot size and be very careful with any minimum or fixed lot sizes.

Differing from the other SNP applications the alternative modes of the PPM resp. PDS are considered by CTM.

14.6 Scheduling in SNP

SNP creates planned orders only outside the SNP production horizon. The SNP production horizon is set in the 'SNP2'-view of the product master. If no entry is maintained in the product master, CTM takes the entry for the PP/DS horizon. The purpose of the production horizons is to prevent the planning result of detailed scheduling from changes caused by medium-term planning on one hand and to allow an overlap between SNP planning and PP/DS planning on the other hand.

Set-up is modelled using fixed resource consumption, sequence dependent set-up is however not possible. The orders are scheduled in SNP according to the production calendar of the location. The properties of the calendars are described in section 11.3.

The scheduling of the orders within and cross buckets is different for each application.

• *Scheduling in SNP Heuristic*
The duration of an order in SNP is a multiple of a day. The choice of the time period (day, week,...) for planning has an influence on the scheduling of the orders as shown in figure 14.8.

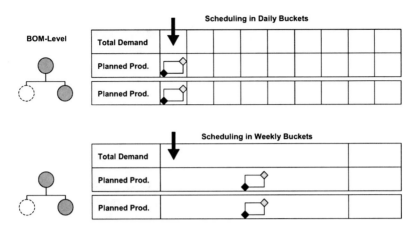

Fig. 14.8. Scheduling in SNP heuristic – receipt in the middle of the bucket

By default the planned orders are scheduled in the middle of the bucket. The 'bucket factor' in the PPM resp. PDS or – if this is initial - the 'period factor' in the 'lot size'-view of the product master determines whether the order is scheduled at the beginning, at the middle or at the end of the time period. Figure 14.9 visualises the impact of these factors. These settings are only valid for the SNP heuristic.

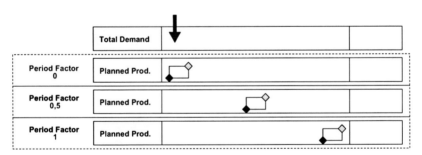

Fig. 14.9. Period factor

The availability time of the planned order is always at 23:59:59, and the time for the dependent demand is at 00:00:00. If the SNP heuristic is used for the production planning of multiple BOM levels, the planned total lead time is less than the sum of each order duration, as figure 14.10 illustrates.

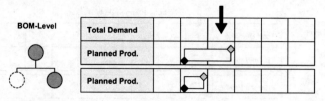

Fig. 14.10. Production scheduling for multiple BOM levels

● *Scheduling in SNP Optimisation*
The order scheduling of the SNP optimiser differs from the scheduling used in the SNP heuristic. Depending on the production time rounding value in the optimiser profile, more or less conservative scheduling is performed. Figure 14.11 shows the respective scheduling results.

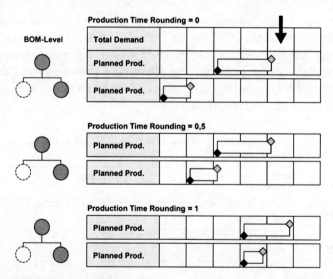

Fig. 14.11. Scheduling by the SNP optimiser

The production time rounding parameter is only effective if the bucket offset in the PPM is initial.

● *Scheduling in CTM*
Like for the SNP optimiser, the scheduling differs from the SNP heuristic as well as displayed in figure 14.12.

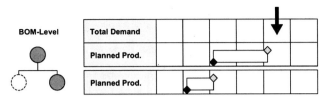

Fig. 14.12. Scheduling in CTM

Because CTM uses pegging, the supplying orders will be in time for the consuming orders.

14.7 Integration to PP/DS and SAP ERP™

14.7.1 Process Implications of Rough-Cut and Detailed Planning

The idea of production planning in SNP is to perform an aggregated, rough-cut medium-term production plan to get an overview about the overall capacity requirements, the requirements for the key components and to trigger procurement for components with long delivery times. The next step towards execution is either a transition to detailed planning in PP/DS or a transfer to SAP ERP™ as shown in figure 14.13.

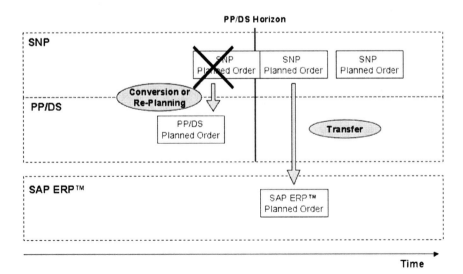

Fig. 14.13. Integration of SNP orders

If the result is transferred to SAP ERP™ the detailed scheduling and planning is usually performed on shop floor level or using legacy systems. For the integration towards PP/DS the process becomes more complicated and the business requirements may differ depending on the roles and responsibilities of the respective planners.

● *Supply Chain Planning vs. Production Planning*
Production planning has traditionally the focus on a local plant, whereas supply chain planning usually has a broader view. Production planning with SNP might tend to either side, depending on the process requirements.

If the supply chain planning is driven by production planning – which is mostly the case if there is either single sourcing for production or there is a dominant plant, the supply network is simple and the distribution planning is not too complex, rough-cut and detailed production planning is often carried out by the same person. In this case instantaneous access to the same objects in both applications has a high priority.

In the opposite case when supply chain planning is driving the planning process, there is usually a separate organisation which is responsible for the most of the supply chain related KPIs like inventories and delivery performance. This is often the case in more complex supply networks with a huge make-to-stock portion, where order lead times are short – e.g. in the consumer goods industry. Here a clear separation between the responsibilities for the whole supply chain and for the manufacturing is desired, so that a distinct hand over of the plan from supply chain planning to production planning takes place. In this case a transparent and traceable processing of the demands has the higher priority – even if several steps and different demand objects are involved.

Other aspects for the integration are the different levels of automation resp. of the possibilities for interactive planning, the frequency, the level of detail resp. the accuracy (i.e. which constraints are taken into account) and whether cross location sourcing decisions have to be made.

Depending on the chosen scenario, the requirements for the interaction between SNP and PP/DS or SAP ERP™ will be quite different. In the following some of the different possibilities are explained.

14.7.2 Integration to PP/DS

The integration between SNP and PP/DS contains a change from an at least daily bucket to a time-continuous planning and involves for production a completely different set of master data. The main parameters are the way of the transformation of the SNP order to the PP/DS order and the use of the SNP and the PP/DS horizon.

● *SNP and PP/DS Horizon*
One of the questions regarding the integration to PP/DS concerns the time horizon where SNP is used and where PP/DS is used. Since procurement can be triggered from SNP as well, the lead time of procurement should not be used as an indication for the horizons. Better indicators for the need of a detailed scheduled production plan are the fixed horizons for production, the set-up times and production cycles.

The scenario does affect the way the SNP horizon and the PP/DS horizon might overlap. The two common ways to integrate SNP with PP/DS are to limit SNP planning (distribution and production) to the medium-term or to use SNP for distribution planning across short- and medium-term and restrict SNP only for production planning to the medium-term as shown in figure 14.14.

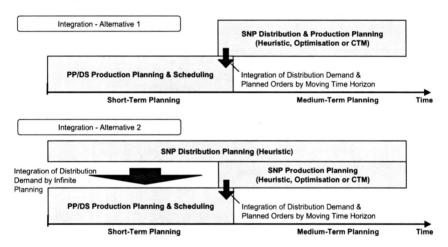

Fig. 14.14. Horizons for SNP – PP/DS integration

The downside of the first alternative is that this way the production is not able to respond to changes in the demand within the short-term horizon. To propagate changes in the demand situation across the supply network to the production, the alternative approach requires the use of the SNP heuristic (because the optimiser and CTM would regard the current production plan as a constraint) with the all its downsides regarding sourcing alternatives and feasible distribution plan.

The indication for the separation between SNP and PP/DS are the SNP horizon (SNP plans outside the SNP horizon) and the PP/DS horizon

(PP/DS plans within the PP/DS horizon). This way it is possible to create an overlap between SNP and PP/DS to smoothen the problems at the borders. The prerequisite for overlapping horizon is however the use of mixed resources to avoid a double consumption of the resource capacity. The problems related to this are often underestimated.

The categories of the PP/DS orders are different from the categories for the SNP orders and are displayed in the key figure 'production (fixed)'. These orders are taken into account by SNP production planning, but are not changed.

• Mixed Resources
The concept of mixed resources is that they have both a bucket capacity for SNP and a time-continuous capacity for PP/DS (see also chapter 14). To avoid a double consumption of the capacity, the PP/DS orders also have a bucket capacity consumption (which needs to be maintained in the PPM resp. PDS) as shown in figure 14.15.

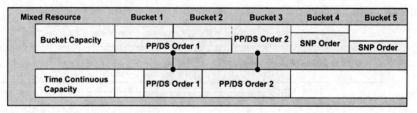

Fig. 14.15. Mixed resource concept

• Integration of the SNP Planned Order to the PP/DS Planned Order
There are basically three options to integrate the SNP production planning result to PP/DS. Note 481906 describes these in more detail.

The first is to use the planned order conversion, i.e. the quantities planned by SNP are converted into orders for PP/DS regardless of the demand situation. The advantage of this alternative is that decisions made in SNP based on a better overview regarding time and network are adhered to. The disadvantage is that PP/DS will not be able to react to short time changes since no production planning run will be required any more.

If SNP is used rather for evaluation and feedback about feasible plans, there is the alternative to perform a PP/DS production planning run within the PP/DS horizon which will delete all SNP orders and create PP/DS orders. In this case the SNP planning result has no impact on the PP/DS planning.

The third option is a complete separation of the SNP planning and the PP/DS planning. In this case SNP planning is performed in an inactive version to get a feasible plan and the result is transferred to PP/DS via DP as planned independent requirements. This alternative has advantages especially in the case that there is an organisational separation between medium-term and short-term planning, e.g. one medium-term planner is responsible for in-house sourcing decisions and there are several local short-term planners to fulfil the requirements.

The only specific functionality is the planned order conversion which is described in the next paragraph.

• Planned Order Conversion

To convert the planning result of SNP into the more detailed model of PP/DS with as less loss of information as possible, the SNP orders are converted into PP/DS orders using a conversion heuristic – e.g. the standard heuristic SAP_SNP_SNGL – either by using the control heuristic SAP_SNP_MULT or with the transaction /SAPAPO/RRP_SNP2PPDS. In this case the selected SNP orders are deleted and PP/DS orders for the same requested quantity and requested date are created. The heuristic allows to influence the lot size and to add an offset to the PP/DS horizon for the order selection.

The PP/DS order is created according to the PP/DS PPM that is chained to the SNP PPM (an entry in the SNP PPM) of the respective order. If no PPM is specified there, the PPM is selected according to the priorities.

• Build-up Inventories

If the business scenario requires to build-up inventories, running a PP/DS heuristic might be counterproductive to the planning result of the SNP horizon. If the production horizon is shorter than the lead time to build-up the inventories, the PP/DS heuristic might delete the orders which have been created by SNP, because no demand exists within the production horizon, as figure 14.16 shows:

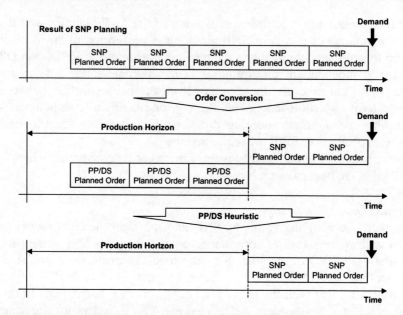

Fig. 14.16. Problem with the build-up of inventories

A possible solution is to change the heuristic to consider demand outside the horizon (by setting the according flag), see also note 481906.

If alternative resources are modelled in SNP for production (because of different properties), these have to be included into different PPMs. As described later on, in PP/DS two possibilities exist to model alternative resources, which have an impact on the integration and the respective planning possibilities.

14.7.3 Integration to SAP ERP™

If no detailed planning (i.e. no PP/DS) is used in SAP APO™, the planned orders are transferred directly to SAP ERP™ – either immediately or periodically according to the settings in the transaction /SAPAPO/SDP110, see also section 25.5. The order number of the SAP ERP™ planned order is received in SAP APO™ ('key completion') but the order category is not affected by the transfer. These planned orders are not fixed in SAP ERP™. They can be converted into production orders in SAP ERP™ without any restrictions, but the conversion can not be triggered from SAP APO™ (see also chapter 18).

15 Detailed Production Planning

15.1 Basics of PP/DS

15.1.1 Order Life Cycle and Order Status

PP/DS focuses on the detailed production planning and scheduling and is therefore mainly concerned with planned orders and production orders. Creating an order contains the steps of calculating the component quantities and the operation durations, and scheduling the operations in the live cache. The prerequisites to create an order are the master data for the products, the resources and the PPMs. In contrast to SAP ERP™ planned orders and production orders (and – if PP-PI is used on SAP ERP™ side – process orders) are not different objects but one object with different categories. Therefore no change in the structure happens when the order type changes.

The order life cycle of a planned order is the conversion to a production order resp. a process order (triggered from SAP APO™ or in SAP ERP™), the release of the production order in SAP ERP™ and its confirmation in SAP ERP™, as shown in figure 15.1.

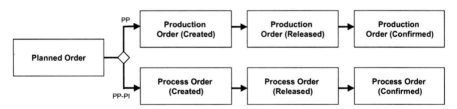

Fig. 15.1. Order life cycle

The order is deleted in SAP APO™ as soon as it is technically completed. There is no order history in SAP APO™.

The properties of an SAP APO™ order are its category, its fixing indicators, its scheduling status and other flags reflecting the order status in

J. T. Dickersbach, *Supply Chain Management with APO*,
DOI: 10.1007/978-3-540-92942-0_15, © Springer-Verlag Berlin Heidelberg 2009

SAP ERP™. Tables 15.1 and 15.2 provide an overview about the order statuses for planned and production orders.

Table 15.1. Order statuses in for planned orders in PP/DS

Order Status	Description	Category
Planned Order Created	Planned Order created by PP/DS heuristic	AI
Output Firmed	Is set for Planned Orders explicitly and is set automatically if a planned order is changed. Production Orders always have the flag 'output firmed'.	AJ (for Planned Order)
Input Firmed	Is set after a manual change of the input quantities and sets the output to firmed.	AJ (for Planned Order)
Date Fixed	Is set if any operation of the order is fixed in the Planning Board.	No Impact
Checked	After ATP Check	AK
Checked & Firmed	After ATP Check of firmed order or multi-level ATP	AL

For the production orders the order categories are listed in table 15.2.

Table 15.2. Order statuses in for production orders in PP/DS

Order Status	Description	Category
Production Order Created	After conversion of a Planned Order in SAP APO™ or SAP ERP™ or creation of a Production Order in SAP ERP™	AC
Released (Production Order)	Release in SAP ERP™	AD
Partially Confirmed (Production Order)	Partial or final confirmation of operations. After the final confirmation of an operation the operation receives the status partial & final confirmation.	AD
Finally Confirmed	After final confirmation of the last operation	AD

The flag 'PP firmed' is set if either the output or the input is firmed or the date is fixed. The scheduling status – fully scheduled, partly scheduled or fully de-allocated – has no impact on the order category (if at least one operation is de-allocated, the order receives the status 'partly scheduled', if all operations are de-allocated, it receives the status 'fully de-allocated').

The categories for PP/DS orders are defined in the general settings with the transaction /SAPAPO/RRPCUST1. Table 15.3 shows the default categories for the most common order types.

Table 15.3. PP/DS categories (extract)

	Receipt Categories	Requirement Categories
Production	AI	AY
Production firmed	AJ	
External Procurement	AG	---
Stock Transfer	AG	BH

An order consists of one or more operation, and an operation consists of one ore more activities. Depending on the setting in the version master, planned orders might be created without any source of supply (i.e. without PPM or PDS). These are dummy orders which do not consume any capacity nor create any dependent demand but suppress the generations of alerts because of missing master data. Therefore we do not recommend to allow these.

15.1.2 Scheduling and Strategy Profile

The durations for the operations of an order are determined using the PDS resp. PPM. These operations are scheduled subsequently and within the same transaction by the scheduler – a COM object – in the live cache. The parameters for the scheduling are defined in the strategy profile with the transaction /SAPAO/CDPSC1. These parameters contain settings regarding the scheduling mode (e.g. find slot, insert operation or infinite), the planning direction (backward, forward, with reverse), a restriction by priorities regarding the alternative modes of the PPM, and whether the validity of the operation and the dependent objects are taken into account. Other settings define the relation of the actual operation to dependent objects, i.e. whether the order internal relationships to other operations and the external relationships to other orders are considered – with the consequence that the dependent objects are probably rescheduled as well.

A strategy profile contains one or more strategies which are processed in their alphanumeric order until the scheduler finds a solution. With more than one strategy it is possible to model scheduling requirements as schedule backwards finite, and if no free slot is available, schedule forwards. Another scenario is that prioritised resources exist in production, e.g. expensive automated machines, which should be loaded with priority, and some older and slower machines as fall back and for peak demands. If these resources are modelled as alternatives with different mode priorities and the first strategy is restricted to high mode priorities only, the scheduler tries to load this resource first. Only if the first strategy does not find a solution, the second strategy is applied to load the resources with lower priority.

Strategy profiles are taken into account each time an order is created or changed. Table 15.4 lists the actions that require strategy profiles and which strategy profile is used.

Table 15.4. Strategy profiles

Action	Strategy Profile	Recommendations
Production Planning Run	Part of the Heuristic	Infinite Planning Do Not Consider Pegging
Capable-to-Promise	Global Parameters	Find Slot
Multi-Level ATP	Global Parameters (Conv. ATP → PP/DS)	
R/3 Integration	Global Parameters	Infinite Planning Backwards with Reserve No Validity Consider Internal Rel.ship No Maximum Constraints No Fix Pegging No Compact Scheduling
Conversion of SNP Orders to PP/DS Orders	Global Parameters	Infinite Planning
Interactive Scheduling in the Planning Board	Overall Profile for Planning Board	Infinite Planning Consider Internal Rel.ship Consider Pegging
Scheduling Heuristic in Planning Board	Part of the Heuristic	
Scheduling Run in Back- ground Planning	Part of the Heuristic	

Orders transferred from SAP ERP™ are the only ones that are allowed to be scheduled into the past (if it is desired to schedule an order interactively into the past, a negative offset has to be maintained in the strategy). This is required for the initial upload of orders with backlog and for confirmations. To avoid an order being scheduled too far into the past, always infinite scheduling should be used for the SAP ERP™ integration. If it can not always be ensured that the PPM allows sufficient breaks, scheduling into non-working times should be allowed as well. Notes 394113 and 417461 describe the recommendations for this. Additionally a hard coded fallback strategy is provided to prevent errors in the integration (check note 460107).

In the interactive planning applications – these are the product view (transaction /SAPAPO/RRP3), the planning board (transaction /SAPAPO/CDPS0) and the product planning table (transaction /SAPAPO/PPT) – it is possible to make changes to the strategy settings. These changes are saved user specifically. If you create a strategy profile, make sure the flag 'active' is set.

15.1.3 Planning Procedure

The planning procedure describes the way production planning reacts to events (e.g. change in a sales order, creation of dependent demand etc.). Some of the possibilities are to
- cover the requirement immediately,
- cover the dependent demand with existing receipts,
- call the production planning heuristic immediately,
- create a planning file entry or
- do not carry out any action.

The planning procedure is defined in customising with the path *APO → Supply Chain Planning → PP/DS → Maintain Planning Procedure* and contains additionally information about the default production planning heuristic, the ABAP class for CTP (see chapter 16.2) and the real quantity (see next chapter). By default the following planning procedures are available:
- manual with check (1): planned orders are only created if there are receipts for the dependent demand.
- manual without check (2): planning is performed only if the product is explicitly selected. No planning file entries are created.
- cover dependent requirements immediately (3): planning is triggered as soon as a planning relevant change happens. This or a similar planning procedure is a prerequisite for CTP.
- planning in planning run (4): planning file entries are created and planning is carried out depending on the control settings of the planning run
- multi-level ATP check (5): should be used for the multi-level ATP scenario.

Note 439596 gives recommendations for the customising of own planning procedures.

15.1.4 Real Quantity

From SAP APO™ 4.0 on the planning mode in the product master determines whether the requested or the confirmed quantity is used for pegging and for production planning. Differing from SAP APO™ 3.1 it is therefore not possible anymore to use the confirmed quantity for pegging and the requested quantity for production planning.

There are different ideas about the significance of the ATP check for the planning from company to company. While many companies want the ATP check to limit the demand signal for production, other companies think that the ATP check promises to the customer the least he can get and therefore still want to meet the requested quantity, even when the customer order is already confirmed with less. So therefore they want to plan for the complete requested quantity. From SAP APO™ 4.0 on it is possible to choose between these two options on locationproduct level.

15.2 Heuristics for Production Planning

15.2.1 Concept of Production Planning and MRP Heuristic

The production planning in SAP APO™ uses two heuristics – one is controlling the production planning – i.e. the planned order creation – per pegging area. A pegging area is a planning segment of a location product, i.e. there can be more than one planning segments per location product. The second heuristic is a MRP heuristic that controls the sequence and the way the production planning heuristic is called.

• *Overview of Production Planning Settings*
The behaviour of the production planning run is controlled by the settings for the MRP heuristic, the production planning heuristic and the planning procedure. Figure 15.2 provides an overview of the settings in the heuristics, the product master and the global settings.

The planning heuristic is either defined in the MRP heuristic or taken from the product master. The default setting for the product master is defined in the global settings with the transaction /SAPAPO/RRPCUST1.

Especially for the area of production planning there are very detailed consulting notes available which are listed in collective note 441102. The following chapter is basically a summary and visualisation of the information in those notes. For more detailed information we recommend to read the mentioned notes.

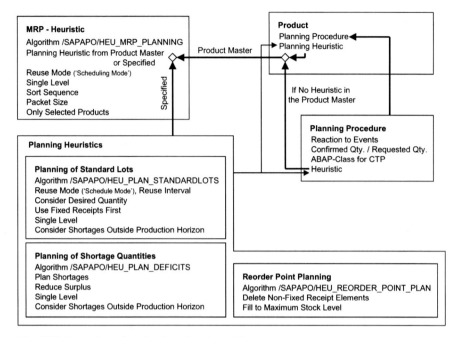

Fig. 15.2. Overview of production planning settings

15.2.2 Production Planning Heuristics

Production planning heuristics create planned orders or purchase requirements. In these heuristics the planning algorithm and its parameters are defined.

In the PP/DS module heuristics are available for different objectives as scheduling, fixing the pegging and more. To see whether a heuristic is suited for production planning the properties of the heuristic can be checked in the table /SAPAPO/HEURFUNC. For production planning heuristics the field PROD_HEUR contains an entry. The PP/DS heuristics are displayed and maintained with the transaction /SAPAPO/CDPSC11.

The standard algorithm set by default in the global parameters for production planning is /SAPAPO/HEU_PLAN_STANDARDLOTS, which is used in the standard heuristic SAP_PP_002. Production planning consists of the steps

1. net requirements calculation
2. procurement quantity calculation and
3. source determination.

Net requirements calculation determines the uncovered requirements by allocating first the fixed receipt elements to the demand and afterwards – depending on the re-use mode – the non fixed receipts as well. Note 448960 describes the net requirements calculation in detail.

The strategy is maintained within the production planning heuristic. Only infinite scheduling modes are allowed. For the use of finite scheduling – which is strongly not recommended – BAdIs have to be applied.

• *Re-use Strategy*
The re-use strategy (one of the basic settings of the production strategy) controls whether 'use earliest receipts first (FIFO)' is applied or 'use timely receipts'. 'Use earliest receipts first' was formerly known as 'use fixed receipts first'. Fixed receipts for net requirements calculation are stocks, production resp. process orders, purchase orders and manually changed planned orders and purchase requisitions.

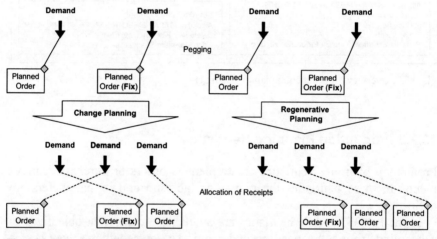

Fig. 15.3. Use earliest receipts first

The re-use strategy controls the allocation of these fixed receipts. For 'use earliest receipts first (FIFO)' the first demand is covered by the first receipt, independent whether the receipt is before or after the demand. The implicit assumption of this method is that the fixed receipt elements are already scheduled as early as possible, so that neither a new – i.e. not fixed – element can be created with an earlier receipt date nor that any non fixed elements are scheduled earlier. Fixing a receipt element in the future – e.g. by changing a planned order manually – might have the implication that more early demands are not covered in time any more as shown in figure 15.3. New receipt elements are created only after the last fixed order.

Note that the dotted lines represent the allocation of the receipt elements to the demand and not the pegging. The MRP in SAP ERP™ uses the same logic.

The alternative is to minimise the lateness with the re-use strategy 'use timely receipts'. Here first the receipts from the past and within the fixing horizon are allocated to the demands in the past and the fixing horizon. Fixed receipts outside the fixing horizon are allocated to the appropriate demands, where a demand is considered as appropriate if it is later than the receipt or at least not earlier as the schedule alert threshold specified in the product master. In the third step the free fixed receipts are allocated to the remaining earlier demands, instead of creating new receipt elements. Production planning prefers to balance the quantities and leaves the resolving of scheduling problems to detailed scheduling steps instead of creating excess coverage.

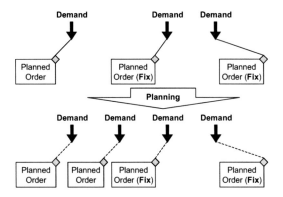

Fig. 15.4. Use timely receipts

Production planning does not consider pegging constraints. If a receipt – fixed or non-fixed – matches a demand regarding the quantity and its availability is before the demand, production planning allocates it independent of the pegging constraints. These constraint violations have to be resolved in scheduling afterwards. The advantage of planning without 'consider fixed receipts first' is that fixing a receipt does not destabilise the plan. When deleting excess coverage orders, always the latest order is deleted.

Production planning heuristics do not use the pegging from live cache but calculate the backlog according to the logic described above. Accordingly the use of backwards pegging does not affect the production planning logic.

Production planning is usually only performed for demands within the production horizon. For some heuristics however – e.g. SAP_PP_002 – it is possible to control that shortages outside the production horizon are considered as well.

• *Planning of Shortage Quantities*
The difference of the planning of shortage quantities with the algorithm /SAPAPO/HEU_PLAN_DEFICITS (heuristic SAP_PP_003) to the planning of standard lots (without 'consider fixed receipts first') is that both for the creation of receipt elements for uncovered demands and the deletion of non fixed excess orders the flags have to be set separately. If none of these flags are set, the heuristic does not do anything at all.

• *Re-use Mode*
After allocating the fixed receipts, the uncovered demands are calculated in the net requirements calculation. Using these demands the procurement quantities are calculated taking the lot size into account and subsequently the source is determined. For the subsequent planning the re-use modes
  • change planning
  • regenerative planning
  • new explosion of planning data and
  • new explosion and regenerative planning
are optional. Using regenerative planning, all non fixed receipts are deleted and new orders are created as calculated before. If change planning is used, the system checks whether dates, quantities and sources of the existing order match the calculated result. Non fixed orders are only deleted if this is not the case. Using the new explosion mode, the existing order is re-created using the current PPM. As long as no planning relevant changes in the product data have been made the order number remains the same. The mode in which production planning is executed depends on both the re-use mode of the planning heuristic and the planning file entry (see next paragraph).

• *Creating Orders in De-allocated Mode*
Applying the BAdI described in note 362208 it is possible to control that new orders are created in de-allocated mode (and by adding an if-statement checking whether 'sy-batch' equals 'X' this behaviour can be restricted to orders created in a background job). This way newly created planned orders are clearly distinguished, which might be an advantage for interactive scheduling of planned orders.

15.2.3 MRP Heuristic

The MRP heuristic is used to control the production planning heuristic in order to perform the production planning in the right sequence of the low level code. The standard MRP heuristic is SAP_MRP_001.

Section 13.3 describes the differences between the single-level heuristic SAP_MRP_001 (recommended) and the not recommended multi-level heuristic SAP_MRP_002.

15.2.4 Net Change Planning and Planning File Entries

Planning file entries mark that planning relevant changes have occurred for a locationproduct and are displayed in transaction /SAPAPO/RRP_NETCH. There are four types of planning file entries
  1. use suitable receipts
  2. delete non firmed receipts
  3. re-explode plan
  4. delete non firmed receipts and re-explode firmed receipts.
The re-explosion of the master data is not performed for orders with fixed inputs, fixed dates and production resp. process orders from their release onwards. The planning file entries are deleted after the execution of a planning heuristic for that locationproduct. Note 557731 describes this topic in detail.

15.2.5 Mass Processing

The usual way to perform production planning is to have background jobs running over night for mass processing and execute production planning manually only for conflict resolution. Running the planning heuristic for all products has the disadvantage that the right sequence – the sequence according to the BOM levels - is not guaranteed, see also note 513827. Therefore it is recommended to use the MRP heuristic instead as a control for the planning heuristic. The MRP heuristic ensures the right sequence – BOM level by BOM level to avoid planning a locationproduct several times. From time to time it is advisable to run the heuristic 'stage numbering' (SAP_PP_020) to re-determine the BOM levels (including stock transfers).

Another setting in the MRP heuristic is the packet size, which determines the number of locationproducts which are planned at the same time. This number should not be too large since in the case of an error the result of the complete packet is rejected, see also note 513827. By default the packet size number is one, but there is some potential to improve the performance by increasing the packet size.

• *Background Planning*
The prerequisite for mass processing is to create the settings for production planning in the background with the transaction /SAPAPO/CDPSB0. These settings define the heuristics and the objects for which they are applied, the version, the propagation range and the time profile (which is not relevant for the production planning heuristics as described in note 457723). If these settings are saved as a variant they can be used for background jobs.

• *Scope of Planning*
In the settings for the background planning it is possible to restrict the production planning run to the selected objects (e.g. locationproducts). Whether production planning is carried out for the components as well (i.e. for the products, for which a dependent demand is created), depends on the planning procedure of the product and on the processing control of the MRP heuristic. The creation of planning file entries is influenced by the planning procedure. If additionally the processing control setting 'only selected products' is not set, these products are planned as well.

The impact of these settings is shown for an example consisting of a finished product, a semi finished product and a raw material in table 15.5 (the raw material is an input to the semi finished product, which is an input to the finished product). Only the finished product is selected for the background planning using the MRP heuristic and the 'plan standard lots' planning heuristic.

Table 15.5. Product selection for planning in the background

	Planning Procedure Automatic		Planning Procedure Manual	
Product	Proc. Control 'Only Selected Products'	Proc. Control -	Proc. Control 'Only Selected Prod.'	Proc. Control -
Finished (Selected)	Orders	Orders	Orders	Orders
Semi Finished (Not Selected)	Planning File Entry	Orders	-	-
Raw (Not Selected)	-	Orders	-	-

The flag 'execute single-level planning' has no impact on the product selection, neither in the planning heuristic nor in the MRP heuristic. By excluding products from the propagation area any planning for these is prevented.

• *Parallelisation*
Usually the nights are short and so is the time window for production planning, so that it is often desired to parallelise the planning run for performance reasons. The best way to parallelise planning runs is to use completely disjunctive work areas, e.g. per business unit. If this is not possible, the risks caused by overlapping parallel planning runs must be clear.

An implication of the parallelisation of the production planning is that each planning job regards the complete requirements and receipt situation of a locationproduct and resolves the shortage independently. If the same locationproduct is processed by parallel planning jobs, there might be double receipts. Another implication is that resource overloads might be generated. If finite planning has to be used, all products which use same resources have to be planned sequentially. Note 513827 provides some recommendations regarding the parallelisation. With increasing parallelisation, there is a general trade off between the decreasing duration for the production planning run and the increasing lock problems.

• *Logging*
The planning log level (no log, normal [i.e. only errors] or detailed) is set per user in the product view (transaction /SAPAPO/RRP3) with the menu path 'Setting → Planning Log'.

• *Cross Plant Planning*
If there are components which are procured from different plants, this has to be taken into account in the design of the planning runs. A strict separation of the locationproducts per plant is possible using the planning procedure, the MRP settings or the propagation area, but would require a sequential planning of the plants.

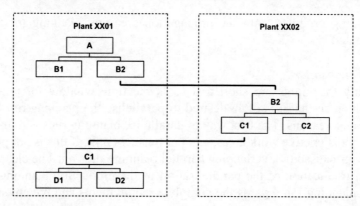

Fig. 15.5. Cross plant planning

If the component's components are procured from the first plant as the material flow in figure 15.5 describes, there are mainly the possibilities to include the planning of that product in plant XX02 in the planning run for plant XX01 (e.g. by not setting the flag for 'selected products only' and automatic planning), or to correct the plan for product C1, D1 and D2 manually on a regular basis.

15.3 Consumption Based Planning

It is possible to perform consumption based planning in SAP APO™ as well. To do so, it is necessary to apply the heuristic SAP_PP_007. A proposition to handle this is to set the heuristic SAP_PP_007 in the product master via user exit if a value for the reorder point is maintained in the material master. Usually this is only used for externally procured materials.

Since it is likely that the current maximum supply is not sufficient to cover all planned demands, a planned shortage in the future is to be expected. To prevent irritating alerts for this, these products have to be excluded from the alert monitor profile. It is however not possible to select the products by procurement type. A proposition for the modelling is to use one or more separate planners for consumption based products to simplify the selection.

15.4 Material Flow and Service Heuristics

Production planning per se might not regard the material flow dependencies between orders, i.e. that the order for the assembly group has to be finished before the order for the finished product starts.

The creation of the right sequence is part of the detailed scheduling. Nevertheless it might be necessary to ensure that the pegging relations are adequate and/or to adjust the orders to the material flow as a starting point for the subsequent detailed scheduling. One reason might be to give the optimiser a better starting situation – the optimiser might be sensitive especially regarding incoherent pegging.

• *Fixing Horizon*
The fixing horizon prevents automatic planning in PP/DS from creating any orders within this horizon. The idea is not to disturb the shop floor or detailed scheduling and execution of the current plan by frequent changes. The fixing horizon is set in the product master. It should cover as well the frozen period for shop floor as the time needed to convert the planned order into a production order. It is nevertheless possible to schedule an order manually within the fixing horizon. FAQs to the fixing horizon are answered in note 441740.

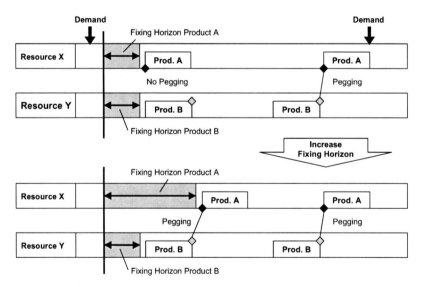

Fig. 15.6. Use fixing horizon to ensure pegging relation

An increase of the fixing horizon for the top level product can be used as well as an additional security to prevent the loss of pegging in cases where many demands are in the past, as shown in figure 15.6.

• *Bottom-up Heuristic*
The bottom up heuristic SAP_PP_009 reschedules dependent demands to the earliest receipt date first for the fix pegged elements, and afterwards for the other demands. The dependent demands are sorted according to their requirement dates – neither priorities nor category differences are taken into account. It is recommended to use forward scheduling in the strategy profile. SNP orders are not considered. Note 560969 describes the properties of the bottom-up heuristic in detail.

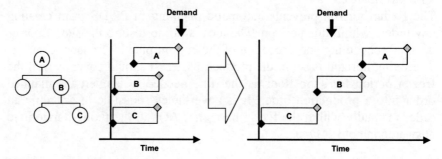

Fig. 15.7. Bottom-up heuristic

The bottom-up heuristic should be executed from the lowest level on which scheduling problems exist. To configure the bottom-up heuristic, the heuristic SAP_PP_009 must be included into the MRP heuristic with the setting to sort the BOM levels in descending order.

• *Top-down Heuristic*
Analogous to the bottom-up heuristic the top-down heuristic SAP_PP_010 adjusts the planned and production orders to the demand date. This might be required to align the orders of lower BOM levels to a scheduled bottleneck.
    Both the bottom-up heuristic and the top-down heuristic should be used with infinite scheduling strategies.

15.5 Tools for Visualisation and Interactive Planning

15.5.1 Product View

The main tool to display the production planning result and perform interactive production planning is the product view, which is called with transaction /SAPAPO/RRP3. Analogous to the receipt and requirements list in SAP ERP™ (transaction MD04), the product view is restricted to one locationproduct at a time. The product view itself has several views, these are

- the 'element'-view, which is an order oriented view where all the orders are displayed in chronological order with the highest level of details,
- the 'period'-view, where requirement and receipt elements are aggregated and
- the 'stock'-view, where the stock types are displayed with the according information regarding sublocation (SAP ERP™: storage location) and version (SAP ERP™: batch).

Further views contain the availability situation (corresponding to the transaction /SAPAPO/AC03, cf. chapter 7), the forecast consumption situation (corresponding to the transaction /SAPAPO/DMP1) and the product master. The element-view and the period-view can be configured to a certain extent in the user settings (menu path '*Settings* → *User Settings*').

• Element-View

The 'element'-view lists each order according to its requirement resp. receipt date with the information about order type, quantities (confirmed and requested), dates (available and start), surplus resp. shortage (derived from the pegging), targets resp. sources of the orders and resources. Additional information are – among others – the order priority, the order status (e.g. fixing) and the conversion indication. Figure 15.8 shows the 'element'-view for a planning example.

XX_HEAVY_250 in XXD1 (Make-to-stock)										
Rqts/avail.	Rqts/av.	Category	Receipt/rqmt. elemt.	Rec./Issue qty	Available	Surplus/shortfall	PP-firmed	Start date	Start time	Target/source
24.03.2003	12:04:04	FC req		100-	17.000	0				
27.03.2003	12:00:00	DEP:ConR1		100-	16.900	0				XXUK/
31.03.2003	12:04:04	FC req.		50-	16.850	0				
17.04.2003	00:00:00	Pl0rd. (F)	33513	21.150	38.000	1.000	✓	17.04.03	16:21:58	/XXD1
20.04.2003	12:00:00	FC req		37.000-	1.000	0				
23.04.2003	20:00:00	PrdOrd (C)	60002587	50.000	51.000	3.000	✓	23.04.03	11:45:00	/XXD1
24.04.2003	19:01:12	Pl0rd.	33510	60.000	111.000	0	☐	24.04.03	09:10:12	/XXD1
24.04.2003	19:35:24	Pl0rd.	33512	2.000	113.000	0	☐	24.04.03	19:01:12	/XXD1
24.04.2003	20:00:00	Pl0rd.	33514	1.000	114.000	0	☐	24.04.03	19:35:24	/XXD1
24.04.2003	22:00:00	SalesOrder	6975/000010/1	1.000-	113.000	0		24.04.03	22:00:00	
24.04.2003	22:00:00	SalesOrder	6976/000010/1	2.000-	111.000	0		24.04.03	22:00:00	
25.04.2003	11:45:00	Pl0rd.	33511	40.000	151.000	0	☐	25.04.03	05:06:00	/XXD1
25.04.2003	23:59:59	SalesOrder	6978/000010/1	10.000-	141.000	0		25.04.03	23:59:59	
26.04.2003	12:00:00	FC req		80.000-	61.000	0				

Fig. 15.8. Element-view of the product view

For a quick analysis of the coverage situation the rows 'available' and 'surplus/shortfall' are quite helpful. The row 'available' simply cumulates all receipt and requirement elements, while the row 'surplus/shortfall' displays the unpegged receipts and requirements. This is the same procedure as used by the alert monitor to calculate the backlog and the excess coverage. The sequence of the columns is changed by drag and drop. To save these settings (this is only possible for the own user), the steps described in figure 15.9 have to be carried out:

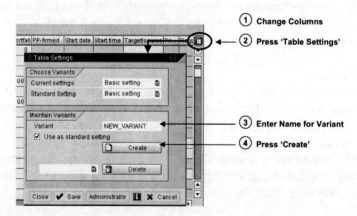

Fig. 15.9. Adopt change of the column sequence

In the 'orders'-view of the user settings (menu path '*Settings* → *User Settings*' or CTR + F10) it is determined whether orders with a quantity of zero – sales orders which are not confirmed to their requested date or production orders which are not yet technically completed – are displayed and whether different stock types are aggregated for display. Using the option for 'enhanced table display' in the 'general'-view of the user settings allows to filter the data and to export it to excel.

Another helpful tool to analyse the planning situation is the display of the pegging network of an order, where all orders (receipts and requirements) which are somehow linked by pegging are displayed using the 'context of order' button, figure 15.10.

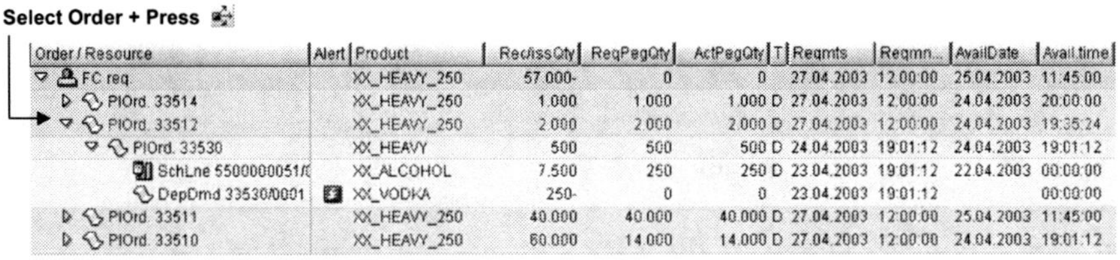

Select Order + Press

Order / Resource	Alert	Product	Rec/IssQty	ReqPegQty	ActPegQty	T	Reqmts	Reqmn...	AvailDate	Avail.time
▽ 🔒 FC req		XX_HEAVY_250	57.000-	0	0		27.04.2003 12:00:00	25.04.2003	11:45:00
▷ 🔷 PlOrd. 33514		XX_HEAVY_250	1.000	1.000	1.000	D	27.04.2003 12:00:00	24.04.2003	20:00:00
▽ 🔷 PlOrd. 33512		XX_HEAVY_250	2.000	2.000	2.000	D	27.04.2003 12:00:00	24.04.2003	19:35:24
▽ 🔷 PlOrd. 33530		XX_HEAVY	500	500	500	D	24.04.2003 19:01:12	24.04.2003	19:01:12
📋 SchLne 5500000051/0		XX_ALCOHOL	7.500	250	250	D	23.04.2003 19:01:12	22.04.2003	00:00:00
🔷 DepDmd 33530/0001	🔳	XX_VODKA	250-	0	0		23.04.2003 19:01:12		00:00:00
▷ 🔷 PlOrd. 33511		XX_HEAVY_250	40.000	40.000	40.000	D	27.04.2003 12:00:00	25.04.2003	11:45:00
▷ 🔷 PlOrd. 33510		XX_HEAVY_250	60.000	14.000	14.000	D	27.04.2003 12:00:00	24.04.2003	19:01:12

Fig. 15.10. Context of order

● *Interactive Production Planning*

Planned orders and purchase requisitions can be created manually within the product view by entering the requested date and quantity into a new row. Manually created or changed orders are fixed. In case of multiple sources, the source for the order is selected from a pop up window where all valid alternatives are listed in the order of their priority. The order types planned order, production order (before release) and purchase requisition are changed by overwriting their availability date and their confirmed quantity with a new requested date and a new requested quantity (production planning and scheduling is still performed afterwards). The usual way to display an order is by double click from the product view (or the planning board). There all information regarding the order structure, the operation dates and the quantities are displayed in the most detailed way.

Using the button 'product heuristic', the production planning heuristic defined in the product master or – if no heuristic is defined there – from the global parameters (transaction /SAPAPO/RRPCUST1) is executed. Using the button 'variable heuristic' any heuristic can be chosen.

Other actions which can be carried out are to trigger the conversion to production orders and to perform an ATP check for the order.

● *Periods-View*

If many orders exist for a locationproduct, the 'element'-view becomes too crowded to provide a clear overview about the planning situation. In this case the 'periods'-view helps to clarify the planning situation by aggregating the quantities of the same order types into daily, weekly or monthly periods. Unlike in SNP no mixed periods are supported. Figure 15.11

shows the order display in weekly periods for the same example as in figure 15.8.

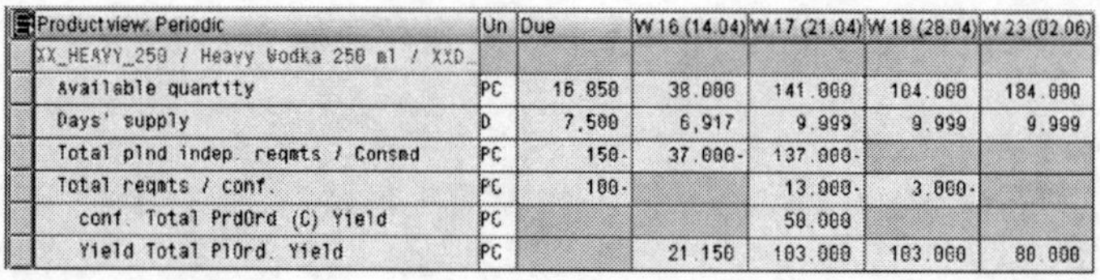

Product view: Periodic	Un	Due	W 16 (14.04)	W 17 (21.04)	W 18 (28.04)	W 23 (02.06)
XX_HEAVY_250 / Heavy Vodka 250 ml / XXD						
Available quantity	PC	16.850	38.000	141.000	184.000	184.000
Days' supply	D	7.500	6.917	9.999	9.999	9.999
Total plnd indep. reqmts / Consmd	PC	150-	37.000-	137.000-		
Total reqmts / conf.	PC	100-		13.000-	3.000-	
conf. Total PrdOrd (C) Yield	PC			50.000		
Yield Total PlOrd. Yield	PC		21.150	103.000	103.000	80.000

Fig. 15.11. Periods view of the product view

In the 'product 2'-view of the user settings the display properties for receipts, customer demands and forecasts are defined. For receipt elements it is possible to choose whether the receipts are aggregated (only one row for all receipt types), partially aggregated (one row per receipt type – in figure 15.11 one for planned orders and one for production orders) or disaggregated (a separate row for each source type – in this case for each PPM) and whether yield and/or total quantity (including scrap) is shown. Interactive planning in the 'periods' view is only possible in the 'disaggregated' rows for the receipts. Analogous settings are possible for the demands and the forecast.

In figure 15.11 only those periods are displayed, for which orders exist. To display all periods, the button 'show/hide zero columns' at the bottom of the screen has to be pressed. The 'orders'-view of the user settings defines whether the 'elements'-view or the 'periods'-view is used as default when entering the product view.

● *Requirements View and Receipt View*
There are two other applications similar to the element view, the requirements view (transaction /SAPAPO/RRP1) and the receipt view (transaction /SAPAPO/RRP4). As their names indicate, they show either only requirements or only receipts, but do have the advantage that it is possible to select more than one locationproduct for display. The objects are restricted by time horizon, location, product number, planner and others.

As in the product view, with CTR+F10 the same user settings are applied. In this case the 'enhanced table display' offers additionally the possibility to sort the elements by any column.

15.5.2 Product Overview

The product overview (transaction /SAPAPO/POV1) provides an overview about the product properties, the days' supply and the alert situation with

the most critical alerts per product – but only one entry per product as
shown in figure 15.12.

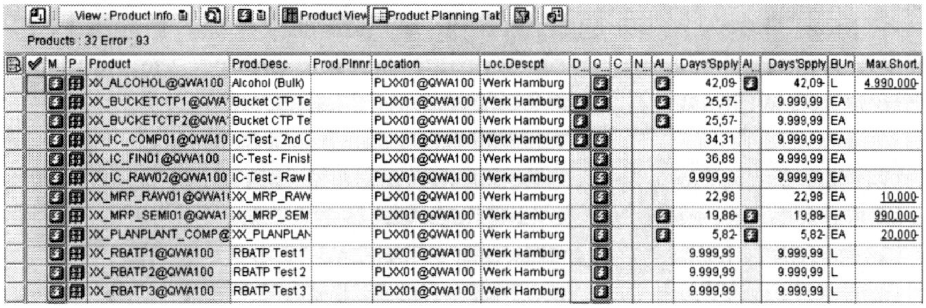

Fig. 15.12. Product overview

The benefit of this transaction is to get an overview about the planning
status of many products at once – and not only about those with alerts.

15.5.3 Product Planning Table

The product planning table includes different charts with the according
functionality, for example the product view, the planning board, the optimiser and the alert monitor, into one application with a common navigation frame. The product planning table is called with the transaction
/SAPAPO/PPT1. As an own feature the product planning table provides a
chart for periodic resource load display and planning.

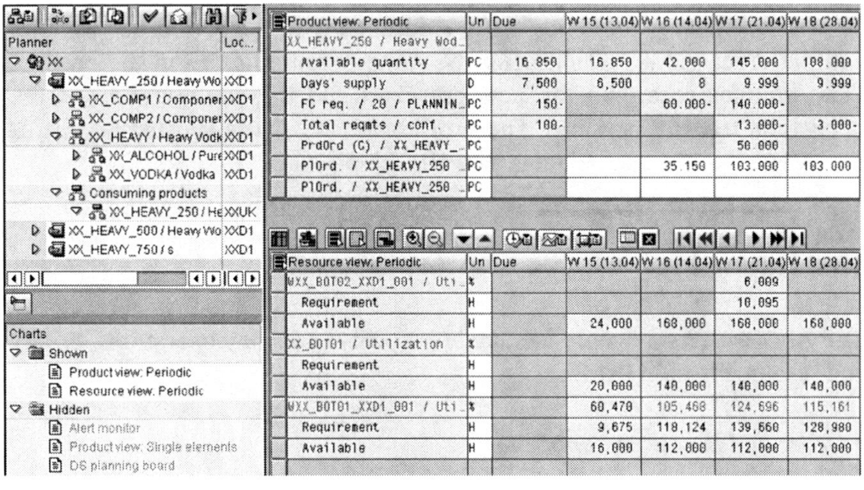

Fig. 15.13. Product planning table

The product planning table is entered with a selection of products, locations, resources and a time horizon. For the navigation between the objects the tree in the top left box is used. Up to three charts are displayed at the same time. The charts are changed using the bottom left box. Figure 15.13 shows a product planning table with the periodic product view and the periodic resource view.

Production view: Periodic	Un	Due	W 15 (13.04)	W 16 (14.04)	W 17 (21.04)	W 18 (28.04)
XX_BOT02_XXD1_001 / Bottling Machine 02 001	k				8,009	
XX_HEAVY_500 / N5000024601PV01 / P1Ord. / Heavy Vodka 500	PC				75.095	
XX_HEAVY_750 / s						
XX_HEAVY_750 / P1Ord.	PC				10.000	
XX_BOT01	k					
XX_BOT01_XXD1_001 / Bottling Machine 01 001	k		19,450	120,741	30,722	
XX_HEAVY_250 / N5000024040P02 / P1Ord. / Heavy Vodka 250	PC				10.000	2.000
Distribtd	PC			1.436	8.564	2.006
XX_HEAVY_250 / N5000024502XXXX / P1Ord. / Heavy Vodka 256	PC					
XX_HEAVY_250 / N5000024501F001 / PrdOrd (C) / Heavy Vodka	PC				50.000	
XX_HEAVY_250 / N5000024501F001 / P1Ord. / Heavy Vodka 250	PC		10.000	80.000	103.000	
XX_HEAVY_250 / N5000024026CRP / P1Ord. / Heavy Vodka 256	PC					
XX_HEAVY_500 / N5000024601PV01 / P1Ord. / Heavy Vodka 500	PC					
XX_HEAVY_750 / s						
N5000024701PV01 / P1Ord.	PC					
XX_HEAVY_750 / P1Ord.	PC				10.000	

Fig. 15.14. Production view within the product planning table

Another speciality of the product planning table is the 'production'-view, where the option exist to adjust the planned production interactively to control the resource load, figure 15.14. The row 'distributed' displays the portion of orders which have their availability date in another period, but do consume capacity in the current period.

Which charts to display, is defined in the user setting (menu path *Settings → User Settings* or CTR + F10). Additionally to the period, the product and the order views, which offer the same possibilities as in the product view, settings regarding the resource chart and the configuration of the navigation tree are possible. The profiles for the charts and the heuristics are assigned in the user settings as well.

15.6 Reporting

SAP APO™ offers a couple of standard reports to display operations, orders and resource load in list form.

• Order and Resource Reporting
With the transaction /SAPAPO/CDPS_REPT the order and resource reporting is called, where

- the orders,
- the operations,
- a production overview,
- the work in process and
- the resource load

are displayed according to the selections. The layout of these lists, i.e. the displayed columns and their sequence, are defined in the layout settings per user or as standard setting. These lists can be exported to other programs, e.g. excel.

The order and the operation lists are self explanatory. The production overview displays the quantities for in-house production in total and according to the order status (e.g. fixed or partially confirmed). The resource load is displayed for the selected period size in the according list. Infinitely planned orders do contribute to the resource load, whereas de-allocated orders do not cause any resource load.

As an alternative orders and operations are displayed in the production list with the transaction /SAPAPO/PPL1.

• *Plan Monitor*

If more sophisticated reports are required, the plan monitor offers the possibility to use a set of predefined key figures of the types 'order quantities', 'order dates', 'stock figures' and 'resources'. Own output figures are defined by performing calculations with the standard ones.

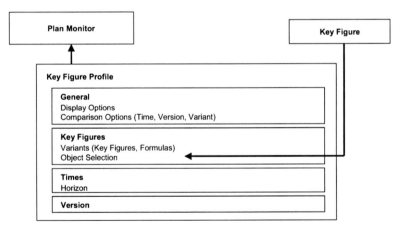

Fig. 15.15. Plan monitor settings

The plan monitor is called as a stand alone tool with the transaction /SAPAPO/PMON. The configuration of the plan monitor is entirely made in the key figure profile with the transaction /SAPAPO/PMONDEF. Like the alert monitor profile (cf. chapter 24), the key figure profile can be assigned to several applications (e.g. planning board and product planning table). Figure 15.15 shows the structure of the plan monitor settings.

The plan monitor displays the information that is defined in variants, where a variant is either a key figure or a self defined formula with selected objects (product, resource, …). The key figure profile defines the version, the time horizon and the display options for the outputs. The definition of the variants themselves is done in four steps:

- Step 1 contains the definition of the variant and its calculation.
- If the variant contains a key figure, in the 2nd step the key figure is selected.
- In the 3rd step the objects are assigned to the key figure with right mouse click. The variant will display the key figure only for the assigned objects.
- Finally in the 4th step the variant is saved by pressing the button 'copy' at the bottom of the screen and saving. As the result, the variant is displayed in the top left box.

Figure 15.16. shows these four steps:

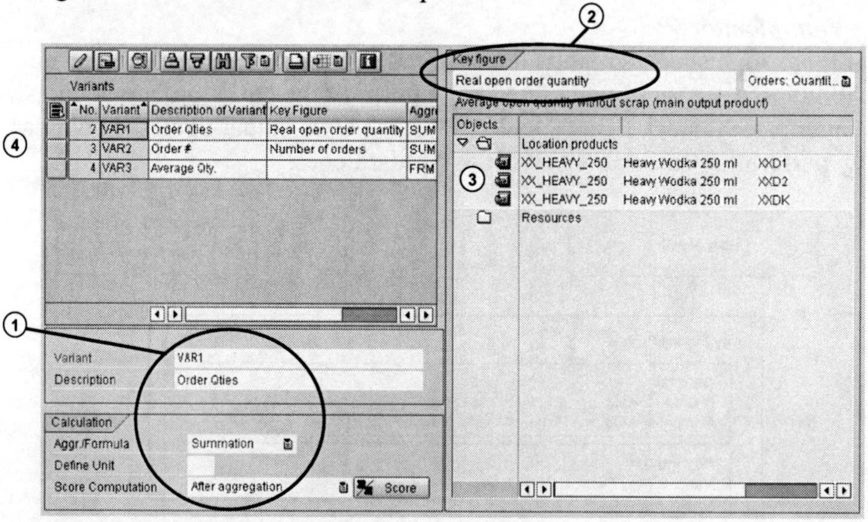

Fig. 15.16. Definition of the key figures for the plan monitor

Before defining a new variant, it is necessary to save the old one. As an example for the calculation of a variant using two key figures, figure 15.17 shows the creation of a variant as the quotient of the standard key figures defined in variant 1 and variant 2 with the according syntax.

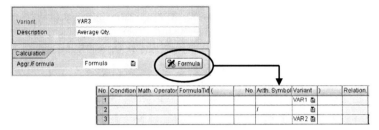

Fig. 15.17. Calculated variant

The plan monitor supports reporting based on durations but not based on quantities. Note 590163 provides additional information for the use of the plan monitor.

15.7 Special Processes for Production Planning

15.7.1 MRP Areas

MRP areas are used in SAP ERP™ mainly if there is a need to separate material for two different lines of businesses, e.g. for normal production and for spare parts. In this case the inventories and the demands should be treated independently, though they are physically often very close.

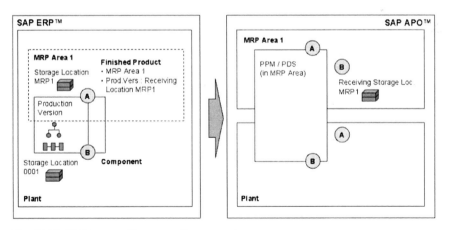

Fig. 15.18. MRP areas without transfer reservation

There are two common scenarios for the use of MRP areas, one is without transfer reservation and the other one with transfer reservation. The process without transfer reservation is shown in figure 15.18. The component is issued as an order reservation and the finished product is received into the MRP area. Differing from SAP ERP™, the MRP area in SAP APO™ is a separate location (of the type 1007). The PPM resp. PDS creates a dependent demand in the plant and a receipt in the MRP area.

The configuration for the process with an explicit transfer reservation is shown in figure 15.19. In this case the component is first moved into the MRP area and the production is modelled completely within the MRP area.

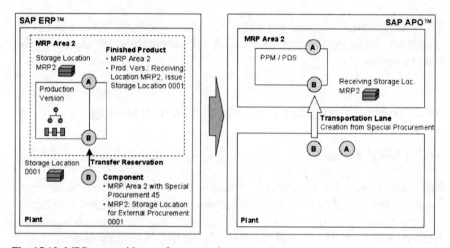

Fig. 15.19. MRP areas with transfer reservation

The stock transfer requisition created in SAP APO™ is transferred to SAP ERP™ as stock transfer reservation. For the usage of stock transport orders to transfer stock between MRP areas further restrictions as described in note 594318 apply.

Probably the most difficult part about planning with MRP areas in SAP APO™ is the set-up and the integration of the master data. Figure 15.20 provides an overview about the required settings and the corresponding objects that are created in SAP APO™ during CIF transfer.

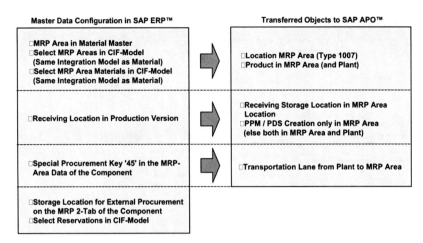

Fig. 15.20. Master data in SAP ERP™ and transferred objects to SAP APO™

Another required setting is that on SAP APO™ side the publication of reservation is maintained in transaction /SAPAPO/CP1 for the target location (i.e. the MRP area).

15.7.2 Production in a Different Location

In some cases the production is modelled in a different plant than the planning and the sales or distribution for the finished product. The special feature of this process is that no transfer is modelled between the plants. In most cases the reason for this is a stricter separation of the responsibilities. In SAP ERP™ this is modelled with the special procurement key '80'. Figure 15.21 shows the modelling in SAP ERP™ and SAP APO™.

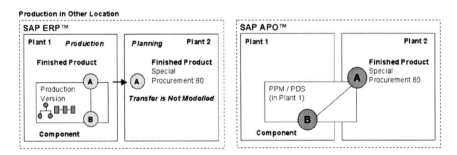

Fig. 15.21. Modelling of production in a different location

A similar process is the reservation from an alternative plant as shown in figure 15.22. This process is however not supported by SAP APO™.

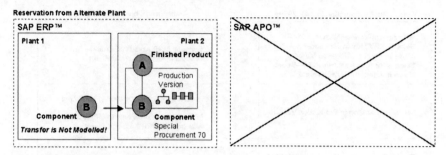

Fig. 15.22. Reservation from alternative plant (not supported)

15.7.3 Capacity Reservations

Though all customers are equal, sometimes some customers are more equal. In this case or if there are agreements based on capacity and not on products (i.e. if it is not yet clear which product will be ordered) it might be required to reserve capacity for special orders. Capacity reservations are maintained using descriptive characteristics (cf. Dickersbach 2005) per time period – e.g. per week – and per resource. The maintenance of the capacity reservations is done in the resource master. Typically customer related characteristics are used, and up to three different descriptive characteristics are possible. At planned order creation the capacity reservations are checked as an additional constraint, and might lead to lateness of the orders in case of missing capacity.

The capacity reservations have a release date – after the release date the capacity reservations are not valid any more. The reason for this is avoiding that capacity reservations originally planned for orders of a certain customer keep the capacity blocked even if these orders are not placed after all.

15.8 Capable-to-Match (with PP/DS Master Data)

The functionality of CTM in general is described in section 10.3, and the motivation and the restrictions to use CTM only for production planning are already mentioned in section 14.5. Using CTM for production planning with PP/DS master data offers the advantage of creating feasible plans in one step, but does have the disadvantage that the results are more difficult to interpret than by an infinite PP/DS heuristic.

Several restrictions exist regarding the master data – e.g. no sequence dependent set-up, no parallel operations and no component assignment to other operations than the first one are allowed. Especially if the production structure is more complicated, it has to be checked in detail whether the modelling is supported by CTM.

Another restriction is that alternative modes are considered in the CTM run – i.e. CTM chooses between the alternatives – but it deletes the alternative modes from the planned order with the consequence that no change of the resource is possible any more – neither manually nor by the optimiser.

CTM does have the advantage of creating a feasible plan based on clear priorities. The downside is that in more complex scenarios the plan might not fulfil the requirements for an optimal or even good plan – e.g. because sequence dependent set-up is ignored. In these cases subsequent scheduling steps have to be performed.

16 Sales in a Make-to-Order Environment

16.1 Process Peculiarities and Overview

Usually ATP is looking at existing or planned receipts. In a make-to-order environment there are no receipts per definition. The idea of an ATP check is therefore not to check whether there are receipts for the requested product but whether there is enough available capacity to produce the product and/or whether the required components are available. In SAP APO™ there are mainly three approaches to tackle these tasks, depending on the business requirements:

- Capable-to-promise (CTP) allows to check for free capacity and optionally for available components as well based on simulative planned orders
- Multi-level ATP (ML-ATP) checks components according to the ATP settings based on infinitely scheduled simulative planned orders
- ATP check against allocations, where the allocations represent an aggregated capacity.

Figure 16.1 provides an overview about the approaches and their properties.

Solution	Bottleneck	
	Component	Capacity
Capable-to-Promise	Check	Check
Multi-level ATP	Check	Check
ATP with Allocations	No Check	Check on Aggregated Level

Fig. 16.1. Solutions for ATP in a make-to-order environment

If neither capacity nor components are a bottleneck, it might be appropriate to use a checking horizon for the ATP check instead.

J. T. Dickersbach, *Supply Chain Management with APO*,
DOI: 10.1007/978-3-540-92942-0_16, © Springer-Verlag Berlin Heidelberg 2009

• *ATP with Allocations*
The ATP check against allocations in general is explained in section 7.5. In this case the idea is to use the allocation quantities as aggregated capacities. This is of course only possible if the production structure allows to define a capacity per product group. It is possible to consider the lead time by combining the allocation check with a product check and using the check horizon for the product check.

• *Combined Sales and Production Planning*
The other two solutions – CTP and ML-ATP – both trigger the creation of a planned order by the ATP check. This means that sales is performing production planning tasks (in the background). The combination of sales and production planning in the ATP check does have impacts on the production side, mainly for the functions for production planning in SAP APO™. These are described in the limitations for CTP resp. ML-ATP.

From a business process side the no major changes regarding the responsibilities and the exception handling are expected. The biggest hurdle is to get the planners' acceptance to pass the control about the planned order creation to sales.

• *CTP for Make-to-Stock*
When CTP or ML-ATP is used, this should be the only source for production planning. It is however possible to apply CTP in a make-to-stock scenario if a part of the demand is planned for with independent requirements. This increases the complexity of the solution and has to be tested very thoroughly.

• *Scheduling*
For both scenarios – CTP and ML-ATP – a subsequent scheduling step is necessary. Even though CTP performs a finite scheduling (either on detailed level or using a bucket capacity), the production sequence will be defined by the more or less arbitrary sequence of the sales orders and will most probably not meet the requirements for a production plan.

• *Time-Continuous CTP vs. Bucket-Oriented CTP*
There are two options how to perform the CTP check, one is the check based on a detailed finite scheduling, and the other one is to check the capacity based on a finite bucket capacity, which does not regard the detailed sequence of the orders within the bucket. We recommend to prefer the bucket-oriented CTP to the time-continuous CTP whenever possible.

This recommendation is resulting from the difficulty for the use of this function is due to the different goals of sales and production. Sales aims to confirm the earliest possible date to the customer and has a strong interest that the confirmed date is met. The targets for the production however are low production costs and therefore few set up times and a high resource utilisation. The more BOM levels are planned, the more these targets become contradictory.

16.2 Capable-to-Promise

16.2.1 Steps Within the CTP Process

There are several steps during a CTP check which involve both ATP and PP/DS functionalities. First in ATP the requested date and the CTP quantity – requested quantity for CTP – are calculated. The requested date for CTP is always the requested date of the sales order, while the CTP quantity might differ from the requested quantity of the sales order depending whether a product check is performed first. PP/DS is called with this information and creates a simulative planned order according to the production restrictions and a simulative demand for the same quantity to prevent any demand element from using this planned order. Note that PP/DS is always able to create planned orders for the requested quantity, even if it might be very late (if production is triggered from a rule, it is possible to inhibit this behaviour in the calculation profile). Now PP/DS returns with the availability date for the CTP quantity and ATP combines the confirmations from the ATP check with the PP/DS result. After saving, the sales order is transferred to SAP APO™, the planned order status is changed from 'simulative' to normal and the simulative demand is deleted. Figure 16.2 shows this procedure.

The transactional simulation is used for the time span which is required to create the necessary planned orders. If the planning task is complex – e.g. because of finite scheduling on multiple resources and planning of several BOM levels – the required time span increases and therefore also the risk increases that the required capacity is used by a parallel planning task (which is triggered by another CTP check). This might result in overloads.

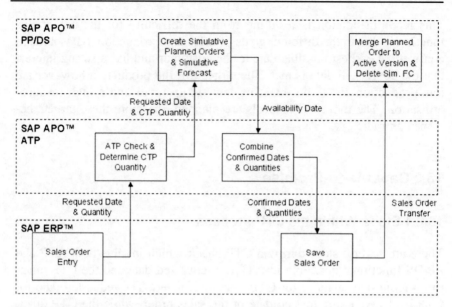

Fig. 16.2. Procedure of the CTP check

16.2.2 Configuration of the CTP Process

The configuration of the CTP process consists mainly of entries in the entities check mode, the check instruction and the planning procedure for all relevant products (finished product and components that have to be checked).

• *Check Mode*
In the check mode the production type has to be defined as 'standard'.

• *Check Instruction/Start of Production*
The check instruction determine that production has to take place. The main option is regarding the start of production. There are several ways to start CTP, which affect the result. Possible ways to start CTP are

- start production immediately,
- start production after product check and
- start production after all basic methods.

Start production immediately basically suits a pure make-to-order scenario, where sales orders are confirmed based on more detailed information than only lead time. Note that there is no fixed link between the sales order and the planned order.

If CTP is used in a make-to-stock environment and any other lot size method than exact lot size is used, this procedure leads to overstocks and is therefore not recommended. Figure 16.3 visualises this effect for a fixed lot size of 40.

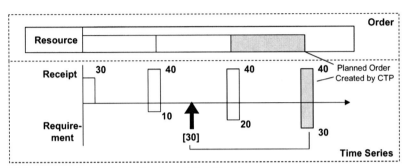

Fig. 16.3. Start production immediately and fixed lot size

The existing supply is not used because 'production immediately' ignores all supplies. With the other options – start production after product check or after all basic methods – first a product availability check is carried out and receipts are created only for the open quantity, figure 16.4.

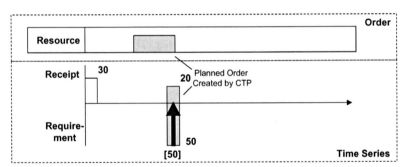

Fig. 16.4. CTP – production after product check/all basic methods

The consequence of this is that a quantity might be confirmed late even though there is enough capacity available to produce in time, figure 16.5. This behaviour can be avoided using the parameter 'late delivery' within the calculation profile in rules-based ATP, where the maximum delay for a confirmation is defined. This possibility applies only when triggering CTP from rules-based ATP.

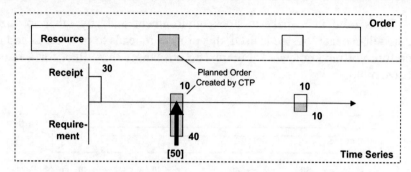

Fig. 16.5. CTP – partial confirmation using production after basic methods

● *Planning Procedure*
Prerequisites for the use of CTP is that the relevant products –the finished product and the components which have to checked resp. planned during the CTP check – have a planning procedure assigned in the product master which triggers production planning for a creation or change of a sales order. The standard planning procedure '3' (cover dependent requirements immediately) is suited for this.

The ABAP class for CTP within the planning procedure has to be maintained only if fix pegging is used (cf. section 19.4).

16.2.3 Problems with Time-Continuous CTP

Until SAP APO™ 4.1 the only way to perform CTP was to use the time-continuous resource capacity from PP/DS. This implied that with each CTP check the planned orders were scheduled finitely with all level of detail. Though there are cases were CTP has been implemented this way very successfully and provides significant benefits (usually in environment with either a very simple production structure or with a very low order volume), this approach has disadvantages which often lead to problems.

● *Scattered Capacity Loading*
Probably the most significant problem is the scattered capacity loading. This problem especially occurs in combination with exact lot size as shown in figure 16.6. The consequence of the scattered capacity loading is that the sales order will be confirmed late although physically there is enough capacity to produce the requested quantity in time. The use of fix or maximum lot sizes reduces this problem (if the lot size is smaller than the usual order quantity), but with the trade off of more planning activities

and therefore decreased performance. A dynamic planned order split is not possible.

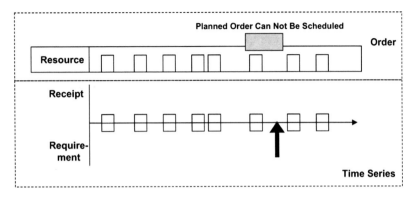

Fig. 16.6. Capacity block

This problem can be reduced by grouping the planned orders on a regular basis, e.g. using the optimiser for fix intervals as shown in figure 16.7.

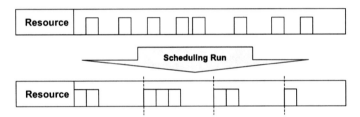

Fig. 16.7. Cleaning of the resources

This is however not a solution to the problem, since sales orders are changing the schedule any time and dependencies to subsequent operations and orders have to be regarded, which gets the more complicated the more BOM levels are included, the more operations exist within an order and the longer the duration of the operations is.

• *Sequence Dependent Set-up and Degrees of Freedom for Optimisation*
When working with sequence dependent set-up the impact of the 'arbitrary' sequence that is provided by CTP will most likely cause huge set-up times which block the resource capacity.

Another problem of the detailed scheduling by CTP is that the degrees of freedom and therefore the potential for a following optimisation of the schedule is limited.

The consequence of all these – the scattered capacity loading, huge sequence dependent set-up and reduced degrees of freedom for optimisation – are low resource utilisation and late order confirmation. Additionally the probability to run into performance problems and scheduling errors increases because of the complex scheduling tasks at each CTP check. Therefore we recommend using bucket-oriented CTP instead whenever possible.

16.2.4 Bucket-Oriented CTP

The approach of bucket-oriented CTP to overcome the problems with time-continuous CTP is to use a bucket capacity for the check. The consequences are that CTP is performing a finite planning and scheduling only on bucket level and no scheduling within the bucket.

• *Bucket Capacity for PP/DS Resources*
The bucket capability of PP/DS resources is a new feature and not linked in any way with the bucket capability of mixed resources. The bucket capacity is aggregated from the time-continuous capacity (or from block planning which is covered in Dickersbach 2005). When an aggregation level changes there is always inaccurateness, therefore the bucket factor in the resource allows deceasing or increasing the bucket capacity to suit the business. Figure 16.8 shows the bucket capacity view of the resource.

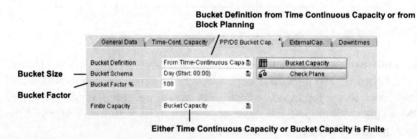

Fig. 16.8. Bucket capacity view of PP/DS resources

The property for PP/DS buckets can be selected for single, multi, single-mixed and multi-mixed resources. The prerequisites are that they are primary resources and calendar resources in all plan masters (PPM, PDS). Depending on the resource and the planning strategy settings, either the time continuous capacity or the bucket capacity is planned finite – never both.

The size of the buckets can be selected in the resource master as well. The most usual (day, week, and month) are provided per default, others can be defined per BAdI.

• *Scheduling Mode*
To support the PP/DS bucket property in scheduling, the strategy profile (transaction /SAPAPO/CDPSC1) provides the scheduling mode 'search for bucket with free capacity'. Additionally the option exists to overrule the resource setting regarding which capacity is the finite capacity (bucket, time-continuous or as defined in the resource). The impact of these settings is shown in table 16.1 for the scheduling mode 'search for bucket with free capacity'.

Table 16.1. Impact of scheduling mode 'search for bucket with free capacity'

	Strategy Settings		
Resource Settings	As Specified in Resource	Time-Continuous Capacity	Bucket Capacity
Time-Continuous Capacity	Infinite	Infinite	Bucket-Finite
Bucket Capacity	Bucket-Finite	Infinite	Bucket-Finite

• *Planning Result of the CTP Check*
The planned orders resulting from the CTP check are scheduled bucket-finitely, but infinitely within the bucket as shown in figure 16.9.

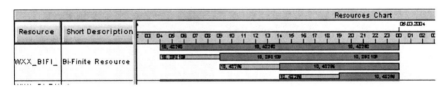

Fig. 16.9. Result of multiple bucket-oriented CTP checks in the planning board

Another advantage of bucket-oriented CTP is that it is intuitively more evident that a subsequent scheduling step is required (though it is required for time-continuous CTP as well).

• *Capacity Consumption and Evaluation*
The capacity consumption for bucket capacity is calculated in a different way than for time-continuous capacity. Especially for sequence dependent set-up there might be significant differences. For the bucket capacity the non-sequence dependent set-up from the plan master (PPM resp. PDS) is

used. The difference between time-continuous and bucket capacity consumption is shown in figure 16.10.

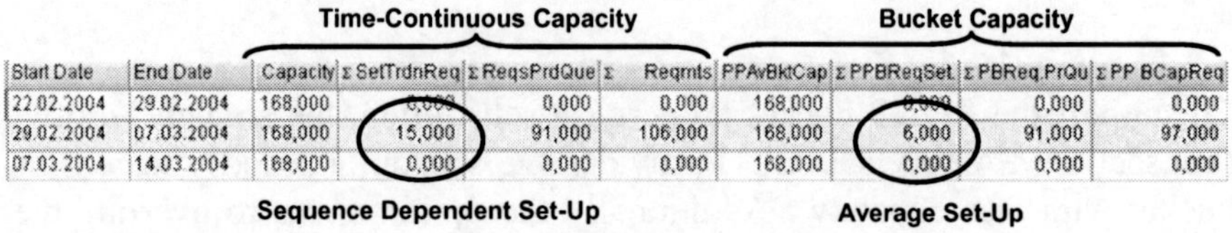

		Time-Continuous Capacity				Bucket Capacity			
Start Date	End Date	Capacity	Σ SetTrdnReq	Σ ReqsPrdQue	Σ Reqrnts	PPAvBktCap	Σ PPBReqSet	Σ PBReq.PrQu	Σ PP BCapReq
22.02.2004	29.02.2004	168,000	0,000	0,000	0,000	168,000	0,000	0,000	0,000
29.02.2004	07.03.2004	168,000	15,000	91,000	106,000	168,000	6,000	91,000	97,000
07.03.2004	14.03.2004	168,000	0,000	0,000	0,000	168,000	0,000	0,000	0,000

Sequence Dependent Set-Up   Average Set-Up

Fig. 16.10. Capacity consumption for time-continuous and for bucket capacity

Both capacity consumptions are displayed with the PP/DS standard reports with transaction /SAPAPO/CDPS_REPT.

It is possible to visualise the bucket capacity consumption in the planning board instead of the 'normal' capacity consumption of multi-resources. To do so, a customising setting the planning board configuration has to be changed as shown in figure 16.11. The configuration of the planning board is explained in section 17.1.

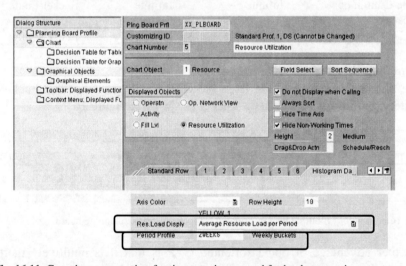

Fig. 16.11. Capacity consumption for time-continuous and for bucket capacity

Note that the period profile has to match the bucket size from the resource master to get sensible results. The period profile is defined in customising with the following path: *APO → SCM → PP/DS → Order View → Detailed Scheduling Planning Board in the Order View → Maintain Period Profiles.*

• *Subsequent Scheduling and Time Horizons*
Due to the fact that the planned orders resulting from CTP are planned
only bucket finite – i.e. there is no finite sequencing – it is necessary to
have a subsequent scheduling step. To avoid problems like overload resulting from the different capacity point of views between CTP and scheduling, the best way is to restrict a horizon to scheduling – e.g. by the planning time fence. Due to the business requirement for flexibility and
responsiveness this might not always be possible.

16.2.5 Interactive CTP

The business scenario for interactive CTP is that there are product allocations on semi-finished product level (cf. section 7.5). In case of late confirmation the planner should be informed via workflow to check the sales
order using interactive CTP.

If a component contains an allocation procedure the product view (cf.
section 15.5.1) is displayed with enhanced options to display different lo-
cationsproducts (user setting 'advanced search for pegging alternatives).
From the product view it is possible either to trigger BAdI
/SAPAPO/EOG_ADDIN1 for posting the stock from a different locationproduct product or to assign a receipt with differing characteristic values to
the request. The update to the sales order is triggered in with the heuristic
SAP_CTP_DLG. As a prerequisite, the indicator for multiple output planning must be set in the product heuristic (cf. section 15.2.2).

16.2.6 Limitations for CTP

Depending on the complexity of the production structure and its modelling
a CTP check can cause significant planning and scheduling load in the system even with the use of bucket-oriented CTP. Therefore it is recommended to use this functionality very carefully and keep the modelling as
lean as possible. Performance problems with CTP increase with the use of
time-continuous CTP, the number of finite resources, the number of
planned orders (due to BOM-levels or lot sizes), the number of operations
within a planned order, the number of automatically planned components
and the number of procurement alternatives. Both from a business and
from a system load point of view CTP is not suited for the production
planning of many BOM-levels.

There are other restrictions which apply to the use of CTP in combination with the following functionalities. CTP should not be used in combination with

- scheduling agreements, since a change in any of the schedule lines causes a new check of all schedule lines,
- periodic lot sizes, since they are not considered during CTP check, which might lead to unfeasible situations after the production planning run and
- planning with the strategies 'planning without final assembly' and 'planning product', since the required capacity to create planned orders from CTP is already blocked with planned orders of a different segment resp. of the planning product, see chapter 5.
- Co-products

These and further restrictions are described in note 426563. Note 426563 answers some FAQs as well. It is technically possible to use CTP in combination with forecast consumption, though the process should be examined carefully.

16.3 Multi-level ATP

16.3.1 Steps Within the Multi-level ATP Process

The basic idea of the multi-level ATP is to confirm a customer request if the components for the product are available in time (i.e. taking the lead time to produce the finished product into account). Like CTP, multi-level ATP is triggered from the check instructions or from the rule and can be used in combination with the basis methods.

Scenarios for multi-level ATP are e.g. business areas where the finished products are assembled to order, e.g. personal computers, and the production capacity is not a bottleneck. In this case the availability of the computer depends on the availability of the components. Figure 16.12 visualises the process for multi-level ATP.

When multi-level ATP is triggered, first the requested date and quantity is determined and passed on to PP/DS. In PP/DS the plan (PPM resp. PDS) for the finished product is exploded and a simulative planned order is created. The strategy profile for this order is the same as maintained for the conversion of the ATP tree in the general settings for PP/DS. The date of the dependent demands for the components is handed over to ATP.

In the second step each component is checked according to the check instructions and the check control. The check instructions are determined by the check mode, which has to be maintained in the product master of each component. The business event is either taken from the customer request from SAP ERP™ or from the check instruction of the finished product (it is possible to maintain a separate business event for multi-level ATP in the check instruction). This way it is possible to choose a different scope of check for the components, e.g. to include the dependent demand as a requirement.

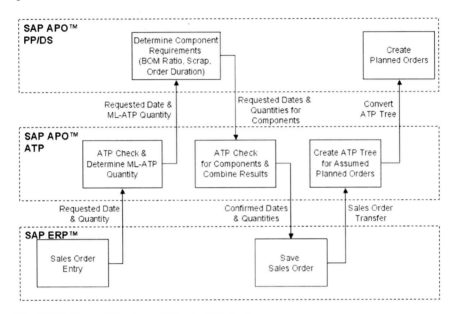

Fig. 16.12. Steps within the multi-level ATP check

If the components have a late availability, the availability of the finished product is calculated in ATP using correlations. Note that the capacity is not considered.

• *Result Screen for Multi-level ATP*
The result of the multi-level ATP is displayed with the same result screen as for rules-based ATP. The availability of the components is displayed as well as shown in figure 16.13.

Product/Location	Material Avail...	Reqmt Quantity	Confirmed Qua...	Cumulated ...	U...	Product avail...	Product Alloc...	Production
▽ XX_PROD1 / XX01 / Item: 000010								
▽ Sched.line: 0001	28.04.2005	10 ■	10	10	EA ■	0		◉ 10
▽ Components of XX_PROD1	28.04.2005	10		10	EA			
XX_COMP1B / XX01	27.04.2005	20 ◉	20	20	EA ◉	20	◉ 20	
XX_COMP1A / XX01	27.04.2005	10 ◉	10	10	EA ◉	10	◉ 10	

Fig. 16.13. Result screen for the multi-level ATP check

According to the availability of the components the sales order is confirmed.

• *ATP Tree and Planned Order Creation*
If the sales order is saved, differing from CTP no planned order is created yet but an 'ATP tree'. The ATP tree contains the information about the required planned order, which was simulated for the confirmation of the sales order. With the transaction /SAPAPO/ATREE_DSP the ATP tree is displayed.

If the first confirmed date for a component is within the scheduling horizon, which is maintained in the global settings for PP/DS (transaction /SAPAPO/RRPCUST1), or the confirmed date of the finished product is within the PP/DS horizon (in the 'PP/DS'-view of the product master), the ATP tree is immediately converted, i.e. planned orders are created in PP/DS. For the conversion of an ATP tree with requirement dates outside the scheduling horizon resp. the PP/DS horizon with the transaction /SAPAPO/ATP2PPDS an offset to the scheduling horizon has to be maintained. The strategy profile for the conversion of the ATP tree is maintained in the global PP/DS parameters. It is recommended to use an infinite strategy.

The dependent demand in the time series for the component (even if no planned order is created yet) to prevent an overconfirmation. The dependent demand is only regarded as a demand element in the ATP time series if the planned order has been checked. This is the case after multi-level ATP.

The planned order is created with the properties 'checked' and 'firmed' and the ATP category AL. The quantity of the planned order can not be changed manually.

• *Subsequent Scheduling*
A subsequent scheduling step is required in any case, not only because capacity restrictions are not considered.

16.3.2 Configuration for Multi-level ATP

The configuration of the multi-level ATP process consists mainly of entries in the entities check mode, the check instruction, the planning procedure for the finished product and the check mode for the components which has to be maintained in the SAP APO™ product master.

• *Check Mode*
In the check mode the production type has to be defined as 'multi-level ATP'.

• *Check Instruction/Re-create Receipts*
In the check instructions the production has to be selected as for CTP. The setting for 're-create receipts' is required to adjust the planned orders to changes in the sales order. This is however only possible in a make-to-order environment. For make-to-stock changes in the sales order cause only an adjustment of the planned orders if the requested quantity is increased or the requested date is brought forward. If a sales order is deleted or its requested quantity is reduced, the according planned orders are neither deleted nor adjusted. In this case the production planning has to be adjusted manually, e.g. using the alert monitor.

• *Planning Procedure*
Because the production planning is done by the conversion of the ATP tree, the planning procedure for the product (in the 'PP/DS' view of the product) should not carry out any actions at the creation or change of sales orders or planned orders to prevent conflicts due to double planning.

16.3.3 Limitations for Multi-level ATP

The conversion of the ATP tree is not a mandatory step. It is also possible to restrict the use of multi-level ATP to the check of the component availability only. In this case no restrictions result regarding production planning.

If however the ATP tree conversion is used, this has severe implications for production planning, because the planned orders are fixed, no lot sizes are supported (only lot-for-lot) and the re-explosion of the PPM is not possible. Therefore it is not possible to adjust the plan with a production planning heuristic, so that the production planning functionality is restricted to the conversion of the ATP tree. Notes 510313 and 557559 provide additional information.

Other restrictions are that sequence dependent set-up, block planning and pegging constraints (including characteristics and shelf life) are not considered. Another system-immanent restriction exists because of the bucket nature of the scheduling for the ATP check. This might lead to delays on each BOM level as described in notes 450674 and 529885.

17 Detailed Scheduling

17.1 Planning Board

The planning board is the central tool for detailed scheduling where operations, orders and the resource load are displayed. Figure 17.1 shows a planning board configuration with the resource chart (Gantt chart) and the order chart. Other available charts are e.g. the operation chart and the resource load chart. Charts are displayed on request using the menu path 'extras', if the according flag is set in the planning table profile. The planning board is called with the transaction /SAPAPO/CDPS0.

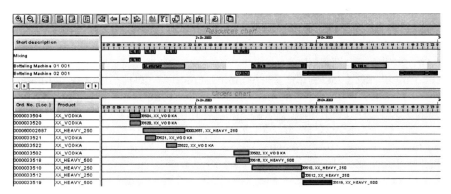

Fig. 17.1. Planning board

The customising of the planning board offers many configuration possibilities. In the example above production orders and planned orders are displayed in a different colour, and properties as fixing and de-allocation are displayed with coloured bars in the middle of the respective object. Generally it is possible to change the layout of the graphical object depending on properties of the according operation, order, product or resource. The information, which is displayed within the objects, is customised as well. By placing the mouse on the object, the information is enlarged as a banner.

J. T. Dickersbach, *Supply Chain Management with APO*,
DOI: 10.1007/978-3-540-92942-0_17, © Springer-Verlag Berlin Heidelberg 2009

In the resource chart three resources are displayed. The working time is marked white and the non-working times grey. If more than one operation is scheduled on a resource at the same time, this is indicated by a thick line underneath the overlapping. Using right mouse click → *Expand Multiple Loading*, the height of the row is adjusted, so that all operations are displayed. For multi resources the 'resource load'-chart provides valuable information. Figure 17.2 shows the resource load for the resource 'Mixing' as in figure 17.1 (with expanded multiple loading).

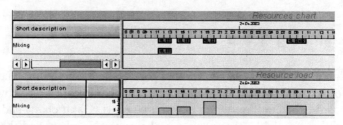

Fig. 17.2. Resource load

Figure 17.2 illustrates that the resource load consumption is independent from the duration of the operation. The maximum capacity is defined as the 'capacity' of the multi resource in the resource master. For single resources the resource consumption of an operation is always 100%.

● *Navigation and Usability*
The initial zoom factor for the planning board is defined in the planning board profile. With right mouse click on the time scale it is possible to switch between hour, day, week, month and year. Interactive zooming is possible by pressing CTR and left mouse button, and draw a window with the mouse. Using right mouse click → *Begin with First Graphical Object* on the resource, the planning board is scrolled to the start of the first object. Via the menu path '*Edit* → *Find*' or CTR + F the order with the entered number is displayed in the planning board.

To highlight all operations of an order, one operation has to be selected and CTR + F6 pressed. With the menu path '*Extra* → *Activity Relationships/ Pegging Relationships* → *Show*' it is possible to display the relationships of an operation or an order.

● *Changes and Transactional Simulation*
For the use of the planning board it is important to keep the concept of the transactional simulations in mind (see chapter 3.8). When calling the planning board, the actual situation is copied into a transactional simulation, which is displayed and in which all planning activities take place. The

orders are not locked, therefore confirmations can be transferred from R/3, but also activities from other planners might change some of the objects. The new plan is written to the active version when saving, and usually the last one to save wins.

There are two consequences of this for the usage of the planning board. The first one is to refresh the planning board frequently to keep the gap between the plan and the reality small – especially for the planning of the near future the feedback from the actual confirmations is essential. The other point which might require some attention is the organisation of the planning process if several planners access one common key resource or different BOM levels of a product are planned by different persons. The propagation range is an assistance from the system side to prevent unauthorised scheduling, nevertheless it does not substitute a clear process design for adjacent or overlapping planning tasks.

If you want to save the result of your scheduling, after pressing the save icon you are asked again whether you want to save, copy or cancel. The right answer is 'copy'.

• Scheduling

Scheduling in the planning board is usually performed by drag and drop. A group of operations is rescheduled at once by pressing 'shift' during drag and drop for the marked objects.

As an alternative to the scheduling of operations by drag & drop the option 'specified date' in the desired date settings of the strategy profile can be used. In this case a pop-up appears for the selected object, where the requested date is entered.

Another option for the interactive planning is to de-allocate and reschedule operations using the icons in the header bar.

• Fixing

Other options for the interactive planning are the fixing of orders or operations using the menu path '*Functions* → *Fix*'. Note that the fixing of operations takes place on operation level, i.e. if the operation of an unfixed planned order (category AI) is fixed, the order will not be fixed on header level. Therefore the category will remain AI and will not change to AJ, which implies that the next production planning run will delete the order although the operation was fixed.

• Planning Board Parameters

The parameters for using planning board are

- the planning board profile, which defines the layout of the planning board,
- the work area, the time profile and the version, which define the content of the planning board,
- the strategy profile, the propagation range, the heuristic profile and the optimiser profile, which define the scheduling possibilities and
- the alert profile.

These entities are grouped in the overall profile.

• Planning Board Profile

The planning board profile is defined with transaction /SAPAPO/CDPSC2 and contains the settings for the selection of the charts, their layout and the layout of the displayed objects. Figure 17.3 gives an overview of the customising structure.

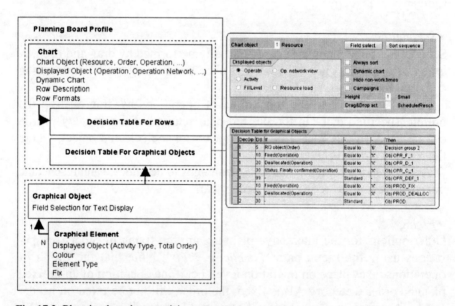

Fig. 17.3. Planning board customising structure

Several charts are available for display in the planning board. The most common are the resource, the operation and the order chart. By using the 'order network' option, pegging is displayed as well. For performance reasons it is advisable to mark as many charts as possible as 'dynamic' – i.e. they are only displayed on request and not by default. The chart format is defined by the row format and the graphical objects. Usually different graphical objects are used for different order types and scheduling statuses, for example planned order and production order, fixed and de-allocated operations. This is defined in the decision table for graphical objects. The graphical object contains one or more graphical elements, which define the shape and the colour of the displayed object. For the graphical element the option exists to prevent its scheduling in the planning board by setting the flag for 'fixed'. If purchase orders are displayed in the planning board (e.g. in the order chart), a separate and fixed object should be used for them. With the decision table for rows it is for example possible to change the colour of the row description in the resource chart depending on resource properties.

Restrictions regarding the number of graphical objects and ways to reduce the number of layer types per graphical element are described in note 374819. It is possible to improve the performance of the planning table by refraining from the display of non-working times and texts as described in note 198852. The runtime of the planning board increases linear with the number of objects (note 379567).

● *Work Area*

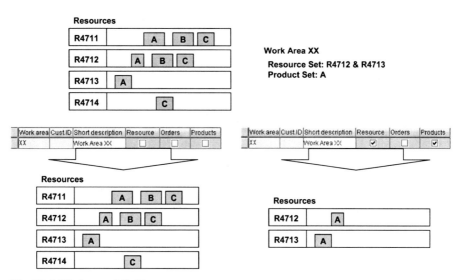

Fig. 17.4. Work area definition

A work area is defined by a set of resources and/or a set of products and/or a set of orders with the transaction /SAPAPO/CDPSC3. Whether the inter-section or the combination is used depends on the flag 'display only selec-ted objects', figure 17.4. If an operation network or an order network is displayed in the chart, this indicator is overruled. Notes 374307, 533824 and 660113 provide additional information on performance and work area in the planning board.

• *Time Profile*
The time profile defines the time horizon for which any objects are displayed. It is possible to distinguish between a display period and a planning period. Naturally the less objects are selected, the better the performance. Usually only an extract of the time horizon is displayed at once on the screen depending on the default zoom factor which is defined in the planning table profile. By defining appropriate segments in the time profile it is possible to improve the performance for the calling of the planning board, because only the objects within the segment are loaded. If the segment is trespassed during scrolling or zooming, the next segments are loaded on that request. Depending on the way the planner works – that is if calling up the planning board is done more often than scrolling in it – using segments might be an advantage.

• *Resource Downtimes: Resolving*
Downtimes are either defined in the resource master or in the planning board by right mouse click on the resource in the resource chart. To our experience, defining the downtimes in the resource master has never caused any problems. If operations had been scheduled on the resource before, they are rescheduled infinitely to the end of the downtime.

17.2 Basics of Detailed Scheduling

17.2.1 Scheduling Strategies

Scheduling strategies define the way operations are scheduled in the system – whether for interactive scheduling in the planning board, for scheduling heuristics or even for production planning heuristics. For production planning the strict recommendation is to infinite planning with as less constraints as possible, which leaves the variety of the different scheduling options to interactive or to background scheduling. For the sake of simplicity the scheduling options are described using the planning board.

Scheduling in the planning board is usually performed by drag & drop (as long as it is not prevented in the chart definition or the customising of the graphical element). The scheduling is performed according to the strategy profile. Forward scheduling has to our experience the most stable behaviour. We recommend using infinite scheduling for at least the dependent objects and consider both internal and external relationships (i.e. pegging) as shown in figure 17.5.

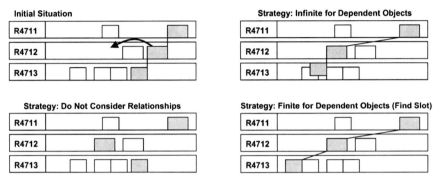

Fig. 17.5. Strategies for interactive scheduling

Using the scheduling mode 'find slot' for dependent objects has the disadvantage that it is absolutely not controllable where the dependent object is scheduled to. Depending on the resource load the planning situation might contain too many constraints, so that the scheduler will not find any solution. Using 'squeeze in' or 'insert' strategies for dependent objects causes many rescheduling actions, which usually lead to undesired disturbances of the plan and a heavy system load. Depending on the actual scheduling task these scheduling modes might be appropriate nevertheless – in this case it is possible to change it user dependent with the icons in the header bar.

Note that the order (resp. the operation) on the top level in figure 17.5 is only rescheduled if there is a maximum pegging resp. maximum activity relationship constraint defined. These constraints should only be applied if there is a necessary and planning relevant technical restriction, since they add significant complexity to planning. In order not to increase the gap between dependent objects, for scheduling backwards the last object should be scheduled, for scheduling forwards the first one. Alternatively the strategy option for 'compact scheduling' can be chosen, which tries to minimise the gaps between dependent objects, figure 17.6. Regarding the restrictions and problems with compact scheduling note 450761 should be considered.

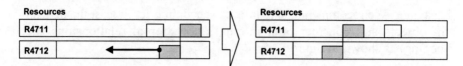

Fig. 17.6. Compact scheduling

For many heuristics the strategy settings have to be applied within the heuristics and the settings in the strategy profile become obsolete. The idea is to guide the user towards the sensible settings.

17.2.2 Error-Tolerant Scheduling

The scheduler applies a kind of 'all or nothing' logic, which means that if one of the selected operations can not be scheduled, scheduling is terminated for all operations. The scheduling heuristics group the activities into packages so that at least the packages will be scheduled, which did not have a failed scheduling attempt in them. Nevertheless there will remain still a lot of operations which will not be scheduled due to one or a few unfeasible problems. Therefore an error-tolerant scheduling is offered. In case that an operation can not be scheduled according to the defined strategy parameters, an emergency strategy is applied which relaxes the scheduling constraints and allows at least the other operations to be scheduled, figure 17.7.

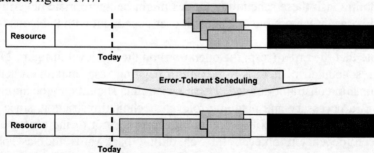

Fig. 17.7. Error-tolerant scheduling

Possible actions in the case of a scheduling error are
- infinite scheduling
- de-allocate
- infinite scheduling & break pegging
- de-allocate and break pegging
- infinite scheduling, cancel pegging & violate order internal relationships
- infinite scheduling, cancel pegging & violate order internal relationships

and can be selected in the planning strategy.

17.2.3 Finiteness Level

Using the finiteness level it is possible to plan in different time horizons with different resources as finite.

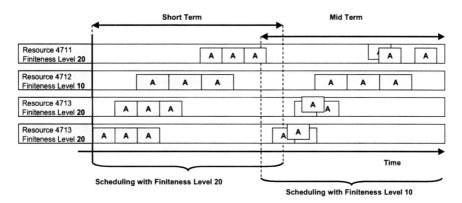

Fig. 17.8. Finiteness in scheduling

The finiteness level is a setting in the resource on one hand and in the strategy profile resp. the optimiser profile on the other hand. Only those resources are scheduled finite which have a finiteness level equal or less than maintained in the profile.

17.3 Scheduling Heuristics

Except from interactive scheduling, SAP APO™ offers a set of standard scheduling heuristics that are included in the planning table using the heu-

ristic profile (transaction /SAPAPO/CDPSC13). The standard scheduling heuristics are
- <u>Schedule Sequence</u>: rescheduling of the selected operations by first de-allocating them and schedule them according to the specified sequence in the heuristic.
- <u>Remove Backlog</u>: operations with backlog are de-allocated and re-scheduled according to the specified sequence in the heuristic.

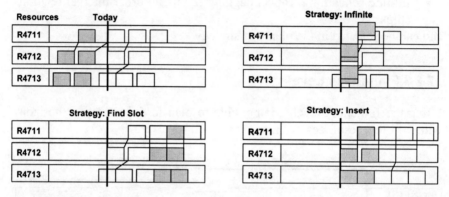

Fig. 17.9. Remove backlog

- <u>Schedule Sequence Manually</u>: selected operations are de-allocated and displayed in the left window of the list. Rescheduling takes place according to the sequence defined by moving these operations per drag & drop into the right window.
- <u>Minimise Runtime</u>: operations are fixed on the selected resource and all connected operations and orders are rescheduled to minimise the total lead time of the according orders.

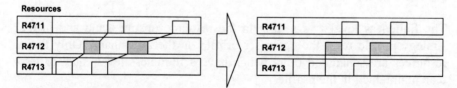

Fig. 17.10. Minimise runtime

- <u>Schedule Operations</u>: schedules already de-allocated operations according to a specified strategy profile

The names of these standard heuristics and the according algorithms are listed in table 17.1.

Table 17.1. Scheduling heuristics

Name/Text	Heuristic	Algorithm
Schedule Sequence	SAP001	/SAPAPO/HEUR_PLAN_SEQUENCE
Remove Backlog	SAP002	/SAPAPO/HEUR_RESOLVE_BACKLOG
Schedule Sequence Manually	SAP003	/SAPAPO/HEUR_PLAN_SEQUENCE_MAN
Minimise Runtime	SAP004	/SAPAPO/REDUCE_LEADTIME
Schedule Operations	SAP005	/SAPAPO/HEUR_DISPATCH

These heuristics contain scheduling parameters which overrule the strategy profile settings (with the exception of SAP005 which has its own strategy profile assigned) and in case of SAP001 and SAP002 the definition of the sequence for scheduling. Table 17.2 lists the available scheduling strategy parameters for these heuristics.

Table 17.2. Parameters for scheduling heuristics

Heuristic	Scheduling Mode	Consider Order Internal Relationships	Consider Fix / Dynamic Pegging	Time Buffers Compact Scheduling
Schedule Sequence	Find Slot / Insert	Always / Prop. Range / Not	Always / Prop. Range / Not	-
Remove Backlog	Infinite / Find Slot / Insert	Always / Prop. Range / Not	Always / Prop. Range / Not	-
Schedule Sequence Man.	Find Slot / Insert	Always / Prop. Range / Not	Always / Prop. Range / Not	
Minimise Runtime	All	Always / In Propagation Range		Time Buffers, Comp. Sched.
Schedule Operations		Strategy Profile		

Some options of the strategy profile are already inhibited for the heuristics by the limited settings. Nevertheless we recommend to test any heuristic carefully regarding the consideration of order internal relationships and especially regarding the fix and dynamic pegging.

We would recommend to test these heuristics thoroughly with the chosen strategy settings for the relevant number of operations before including them into a regular planning process (e.g. production planning in de-allocated mode and schedule operations afterwards with the heuristic). If the heuristics do not meet the expectations, using the sequence optimiser should be considered as an alternative (cf. section 17.5).

• *Multi-level Scheduling Framework*

The idea of the multi-level scheduling framework is to allow heuristics to schedule an activity network instead of individual activities. The approach is to select the activity network from live cache and control the scheduling by heuristics. Still each activity will be scheduled only once, which implies that no optimisation is performed. The advantage of the multi-level scheduling framework is that it allows heuristics to process scheduling problems of a higher complexity than it is possible with the single-level heuristics. In comparison to the PP/DS optimiser the heuristics have the advantage of a better understandability of the solution and the possibility to apply customer specific scheduling logics via BAdI.

The following examples help to understand the motivation for the multi-level scheduling framework. Figure 17.11 shows a scheduling problem of medium complexity:

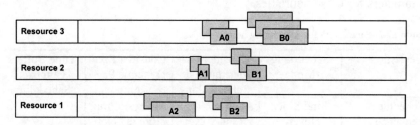

Fig. 17.11. Scheduling problem of medium complexity

The material flow is unidirectional from resource 1 to resource 3 and the bottlenecks (resource 1 and resource 3) are clearly visible. The scheduling approach with single-level heuristics would be:

      Step 1: Schedule sequence on resource 1.

      Step 2: Multi-level bottom-up heuristic.

      Step 3: Schedule sequence on resource 3.

But looking at a more complex scheduling problem where the material flow is not correlated to any resource order and the bottlenecks are moving as shown in figure 17.12 (e.g. in high tech industries), no successful scheduling approach with single-level scheduling heuristic is possible.

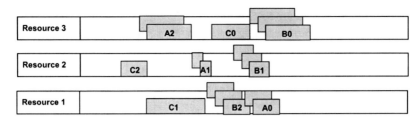

Fig. 17.12. Scheduling problem of high complexity

In this case the multi-level scheduling framework becomes an advantage, because by reading the activity network from live cache it becomes clear which activities have to be scheduled first. Figure 17.13 shows this approach.

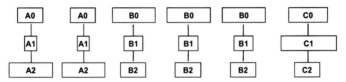

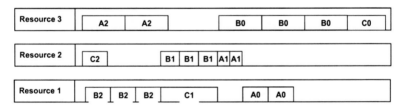

Fig. 17.13. Multi-level scheduling framework – approach

Per default only the multi-level scheduling heuristic SAP_DS_01 (only forward scheduling) for stable forward scheduling is provided. The scope of this heuristic is to schedule backlog to a feasible plan again without changing the sequence of the operations across the resources.

17.4 Sequence Dependent Set-up

In many production processes the set-up for an operation depends on the predecessor operation. In discrete manufacturing there are often several parameters of a resource which have to be adjusted depending on the products which are planned. Sequence dependent set-up is modelled in SAP APO™ by assigning a set-up group resp. a set-up key to the operation and defining the set-up duration between the set-up groups resp. keys in a set-up matrix. One or more set-up keys are assigned optionally to a set-up group to refine the set-up duration for certain set-up groups. The set-up group or – if set-up keys are used - the set-up key is assigned to the PPM on operation level. Though there is no integration, the set-up group corresponds to the set-up group category in SAP ERP™ and the set-up key to the set-up group key in SAP ERP™. The set-up matrix defines the set-up duration between the set-up groups resp. the set-up keys and is assigned to the resource. Figure 17.14 gives an overview about this structure.

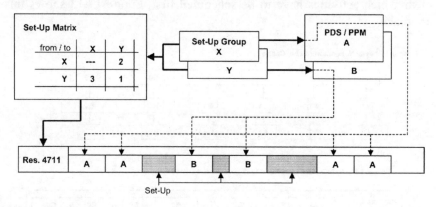

Fig. 17.14. Calculation of sequence dependent set-up

Note that the set-up matrix has to be assigned to the resource both version independent (since the PPM consistency check looks here whether the set-up group of the operation is included in the set-up matrix of the resource) and version dependent (this is where the scheduler gets its information from). To use the set-up duration from the set-up matrix, the flag for sequence dependent set-up has to be set in the set-up activity.

The sequence of the activities is calculated using the start date of the production activity. Before SAP APO™ 4.0 the sequence was calculated using the start date of the set-up activity, which could lead to recursions and cause unexpected results as described in note 445899. The calculation of the dynamic set-up durations requires a clear sequence of predecessors and successors, which is the reason that sequence dependent set-up is not allowed for multi resources and which is not always guaranteed with infinite planning. In this case planning in de-allocated mode should be examined.

There are other recommendations regarding the modelling in combination with sequence dependent set-up in note 445899. A rather important one is not to assign any input components to a dynamic set-up activity, since a change in the sequence leads to new material requirement dates. This is controlled by the customising setting in SAP ERP™ with the path *Integration With Other mySAP.com Components → APO → Application Specific Settings and Enhancements → Settings and Enhancements for In-House Production → General Settings for Manufacturing Orders → Assign Components to an Operation Segment.* The set-up group and the set-up matrix in SAP APO™ are location dependent. Regarding the set-up groups and the set-up keys there is a structural difference between SAP ERP™ and SAP APO™ as well – in SAP ERP™ it is possible to assign a set-up key to more than one set-up group. For reasons of PPM integration the SAP ERP™ modelling should consider the SAP APO™ possibilities. Figure 17.15 shows the interdependencies of the relevant entities in SAP ERP™ and in SAP APO™.

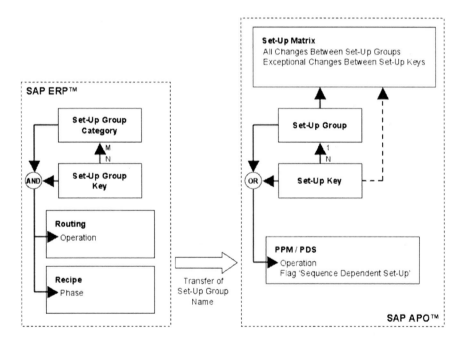

Fig. 17.15. Entities for set-up in SAP ERP™ and in SAP APO™

Another difference between SAP ERP™ and SAP APO™ is that the maintenance of the set-up group categories and keys is customising in SAP ERP™ (transaction OP18 resp. OP43) and master data in SAP APO™ (transaction /SAPAPO/CDPSC6). The set-up matrix is another master data object in SAP

APO™ and is maintained with the transaction /SAPAPO/CDPSC7. Subsequently the SAP APO™ entries are not transported to other systems (quality assurance, productive system).

To have the entry for the set-up group resp. set-up key transferred from SAP ERP™ into the according field in the operation of the PPM resp. PDS, it is sufficient to assign a set-up group category and – optionally – a set-up group key with the same name as the corresponding object in SAP APO™ to the operation of the routing. If a set-up key is assigned to the operation of the routing as well, only the set-up key is assigned to the operation of the PPM resp. PDS. For recipes the set-up group categories resp. keys are assigned to the phase representing the set-up and transferred the same.

In some cases set-up matrices can become rather huge if the set-up depends on multiple criteria. Therefore SAP APO™ offers the option to generate a set-up matrix based on characteristics.

17.5 Sequence Optimisation

17.5.1 Optimisation as Part of the Planning Process

The scope of the optimiser in PP/DS is the scheduling of orders and not the creation of orders like with the SNP optimiser. To our experience the ideas of an optimal schedule do differ, so therefore we recommend to define the expectations first before starting to implement the optimiser. For example some of the standard targets as lead time minimisation and set-up minimisation can be opposite, and whether one rather wants to produce early to have some time buffers or rather as late as possible depends on the optimisation horizon as well as on the business.

To our opinion the most important issue for the implementation of the optimiser is to fit its usage into an integrated planning procedure consisting of production planning, optimisation and interactive scheduling. The approach to load the complete planning scope into the optimiser, press the 'optimise' button and expect all problems to be solved is tempting but in many cases does not provide the expected results. Generally the slicing of the planning tasks in regards of production area and time horizon is one of the more difficult but nevertheless most important tasks. Another point of attention should be the expected interaction between the optimiser result and interactive planning, since the optimiser might provide with each planning run results which are similarly good regarding its objectives but totally different in detail.

A trade off between production resp. set-up costs and storage costs can not be modelled with the standard settings.

The design of an appropriate scenario always depends on the individual requirements, and this is where the project focus should be. An analysis whether the optimisation algorithms might be tuned or how the weights for the target function are optimised is not that important.

The optimiser can be a valuable tool to reduce sequence dependent set-up times, to reduce lead times or – in case of inevitable lateness – to schedule orders according to the priority of the demands. And the optimiser might have an important role in creating feasible plans at all. On the other hand, the downside of optimisation is that its results are not always easy to understand.

17.5.2 Optimisation Model and Scope

The result of the optimiser depends – except from the weights and the algorithm – on the time horizon, the selected resources and order types, and on the runtime for the optimisation. The optimisation problem for sequencing belongs to the more complex problems which are not solved by linear approaches. Therefore the optimiser applies a heuristic approach, which does not guarantee the optimal solution, and does not even provide necessarily the same results under the same circumstances. But to our opinion this is not an issue.

Optimisation in the background is included into the background production planning definition (transaction /SAPAPO/CDPSB0) or is executed with separate variants in transaction /SAPAPO/OPTB0. For interactive optimisation it is possible to call the optimiser from the planning board, the production planning table or the product view.

● *Data Model*
The optimiser uses its own data model. The data is read from the live cache, converted into the optimisation data model and written back to the live cache after optimisation as shown in figure 17.16. If the optimiser is called in the interactive mode from a transactional simulation, the data is written back to that transactional simulation.

The scope of the optimisation is defined by the selected resources, the time horizon and the selected order resp. operation types. If the optimisation is executed in the interactive mode, the data selection of the planning board resp. the product planning table is taken.

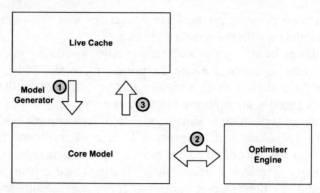

Fig. 17.16. Structure of the PP/DS optimiser

• *Pegging*

After reading the plan from live cache the optimiser performs an internal kind of pegging which is not changed during the optimisation run. Because the pegging is not changed during optimisation, the initial situation has a very big impact on the result of the optimisation. Pegging constraints like maximum pegging length and shelf life are considered by the optimiser.

As shown in figure 17.16, the optimiser reads the planning situation from live cache and transforms it into the core model for optimisation. The order assignments which are defined by the pegging relationships are kept during the optimisation run. Consequently the optimisation problem becomes the more complicated, the more receipts are pegged to one requirement resp. the more requirements are pegged to one receipt. It is recommended to use the over- and underdelivery tolerances for pegging and the pegging strategy 'FIFO' to keep the pegging relationships simple, see also note 532979.

• *Scope of the Optimisation*

The pegging constraints and the order internal constraints – both within the optimisation time horizon and to orders outside the horizon – are considered. Hence by selecting the resources for optimisation, the constraints for the solution are also defined. Figure 17.17 visualises the potential for the optimisation depending on the selection of the optimisation scope.

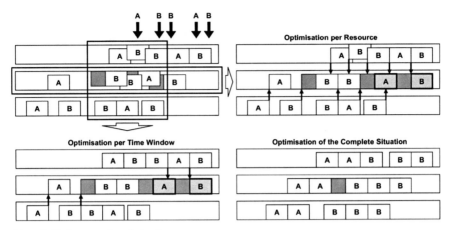

Fig. 17.17. Impact of optimisation scope

In both cases, when the time window for optimisation is too small and when the depending orders on different resources are not selected, the result of the optimisation run is sub-optimal as the late orders (displayed in grey colour) show.

Pegging constraints and shelf life is considered by the optimiser but does increase the complexity of the optimisation problem. Therefore these constraints should only be used if necessary. Since planning situations do not always have solutions which meet the demand in time, the demand on the top level is considered as a soft constraint – no matter whether it is a dependent demand or a customer demand. It is however only possible to prioritise customer demands in the optimiser profile (using the delivery priority).

● *Alternative Sources of Supply*
As the PP/DS optimiser does not create any orders, it does not change sources either, i.e. alternative plans (PPM resp. PDS) are not taken into account. To use the option of scheduling operations to alternative resources, they have to be modelled as alternative modes within one plan.

17.5.3 Optimisation Controls Within the Optimisation Profile

The optimisation profile is maintained with transaction /SAPAPO/CDPSC5 and contains all the relevant settings for optimisation, as

- the optimisation horizon and the start of the schedule,
- the weights of the target function, the optimisation algorithm and the runtime,
- the order selection,
- the backwards pegging,
- the use of infinite resources,
- the prioritisation and
- the backward scheduling.

Some of these settings have already been explained in the previous paragraph. The other ones are described in the next paragraphs.

• Optimisation Criteria
The target function for the optimiser contains the criteria
- lead time (calculated as the duration between start schedule and the end of the last operation),
- set-up times,
- set-up costs,
- average lateness and
- maximum lateness.

The relative impact of these criteria is determined by the weighting factors. Each criterion might tend to favour a different solution. For example an optimal solution regarding the lead time might increase the set-up times and vice versa, figure. 17.18.

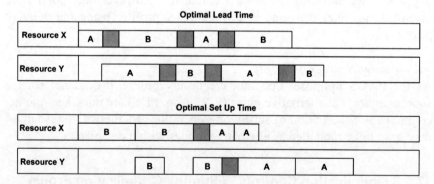

Fig. 17.18. Trade off between lead time and set-up time optimisation

Except for special cases, a more or less arbitrary mix between lateness, lead time and set-up times proves to be appropriate. These settings might be tuned during usage, and we would not recommend focusing too much on the settings of the weighting factors.

• *Optimisation Algorithms*

For PP/DS optimisation genetic algorithms are used. Genetic algorithms are the method of artificial intelligence for optimisation. The basic idea of genetic algorithms is to imitate the evolutionary process by starting with a set of initial solutions, combine and mutate them and select the next generation of solutions for the following iteration step according to the fitness (i.e. the quality of the solution) of the members. Its main advantage is the comparatively low aptitude to local minima. Using the APEX interface it is possible to link other optimisation algorithms for special problems, e.g. for scrap minimisation in the metal, paper and wood industries.

• *Backward Pegging*

As mentioned before, the existing pegging has a severe impact on the optimiser. If no pegging relationship exists because of lateness, the optimiser is able to perform a 'backward pegging' (independent of the settings in the product master) to create relationships for the core model. The range for backward pegging is defined in the 'basic settings' view of the optimiser profile. Since the optimiser does not create any new assignments, without an existing assignment in the core model the orders are scheduled independently of each other. Figure 17.19 illustrates this behaviour.

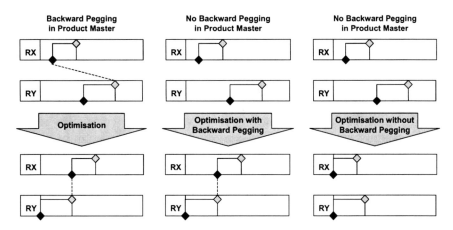

Fig. 17.19. Backward pegging

• *Order Selection*

Except by the resources and the optimisation horizon, orders can be excluded from the optimisation scope depending on their order type (external procurement, transport, ...) and whether they are partially outside the horizon. Fixed operations are always considered as fixed by the optimiser, and whether released production orders are fixed or not, is determined in the optimiser profile. The scheduling status (scheduled, de-allocated or both) is a further selection criterion to define the optimisation scope.

Pegged orders for external procurement are selected as well, if defined in the optimisation profile. If they are not selected, the dependency via the pegging relation is regarded nevertheless. To take the planned delivery time for purchase requisitions into account, the according flag has to be set in the 'order selection' view of the optimiser profile. Purchase orders are considered as fixed.

The fixing horizon is not taken into account by the optimiser. To prevent rescheduling within the fixing horizon, the optimisation horizon and the start of the optimised plan have to be outside the fixing horizon (see also note 517426).

• Finite and Infinite Planning
Resources are considered as finite unless the button 'schedule according to the settings in the master data' is set in the 'order processing' view of the optimiser profile. In this case the setting of the resource master is used. If sequence dependent set-up is used, the resources are regarded as finite nevertheless (see also note 435160).

• Finite and Infinite Planning: Calendar Resources
The modelling of the goods receipt time as an activity on the handling resource of the location is a problem for the optimiser, since all orders of a location with goods receipt have an activity on the same handling resource. If the handling resource is selected, a lot of activities are included into the optimisation model which are not necessary from a problem point of view. On the other hand, if the handling resource is not selected, the goods receipt activity is regarded as fixed and the scope for the optimisation is restricted in an undesired way. But if a calendar resource is used as handling resource, the goods receipt activities do not consume any capacity and therefore number of activities for the optimiser does not increase.

• Slicing of the Optimisation Runs
The supply dates are usually regarded as hard constraints, whereas the demand dates are soft constraints – there has to be a valve for an unfeasible planning problem. If the bottleneck which really requires optimisation is somewhere at the beginning of the product flow, this way it is possible to use the optimiser only for the bottleneck and adjust the orders on the following resources with scheduling heuristics.

If the bottleneck is however at the end of the material flow, it is possible to ignore the supply dates of orders on not selected resources with the setting 'do not consider upstream dependencies' in the optimiser profile.

• *De-allocation of Late Orders*
It is possible to control in the optimiser profile that orders are de-allocated in order to keep the due dates. The advantage of this behaviour is that it becomes more transparent where the capacity problems are and it is easier for the planner to implement an action to this.

• *Prioritisation*
It is possible to prioritise demands according to their delivery priority, orders according to their order priority and status (if already begun or confirmed) and modes according to their mode priority. Figure 17.20 shows the impact of the prioritisation of the sales orders for product B versus the forecast for product A.

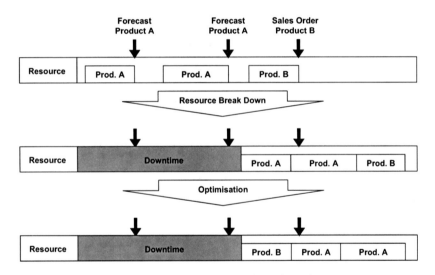

Fig. 17.20. Prioritisation of sales orders vs. forecasts in optimisation

• *Backward Scheduling*
The planning direction of the strategy profile is not relevant for the optimiser. During optimisation, the operations are scheduled as early as possible (beginning at the 'start schedule' date). Setting the flag 'backward scheduling', the optimiser moves the operations as close as possible to the demand date – optionally with a change of the resources, figure 17.21.

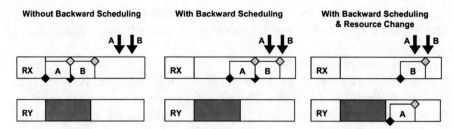

Fig. 17.21. Backward scheduling

The backward scheduling step is performed after the optimisation has finished.

• *Example*

The impact of the parameters for optimisation might become clearer looking at the following example. In figure 17.22 a non-feasible planning situation is shown with overload on resource X (orders for A and for B) and lateness for the orders C and C2.

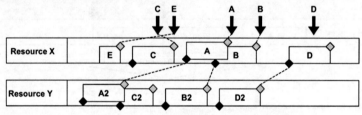

Fig. 17.22. Example – initial situation

The desired optimal solution might be to produce in time, but as late as possible, figure 17.23.

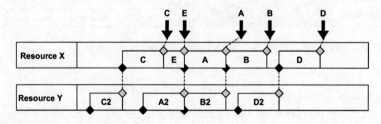

Fig. 17.23. Example – desired solution

Without backward scheduling the optimiser tries to produce as early as possible, figure 17.24.

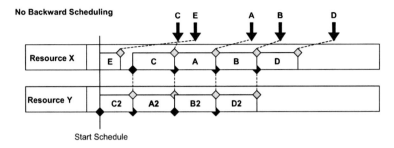

Fig. 17.24. Example – without backward scheduling

Another prerequisite here is to use backward pegging, since the optimiser does not create pegging relationships. If not, in this example the input C2 is not recognised as necessary for order C and will not be scheduled in time, figure 17.25.

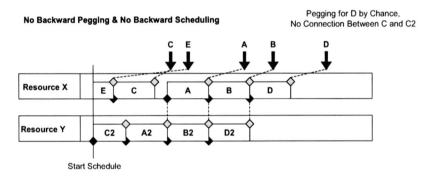

Fig. 17.25. Example – without backward pegging

17.5.4 Handling and Tools for Optimisation

The logs for the optimiser are displayed with transaction /SAPAPO/OPT11. In the transaction /SAPAPO/COPT00 it is defined how long the logs are kept before deleting them.

If the optimisation run is terminated without solution, the first guess is to increase the runtime. If the information from the logging does not help, for the next step we recommend to restrict the optimisation horizon to identify the objects which cause the problems.

- *Volume and Data Modelling*

There are experiences that the optimiser is able to process about 200,000 operations. Nevertheless it is recommended to make the model as simple as possible in order to reduce the run time and improve the quality of the solution.

- *Technical Settings*

The optimisation set up, that is the RFC destination to the optimisation server and the logging, is defined in the optimiser server master data with the transaction /SAPAPO/COPT01. As a prerequisite the RFC destination (as TCP/IP connection) with the path for the optimisation executable file must exist. The number of users per optimiser is limited in the optimiser server master data as well. With the transaction /SAPAPO/OPT03 the users for the respective optimiser applications are displayed and administrated. These technical settings apply to all optimisers, i.e. the PP/DS optimiser, the SNP optimiser and the CTM planning engine. The version of the optimiser is displayed with the transaction /SAPAPO/OPT09.

18 Production Execution

18.1 Planned Order Conversion

The conversion of planned orders into production orders is either triggered in SAP APO™ or in SAP ERP™. The conversion itself is in any case performed in SAP ERP™. To trigger the order conversion in SAP APO™, the two alternatives exits: Either with the transaction /SAPAPO/RRP7 or by setting the conversion indicator interactively for the individual order.

Since the planned order and the production order are different objects in SAP ERP™, at the time of the order conversion the planned order is deleted in SAP ERP™ and a production order is created, including the BOM explosion, the explosion of the routing and a scheduling of each operation. If the conversion is triggered in SAP ERP™, simply the information about the deletion of the planned order and the creation of the production order is transferred, so that the scheduling of the planned order is lost, figure 18.1.

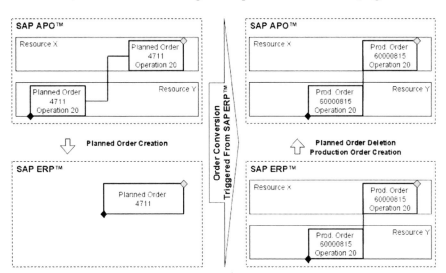

Fig. 18.1. Order conversion triggered from SAP ERP™

J. T. Dickersbach, *Supply Chain Management with APO*,
DOI: 10.1007/978-3-540-92942-0_18, © Springer-Verlag Berlin Heidelberg 2009

Triggering the order conversion from SAP APO™ has the advantage that the connection between the deleted planned order and the new production order is considered, so that the production order is matched with the planned order and the operation dates are kept as shown in figure 18.2.

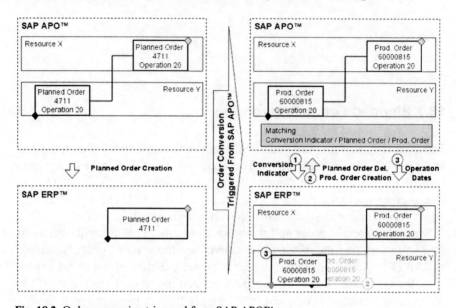

Fig. 18.2. Order conversion triggered from SAP APO™

The operations are scheduled in SAP ERP™ but the scheduled operation dates are overwritten by the operation dates from SAP APO™. Therefore, if scheduling is performed in SAP APO™ with planned orders, the order conversion has to be triggered from SAP APO™ to prevent the loss of the scheduling result.

- *Opening Horizon*

The conversion of the planned order usually has to take place some time before the scheduled start to provide enough time for the preparation tasks as the printing of the order papers and the transport of the components to the resource. This time buffer is modelled by the opening horizon in the 'PP/DS'-view of the product master.

The conversion of planned orders into production orders is triggered from SAP APO™ by setting the conversion indicator either individually per order or automatically in the mass conversion run with the transaction /SAPAPO/RRP7. Triggering the conversion from SAP APO™ is only possible for PP/DS orders, not for SNP orders. In the mass conversion run the planned orders are selected according to product, location and production planner. Only those orders are selected, which have their start date within the opening horizon, figure 18.3.

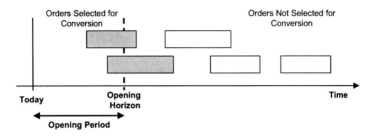

Fig. 18.3. Opening horizon

By maintaining an offset to the opening horizon, the order selection is extended according to formula 18.1:

$$\text{Today} + \text{Offset} \geq \text{Start Date} - \text{Opening Period} \qquad (18.1).$$

- *Conversion Rule*
For the conversion of planned orders the conversion rule is taken into account which defines
  - whether an ATP check is performed with the order conversion,
  - whether only those orders are converted which are pegged to customer orders (requirement check) and
  - whether a customer specific logic as defined in a BAdI is applied.

The conversion rule is defined with the customising path *APO → Supply Chain Planning → PP/DS → Transfer to Execution* and is assigned to the global parameters (transaction /SAPAPO/RRPCUST1) or to the locationproduct master.

18.2 ATP Check and Batch Selection

It is common to perform an ATP check either at the creation or at the release of a production order. If the batch management for some components is active, in some cases a batch selection and determination is performed for these. These two activities are independent of each other.

- *ATP Check*
The availability of the components of an in-house order is checked in SAP APO™ either by triggering the ATP check for the order in the product view

or at the time of its conversion (if defined in the conversion rule). The business event is by default PP, it is maintained in SAP APO™ in the global settings with the transaction /SAPAPO/RRPCUST1.

• *Batch Determination*
The batch determination process in SAP ERP™ is independent of anything that is done in SAP APO™. Fix pegging for example does not have any influence on batch determination in SAP ERP™. On the other hand it is possible to reflect the result of the batch determination in SAP APO™ by creating fixed pegging edges between elements with the same batch number, but this is an offline process using a separate heuristic. Section 19.4 about fixed pegging describes this in more detail.

18.3 Production Order Handling

The production order cycle contains the statuses created, released, partially confirmed, finally confirmed and technically completed. These statuses are transferred to SAP APO™ and are displayed in the order details. The change in the order status and the confirmation – especially deviations in the confirmed quantity – has the impacts as listed in table 18.1. Note that the category for the 'dependent demand' (AY) changes to 'order reservation' (category AV).

Table 18.1. Production order cycle

Order Status in SAP ERP™	Order Category	Receipt Quantity	Order Reservation
Created	AC	Full Order Qty.	Full Reservation Qty.
Released	AD	Full Order Qty.	Full Reservation Qty.
Finally Confirmed (Some Operations)	AD	Adjusted Qty.	Confirmed Operation: Full Reservation Qty.
			Following Operation: Depending on Propagation
Finally Confirmed (All Operations)	AD	Adjusted Qty.	Full Reservation Qty.
Technically Completed	-	Order is Deleted	Reservations are Deleted

Starting with the conversion of the planned order to a production order the SAP ERP™ system becomes the more the master for changes of the production order the more execution is concerned. Depending on the order status the following options exist in SAP APO™ as listed in table 18.2.

Table 18.2. Production order changes in SAP APO™

	Created	Released	Partially Confirmed
Scheduling of Operations	Yes	Yes	No
Change the Order Quantity	Yes	No	No
Change the Component Quantity	Yes	No	No
Delete/Add Component	Yes	No	No

A major difference in the scheduling of production orders between SAP APO™ and SAP ERP™ is that no reduction of the buffer times is possible in SAP APO™ by different scheduling strategies. Unlike in SAP ERP™, the buffer times in SAP APO™ are modelled related to the resources and not to the orders.

The transport times between the operations to move the goods from one resource to another are calculated in SAP ERP™ using a transport matrix. These durations are transferred to SAP APO™ as relationship constraints, but are not scheduled according to the factory calendar. This might lead to conflicts if a transport time between two operations is required, but the first operation ends on Friday evening and the second operation starts on Monday morning. In this case the minimum time constraint between the operations is respected, nevertheless the required time for transport is not planned this way. The notes 380141 and 321956 provide more information about the integration of transport times.

• *Confirmation*
The expected production quantity of an operation depends on the order quantity and the scrap. The confirmed quantity of an operation might however differ from the planned quantity. Depending on the production process the requirements regarding the adjustment of the subsequent operations in case of over- or underconfirmation differ. In a typical assembly scenario an increased scrap in the first operation causes that less units are processed in the following operations, therefore the duration for these operations is shorter than planned and less components are required. In other scenarios, e.g. for highly automated production processes where the work pieces are attached to a carrier belt, the duration of the process depends on the length of the belt and not on the scrap. Whether the duration and the component demand of the following operations are adjusted, is defined in SAP APO™ with the transaction CFO1 per plant and production scheduling profile. The production scheduling profile is maintained in the 'work scheduling'-view of the material master. Figure 18.4 shows the effect of the according flag.

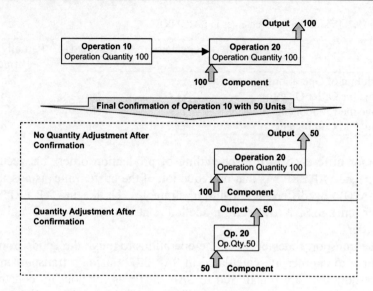

Fig. 18.4. Operation and quantity adjustment after confirmation

Note that the receipt quantity of the output quantity is adjusted in both cases. These settings are only valid for PP and not for PP-PI.

19 Modelling of Special Production Conditions

19.1 Alternative Resources

Alternative resources can be modelled either as alternative modes within one PPM resp. PDS or as alternative PPM resp. PDS. Since the selection of the plan takes place during production planning and not in scheduling, changes of the plan in scheduling are not supported by the scheduling applications but have to be performed interactively. Therefore alternative resources should be modelled as alternative modes if the resource selection is supposed to take place during scheduling.

If alternative resources are modelled using alternative PPM resp. PDS (i.e. alternative production versions), the possibilities to change the resources in PP/DS are limited. The advantages and disadvantages resp. the properties of the different ways of modelling alternative resources – either using alternative modes within one plan or using alternative plans with only one mode – are listed in table 19.1.

Table 19.1. Properties of alternative modelling approaches

Resource Change	Alternative Mode	Alternative Plan (PPM or PDS)
Planning Board	Drag & Drop	Plan Change in Order
Optimiser	Automatically	Not Possible
Product View	Not Possible	Plan Change in Order
Production View in the Product Planning Table	Manual Resource Loaded	

Generally the use of alternative resources in PP/DS is easier having alternative modes instead of alternative plans. Regarding the integration to SNP (where alternative resources can not be modelled using alternative modes) the use of separate plans in PP/DS as well allows to keep the resource selection of SNP.

J. T. Dickersbach, *Supply Chain Management with APO*,
DOI: 10.1007/978-3-540-92942-0_19, © Springer-Verlag Berlin Heidelberg 2009

• *Alternative Modes*

Alternative resources can be modelled using alternative modes. From an integration point of view there are basically two ways to create alternative modes in SAP APO™, these are resource classification and alternative sequences.

• *Alternative Modes Using Resource Classification*

If the duration of the operation is the same, resource classification is the easiest way to model alternative resources. Both the work centers and the operation of the routing are classified with the same class (class type 19 [resource]). To enable the assignment of a resource class to a routing, the transaction OPCA has to be entered within the routing maintenance. During the PPM or PDS transfer the CIF reads the resources which have the same evaluation as the operation and transfers them as alternative modes, figure 19.1. The classification is needed only on the SAP ERP™ side, so no classes or characteristics have to be transferred to SAP APO™. All modes will have the same priority and the same duration.

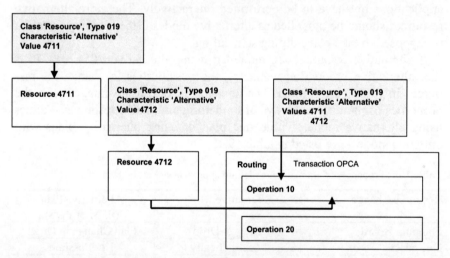

Fig. 19.1. Alternative resource modelling using classification

A proposition for the naming convention is to use the work center keys as characteristic values. Figure 19.2 shows the correspondence of a classified routing and the according PPM resp. PDS.

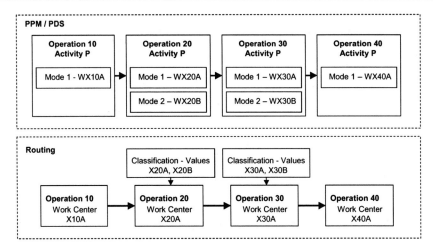

Fig. 19.2. Alternative resources in routing and PPM/PDS using resource classification

• *Alternative Modes Using Alternative Sequences*
Alternative sequences in the routing are transferred as alternative modes. The prerequisite is that the number of operations is the same in each alternative sequence – if necessary, dummy operations have to be included. As described in note 357178, for the PDS it is however required to implement the BAdI CHANGE_EWB_STRUCTURES. For the PPM the user exit CIFPPM01 has to be implemented (see note 217210).

A limitation in this modelling is that the change of the resource in SAP APO™ is transferred to the production order in SAP ERP™, but not as a change of the sequence. The consequence is that the texts and PRTs might be wrong in the production order.

• *Mode Linkage*
The mode linkage determines the dependencies between adjacent operations regarding their modes. By default the mode linkage is '0', which means that there are no restrictions regarding the mode selection. The possible production paths are shown in figure 19.3.

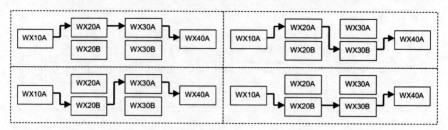

Fig. 19.3. Mode linkage – no mode restrictions

If the duration of the operation depends on the selected resource, alternative sequences have to be maintained in SAP ERP™, see figure 19.4. Since the alternatives are transferred as alternative modes (and not as alternative activities), the number of operations in the alternative sequence have to be the same as in the original sequence. If the production process on the alternative resource requires fewer steps, dummy operations have to be created. The lot size interval of the alternative sequence has to be the same as in the basic sequence. If the alternative sequences have different priorities, a proposition is to use a user field in the sequence header to maintain the mode priority and transfer it via user exit.

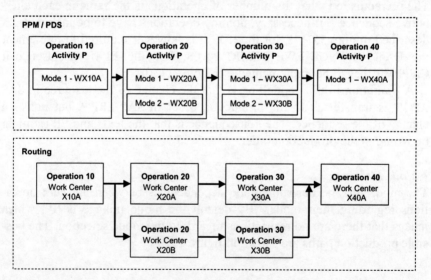

Fig. 19.4. Alternative resources in routing and PPM/PDS using alternative sequences

Except that different durations might be maintained per mode, the main difference to the modelling with resource classification is that the mode linkage in the PPM resp. PDS relationships is set to identical mode per default. Figure 19.5 shows the impact of this on the possible production paths.

Fig. 19.5. Mode linkage with identical mode number

It is even possible to combine resource classification with alternative sequences. To receive sensible results, the mode combinations have to be exploded via user exit and mode linkage '3' (identical mode number) has to be used.

● *Alternative Resources in CTM*
CTM uses alternative resources to schedule the order, but the information of the alternatives is removed from the order after scheduling with the consequence that subsequent optimisation resp. manual scheduling can not select alternative resources.

19.2 Modelling of Labour

Labour can be modelled using secondary resources. A proposition is to use a multi resource as secondary resource and model the number of workers as capacity. The ratio of machine time and labour time of an operation should be used as the fix capacity consumption, while the duration on the secondary resource (the labour) is the same as on the primary resource (the machine).

In many cases workers are not required to give their full attention to one machine only but a pool of workers operate a couple of machines which require more attention at certain moments, e.g. when feeding a new input batch or at set-up. The modelling of workers as a pool has the additional advantage that the planning is not at the level of individual workers which is apt to cause instability and unnecessary constraints.

The SAP entity to model a pool of workers is the use of a capacity in SAP ERP™. This capacity can be assigned to many work centers, figure 19.6. The capacity is transferred as a resource to SAP APO™ as well and should be modelled as a multi or a multi-mixed resource.

Given the example of a work center that requires two hours of machine time and one hour of labour for the basis quantity, the transfer to the 'plan' in SAP APO™ would not take the different standard values into account.

Figure 19.6 shows this case in the left picture (the machine is relevant for scheduling).

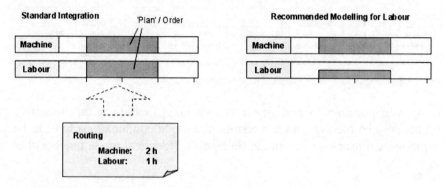

Fig. 19.6. Modelling for labour

To consider the different requirements for machine and labour we recommend to adjust the capacity consumption of the labour operation to the ratio of the labour duration divided by the machine duration. This is for most cases a better representation of the work load, since the standard value of one hour is distributed across the machine duration. However, the logic has to be implemented via user exit.

In many cases labour capacity is not modelled at all. If labour is modelled, it is modelled as a total. It is not a common approach to perform a more detailed modelling.

19.3 Overlapping Production

In many cases – especially for commodities – production orders have a high order quantity and the next production step will start (using the already manufactured products as an input) before the first production step is finished, i.e. before the whole order quantity is produced. Two cases have to be considered for the modelling of overlapping production: Overlap between operations of the same order and overlap across orders.

● *Overlap for Orders: Continuous Flow*
Continuous flow is characterised by the output rate resp. the consumption rate which are calculated by the duration of the operation, the quantity and the offset. The total of the consumed quantities has to be less than the total of the produced quantities, figure 19.7.

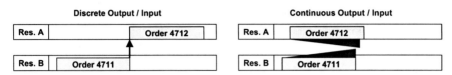

Fig. 19.7. Continuous flow between orders

Continuous flow is considered in production planning but is not supported by detailed scheduling. The implication is that a consistent handling is only possible using the PP heuristic for continuous input/ output SAP_PP_C001 and the bottom-up heuristic for continuous input/ output SAP_PP_008.

In the case of multiple consuming orders (and alternative resources), interactive scheduling will lead to a cumulation of the consuming order at the start of the supplying order. The PP/DS optimiser on the other hand creates start-start and end-end relationships, and therefore the consuming order can not end before the supplying order. The consequence of this behaviour is that the consuming orders are cumulated at the end of the supplying order.

Continuous flow is configured by setting the consumption type in the PDS resp. the PPM to 'C'. The heuristic SAP_PP_C001 for planning with continuous input/output has to be used.

• *Overlap for Operations: Minimum Relationship*
The overlap within one order does not need to calculate the quantities and rates for production and consumption but can be modelled using a start to start relationship between the operations and a minimum offset to account for the first send-ahead quantity to be produced.

19.4 Fixed Pegging and Order Network

Sometimes production processes require a one to one relationship between orders of different BOM levels, and a modelling of the complete production as one order is not possible for legal, technical or costing reasons. Examples for this are common in pharmaceutical production, in metal industries or in discrete manufacturing, e.g. because chemical reactions are part of the production process and batch properties have to be taken into account. The requirements regarding this kind of 'order network' aim to handle this construct as one object in planning and scheduling.

An option to model such behaviour is to use fixed pegging, which has some restrictions as described in section 3.7 (see also note 704583 and 698427).

• *Document Changes*
Up to SAP APO™ 4.1 fixed pegging did have the disadvantage that the fixed pegging arc got lost in most cases. With SAP APO™ 4.1 however it will be kept for most of the changes as shown in figure 19.8

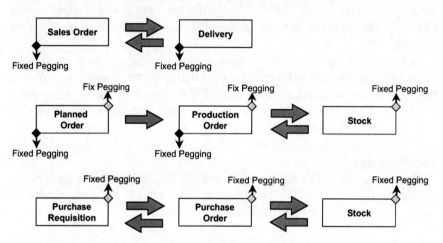

Fig. 19.8. Fixed pegging during document type changes

Note 698427 lists all supported document type changes for which the fix pegging arc is kept. Not supported at all are forecasts and forecast consumption, scheduling agreements and REM backflush, since these are not sensible from a process point of view. The stock transfer orders have two nodes – fixed pegging is kept for the supply node at the target location, but not for the requirement node at the source location.

• *Handling of Fixed Pegging*
The prerequisite to create fixed pegging is that either in the global settings for PP/DS or in the 'demand'-view of the product fixed pegging is activated. The meaning of this setting differs from the previous releases: This setting only allows that fixed pegging can be created but does not cause fixed pegging. The fixed pegging arcs are created either manually (e.g. in the product view) or via heuristic. Two heuristics are available as a standard, SAP_PP_019 to create fixed pegging and SAP_PP_011 to delete fixed pegging. The fixed pegging is created either based on the existing dynamic pegging, on the batch information or on user defined settings.

To increase the transparency of the pegging relationships and alternative nodes for pegging the pegging overview (transaction /SAPAPO/PEG1) was developed, figure 19.9.

Recpt Element	R	Reqmt Element	R	Receipt Qty	Reqmts Qty	DynPegQty	FixPegQty	ActPegQty
PlOrd. 48297		SalesOrder 12080/00		11	11-	11	0	11
PlOrd. 48297		SalesOrder 12083/00		11	0	0	0	0
PlOrd. 48299		SalesOrder 12083/00		30	30-	25	5	30
PlOrd. 48299		SalesOrder 12080/00		0	11-	0	0	0
PlOrd. 48299		SalesOrder 12082/00		0	20-	0	0	0
PlOrd. 48299		SalesOrder 12078/00		0	10-	0	0	0
PlOrd. 48295		SalesOrder 12078/00		10	10-	10	0	10
PlOrd. 48295		SalesOrder 12083/00		10	0	0	0	0
PlOrd. 48298		SalesOrder 12082/00		20	20-	20	0	20
PlOrd. 48298		SalesOrder 12083/00		20	0	0	0	0

Pegging Overview For XX_BUCKETCTP1@QWA100 In PLXX01@QWA100(Make-to-sto

(Tabs: Elements | Periods | Quantities | Stock | Pegging Overview | Product Master | ATP)

(Buttons: Alternative Reqmts | Alternative Recpts | Qty | Qty)

Fig. 19.9. Pegging overview

The existing pegging relationships are displayed as tupels and alternative requirements and/or receipts are displayed per pegging relationship on demand. The fixing of the pegging is done in interactive mode.

• *Fixed Pegging and Batch Determination*
If two orders are completed at about the same time, it could happen that a switch takes place, i.e. the batches from the production orders are not assigned to the production orders they had been previously pegged to.

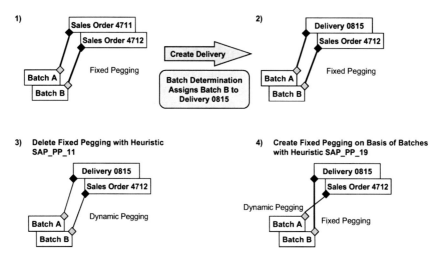

Fig. 19.10. Synchronisation of the fixed pegging and the batch determination

Nevertheless the execution system has to ensure that an appropriate batch is assigned to the next production order, and from a planning point of view it is irrelevant whether there has been a switch of batches or not. The batch determination in SAP ERP™ is the leading process. The adjustment of the fix pegging to represent the batch determination is possible using the heuristic SAP_PP_019 as an offline process step.

• *Fixed Pegging and ATP*

It might be desired to consider the fixed pegging relationship in ATP as well. In the following example the sales order has a fixed pegging relationship to a production order and to the batch 'B' because some minor modifications have been made for the customer. There is sufficient stock available in other batches, figure 19.11.

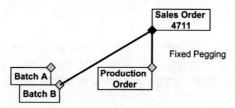

Fig. 19.11. Consideration of fixed pegging in the ATP check

For the creation of the delivery the scope of the check is restricted to stock only. Therefore the desired ATP check result is a confirmation of the requested quantity minus the quantity of the production order.

The scope of the check (i.e. the ATP categories) within the check control is considered. The prerequisite for the consideration of the fixed pegging in the ATP check is that the production type within the check mode is set to 'characteristic evaluation' and that the ABAP class for the CTP scenario in the planning procedure of the product is either /SAPAPO/CL_RRP_FIX or /SAPAPO/CL_RRP_FIX_ONLY, figure 19.12.

Check Mode: Production Type 'Characteristic Evaluation'   **Planning Procedure: ABAP Class for the CTP Scenario**

Maintain Check Mode	
Assignment Mode	Assign customer requirements to planning with assembly
Production Type	Characteristic Evaluation
Check mode text	Warehouse consumpt.

Planning Procedure		
PP Plng Proced.	Heuristic	ABAP Class for the CTP Scenario
6	SAP_PP_002	/SAPAPO/CL_RRP_FIX
H	SAP_PP_002	/SAPAPO/CL_RRP_FIX_ONLY

Fig. 19.12. Prerequisites for the consideration of fixed pegging in the ATP check

The ABAP Class /SAPAPO/CL_RRP_FIX_ONLY causes that the confirmation is done only with receipt elements that are fixed pegged, the class /SAPAPO/CL_RRP_FIX allows both fixed and dynamically pegged receipts.

19.5 Push Production

Push production is a functionality that is mostly required in the process industries, e.g. if bulk – i.e. a semi-finished product – is created with fixed or minimum lot sizes and needs to be processed further due to technical or storage capacity reasons.

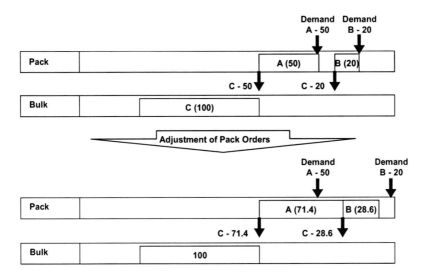

Fig. 19.13. Example for push production

In this case the push production functionality provides a list of products that have this bulk as input and allows creating orders for these interactively. The push production functionality is called from the product view via the menu path '*Goto*' or with transaction /SAPAPO/PUSH.

Part VI – External Procurement

20 Purchasing

20.1 Purchasing Overview

20.1.1 Process Overview

The focus for this book is on operative procurement and not on strategic procurement. Strategic procurement selects the suppliers and negotiates the conditions and quantities. As a result a contract or a scheduling agreement might arise, and often there are costs of scale to be calculated. Operative procurement on the other hand is usually carried out by the planner and allows to choose between alternative suppliers.

Since sometimes components do have a long lead time, their procurement has to be triggered quite early. Another question is triggering procurement based on a feasible plan to save storage costs or to have the components early enough in place, if the production must take place in advance.

20.1.2 Order Life Cycle and Integration to SAP ERP™

The main objects for external procurement are the purchase requisition and the purchase order. Similar to production planning, SAP APO™ creates only the objects for planning, i.e. purchase requisitions, but it is able to trigger their conversion into purchase orders. The creation of purchase orders is technically performed in SAP ERP™. Figure 20.1 shows two alternative order life cycles for external procurement, depending whether the conversion is triggered from SAP APO™ or from SAP ERP™.

J. T. Dickersbach, *Supply Chain Management with APO*,
DOI: 10.1007/978-3-540-92942-0_20, © Springer-Verlag Berlin Heidelberg 2009

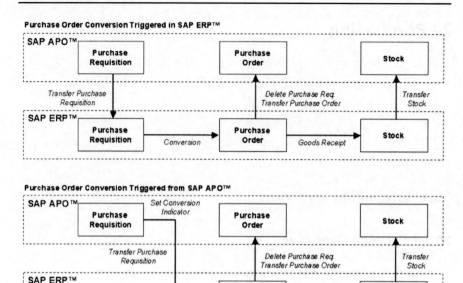

Fig. 20.1. Order life cycle

Differing from in-house production, where it is recommended to trigger the conversion of planned orders to production orders from SAP APO™, the conversion functionality for external procurement offers less functionality than SAP ERP™. By triggering the conversion from SAP APO™, only a one to one relationship between purchase requisition and purchase order is achieved, whereas the purchase order conversion in SAP ERP™ often combines several purchase requisitions (for different materials) for the same supplier to one purchase order. The main advantage triggering the conversion from SAP APO™ is that the purchase requisitions do not have to be transferred to SAP ERP™, which reduces the load for the core interface.

• *Differences Between SNP and PP/DS*

Purchase requisitions are created for external procurement in SNP as well as in PP/DS planning. Both PP/DS and SNP purchase requisitions have the category AG and are displayed by default as 'planned distribution receipts' in the SNP planning book. Both kinds of purchase requisitions can be converted into purchase orders from SAP APO™. For SAP ERP™ it does not make a difference whether the purchase order was created in a SNP or in a PP/DS planning run.

The transfer settings of the according planning application are considered. If the purchase requisition is created by a SNP application, the source of supply is not displayed in the order until it is transferred to SAP ERP™.

● *Scheduling*
The purchase requisition resp. the purchase order contains activities for goods receipt and transport, if the according durations are maintained in the master data (like the stock transfer order).

For the scheduling of these activities in PP/DS it is necessary that the handling resource and the transport resource are maintained. For the initial scheduling – i.e. the scheduling at the time of the order creation – of the purchase requisitions the planned delivery time is considered as well. The planned delivery time is taken from the procurement relationship, or – if no procurement relationship is assigned – from the 'procurement'-view of the product master. The planned delivery time is however not an activity and is therefore not scheduled or otherwise considered at any order changes.

The opening period is used to select the purchase requisitions which are due for conversion. The logic is the same as for production planning: all orders are selected, for which

$$\text{Today} + \text{Offset} \geq \text{Requirement Date} - \text{Opening Period} \qquad (20.1)$$

applies. Figure 20.2 shows the relevant dates, activities and other entities for the scheduling of the purchase requisitions.

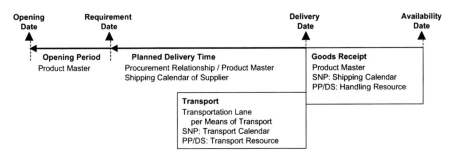

Fig. 20.2. Scheduling of purchase requisitions

The planned delivery time is used to determine the earliest delivery date. If the requested delivery date (the requirement date minus goods receipt) is before today plus the planned delivery time, the latter is used – see figure 20.3. If the transport duration is maintained, the maximum of planned delivery time and transport duration is used instead of the planned delivery time.

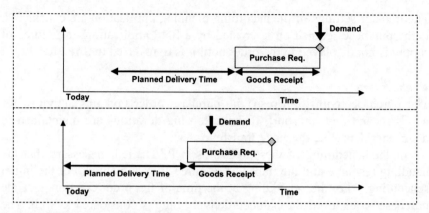

Fig. 20.3. Scheduling and planned delivery time

During manual scheduling the planned delivery time is no constraint. It is assumed, that if the planner violates the planned delivery time by manual scheduling, this is justified. There is however the risk that a purchase requisition is rescheduled automatically, e.g. because a pegged planned order is rescheduled with the setting 'consider pegging' in the strategy profile. In this case the only indication to check whether the planned delivery time is violated, is using the alert for 'opening date in the past'. For the consideration of transport and goods receipt durations rescheduling in PP/DS it is required that transport resources resp. handling resources are maintained.

To take the shipping calendar of the supplier into account, a single resp. multi resource has to be assigned to the transportation lane with an according factory calendar. In this case – like for stock transfer orders – an additional activity for transportation is created and scheduled, and the maximum of the scheduled transportation duration and the planned delivery time is used.

For the transfer of purchase requisitions from SAP APO™ to SAP ERP™ it is necessary that the material short text in the SAP ERP™ material master is maintained in the logon language of the RFC connection. If the material has no short text in that language, SAP ERP™ does not create any purchase requisition.

● *Conversion*

If the conversion of purchase requisitions into purchase orders is triggered from SAP APO™, the conversion indicator has to be set either individually per order or automatically in the mass conversion run with the transaction /SAPAPO/RRP7 – independent whether the purchase requisition has been created by a SNP or a PP/DS application.

In the mass conversion run the purchase requisitions are selected according to product, location and purchasing group (the purchasing group is transferred from SAP ERP™ into the 'SNP2'-view of the product master and is only used in this transaction as a selection criterion). Regarding the horizon, the purchase requisitions are selected according to the condition described in formula 20.1. By maintaining an offset to the opening horizon, the order selection is extended.

• *Purchase Orders*
Purchase orders represent agreements with the suppliers. Differing from the purchase requisition, the purchase order is not merely an object for planning, but has usually a physical counterpart as well – for example a fax at the supplier. Since it is an object that has to be controlled by execution, no changes are allowed in SAP APO™ – neither dates nor quantities. Changes of the purchase order in SAP APO™ are inhibited by the system, with the exception of the scheduling in the planning board, where appropriate customising of the graphical elements is necessary to prevent this (cf. section 17.1). The purchase order in SAP ERP™ is not affected by the change in SAP APO™ anyhow. If a purchase order is deleted manually in SAP ERP™, a new purchase requisition is created for the deleted amount automatically. This new purchase requisition is transferred to SAP APO™.

Purchase orders have the category BF and are displayed in the SNP planning book by default in the key figure 'distribution receipt (TLB confirmed)'.

20.2 Suppliers and Procurement Relationships

Usually the supplier selection is performed in SAP APO™ during the production planning run in SNP or PP/DS when the purchase requisition is created. The prerequisites for the supplier selection are the master data for the supplier and the procurement relationship.

Both are transferred from SAP ERP™, where the info record in SAP ERP™ corresponds to the procurement relationship. Like the info category in the info record, different kinds of procurement relationships exist for standard procurement, consignment and subcontracting. The transfer of an info record, a contract or a scheduling agreement triggers the creation of a procurement relationship and an according transportation lane from the supplier to the plant as well. Figure 20.4 provides an overview about the master data correspondences between SAP ERP™ and SAP APO™.

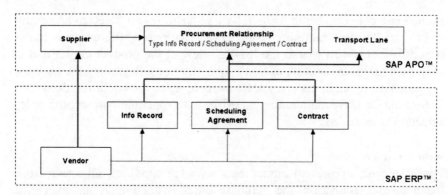

Fig. 20.4 Master data correspondences between SAP ERP™ and SAP APO™

The procurement relationships are maintained with the transaction /SAPAPO/PWBSRC1 and are defined per target location, supplier and material (and procurement type, as described in the following paragraphs) and contains the information about the planned delivery time, the costs (cost scales are considered) and the validity period. The type of the procurement relationship depends on its correspondence to an info record, a scheduling agreement or a contract.

Note that the maintenance of a postal code might be required for the transfer of the vendor to SAP APO™ (depending on the SAP ERP™ customising).

20.3 Supplier Selection

Purchase orders have to be assigned to a supplier, but for purchase requisitions the supplier assignment is not mandatory – they may exist without any assignment to a supplier. If a valid procurement relationship exists in SAP APO™, the supplier is selected and the purchase requisition is assigned during the planning run – i.e. when the purchase requisition is created. This assignment is transferred to SAP ERP™. Purchase requisitions without source are transferred as well; in this case the supplier selection has to take place in SAP ERP™. In case of procurement alternatives, the procurement relationship is selected according to the

- ability to meet the requested date,
- the priority and
- the cost

(in this order). The procurement priority is taken from the assigned transportation lane. Costs of scale are considered and are displayed in SAP APO™ in the procurement relationship.

The source (the procurement relationship) of the purchase requisition can be changed interactively, which triggers a new scheduling. In this case the different planned delivery time is taken into account and the availability date of the purchase requisition is adjusted if necessary. Restrictions by validity periods and lot size are considered. The validity of the procurement relationships is described in detail in note 547328.

Whether procurement relationships are valid for the requested date and quantity and the sequence in which procurement relationships are considered for the selection (priorities and costs) can be checked by creating a purchase requisition in interactive planning (PP/DS or SNP). The procurement alternatives are listed in their sequence for selection.

To display purchase orders and purchase requisitions per supplier, the supplier can be used as selection criterion for the source in the receipt view with the transaction /SAPAPO/RRP4.

• *Quota Arrangements*
Often a product is procured by more than one supplier on a regular basis to reduce the risk of procurement shortfall and other dependencies from the supplier. To split the procurement orders among different suppliers, quota arrangements are used to overrule priorities and costs. The split is performed according to the ratio in the quota arrangement.

Quota arrangements are created with transaction /SAPAPO/SCC_TQ1. For procurement the inbound quota arrangement per procurement relationship is relevant. Figure 20.5 shows the maintenance for quota arrangements. In this example 20 percent of the demand is assigned to the first supplier and 80 percent to the second supplier.

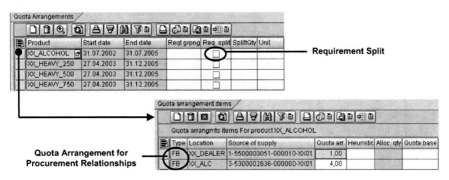

Fig. 20.5. Quota arrangements

To enable the requirement split within one demand, it is necessary for the PP/DS heuristics to use the production planning heuristic SAP_PP_Q001 (with the algorithm /SAPAPO/HEU_PLAN_QUOTA), e.g. by assigning it to the product master in the 'PP/DS'-view.

Depending on the application and on the quota arrangement settings, either the individual demands are split or each demand is assigned entirely to one supplier. In this case the selection of the supplier is based on the history and the assignment of the other purchase requisitions.

For PP/DS it is additionally possible to define a production planning heuristic per procurement relationship. If the heuristic SAP_PP_Q001 is maintained in the product master, the defined production planning heuristic is applied to of the requested quantity per quota.

20.4 Scheduling Agreements

Scheduling agreements are used if products are procured for considerable quantities with a high frequency. Especially in the automotive industry they are a common way of procurement. The principle is to have one object – the scheduling agreement – with a target quantity and the according conditions, to plan the receipts as 'schedule lines' (corresponding to the purchase requisitions) and to send the orders – the 'releases' to the supplier with a reference to the scheduling agreement.

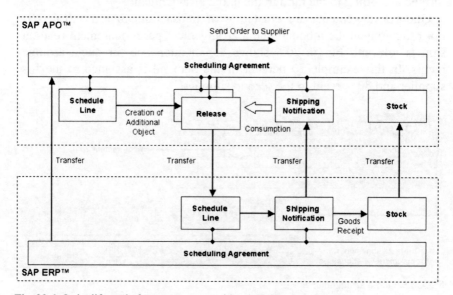

Fig. 20.6. Order life cycle for procurement with scheduling agreements

The releases are created for a defined horizon and are updated in defined intervals. Additionally to the operative releases, it is possible to send the supplier forecast releases to inform him about the planned requisitions in the farther future. This way the scheduling agreement is an object which supports the collaboration with the supplier.

The scheduling agreement is created in SAP ERP™ (transaction ME31L, type 'LP' is mandatory) and transferred to SAP APO™ as a procurement relationship of the type 'scheduling agreement'. Note that each item has to be marked for external planning in the additional data of the scheduling agreement in SAP ERP™ (SHIFT+F5), figure 20.7.

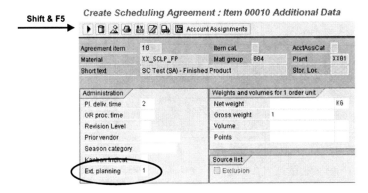

Fig. 20.7. External planning indicator within the scheduling agreement

If a scheduling agreement is created in SAP ERP™, additionally an info record is created as well. Make sure either to exclude this info record from the integration to SAP APO™ or to deactivate the according procurement relationship. Differing from the standard procurement, both the planning and the execution of the scheduling agreement is done in SAP APO™. The order life cycle for scheduling agreements is shown in figure 20.6. The schedule lines are created as the result of a production planning run and are not transferred to SAP ERP™. For the schedule lines, the releases are created in SAP APO™ and sent to the supplier. Only the releases in SAP APO™ are transferred to SAP ERP™, but as schedule lines. Note that after the creation of releases the schedule lines are still visible and active for pegging and planning. The releases are neither relevant for production planning nor for pegging. Depending on the SAP ERP™ customising, it is necessary to create shipping notifications to enable the goods receipt. Shipping notifications are created in the schedule line overview with the transaction ME38 with the path '*Item → Confirmation → Overview*'. The prerequisite to create shipping notifications is that the according confirmation control key is maintained in the item detail of the scheduling agreement. Shipping notifications are transferred to SAP APO™ and reduce both schedule lines and releases.

• *Release Creation*

The scheduling agreement in SAP APO™ and the customising setting – the release creation profile – is shown in figure 20.8. The decisions whether forecast releases are used and whether the releases are written into the live cache after creation or after sending are made in the scheduling agreement view of the procurement relationship.

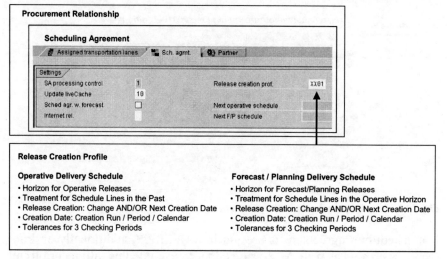

Fig. 20.8. Structure of scheduling agreement settings

The release creation profile defines the horizon for which the releases are created and the necessary events to create the releases. The relevant events are the creation date and the change of the schedule lines. The creation date might be limited to certain periods, or each release creation run might be a valid creation date. The other event is the change of the schedule lines, where it is possible to define tolerance levels for the increase or the decrease of the quantity to prevent the creation of the event for only small changes. By defining whether one event is sufficient to create a new release or both events are required, many options exist to control the sending of new releases to the supplier.

The release creation profile contains a different set of parameters for operative releases and for forecast releases. The release creation profile is maintained with the transaction /SAPAPO/PWB_CM1. The releases are created with the transaction /SAPAPO/PWBSCH1, or – for interactive planning – in the 'periods'-view of the product view. If the releases are created in interactive planning, the release profile is not taken into account.

With the creation of a release a trigger for sending a notification to the supplier is created. These triggers are either immediately sent or processed afterwards with the transaction /SAPAPO/PWBSCH2. As a prerequisite the trigger type and the medium for the sending of the notification have to be configured and the communication data for the supplier has to be maintained in SAP APO™.

The status and the history of a scheduling agreement is displayed with the transaction /SAPAPO/PWBSCH3. Figure 20.9 shows an example where both operative and forecast releases are used.

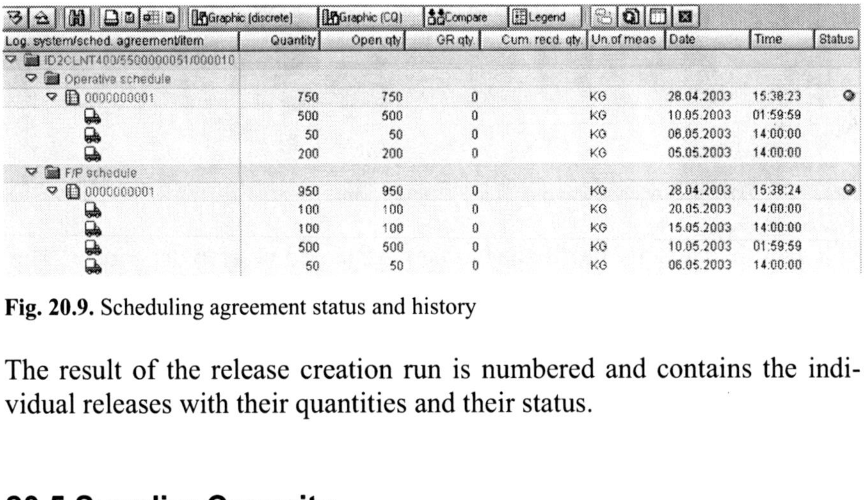

Log. system/sched. agreement/item	Quantity	Open qty	GR qty	Cum. recd. qty	Un.of meas	Date	Time	Status
ID2CLNT403/5500000051/000010								
Operative schedule								
0000000001	750	750	0		KG	28.04.2003	15:38:23	◉
	500	500	0		KG	10.05.2003	01:59:59	
	50	50	0		KG	06.05.2003	14:00:00	
	200	200	0		KG	05.05.2003	14:00:00	
F/P schedule								
0000000001	950	950	0		KG	28.04.2003	15:38:24	◉
	100	100	0		KG	20.05.2003	14:00:00	
	100	100	0		KG	15.05.2003	14:00:00	
	500	500	0		KG	10.05.2003	01:59:59	
	50	50	0		KG	06.05.2003	14:00:00	

Fig. 20.9. Scheduling agreement status and history

The result of the release creation run is numbered and contains the individual releases with their quantities and their status.

20.5 Supplier Capacity

In some cases the procurement quantity of a supplier is limited – either by the physical capacity or by the volume of the contract. The restriction might apply to a single product or to a set of products. This kind of capacity restriction is modelled in SAP APO™ using a transport resource for the capacity and the capacity consumption in the 'product specific means of transport'-view of the transportation lane as resource load.

The general logic of the transport resource capacity consumption is described in section 11.3. Since these capacity restrictions are usually defined per week, month or year, a peculiarity of the modelling is to avoid the daily capacity definition that is provided in the resource master. Figure 20.10 shows how to define a capacity for other than daily buckets using the capacity profile.

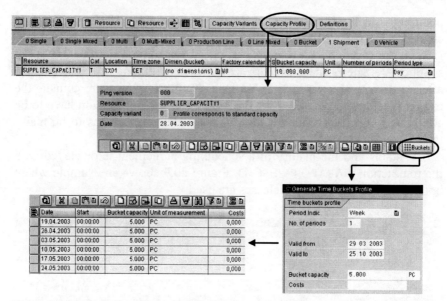

Fig. 20.10. Transport resource to model the supplier capacity

Note that the capacity profile has to be maintained version dependent. The required system steps for using transport resources as supplier capacities are:

- Assign transport methods to the transportation lane. If more than one contract exists for one supplier, for each disjunctive capacity restriction a different transport method has to be assigned.
- Create a transport resource for each contract and assign it to the according transportation lane and transport method.
- Maintain the capacity consumption for each concerned product in the 'product specific means of transport'-view of the transportation lane. One transport resource (i.e. one capacity restriction) can be loaded by more than one product, and the factor for the capacity consumption might be different per product.

The scheduling of the orders is performed according to the strategy profile. If the capacity of the current period is not sufficient, other periods are searched, e.g. backward with reverse until a bucket is found with enough free capacity for the entire order. Note that no order split is performed. To prevent that large orders are not scheduled because they exceed the maximum bucket capacity, a suitable maximum lot size has to be defined in the product master.

The consumed capacity is displayed per transport resource in the 'resource load' report with the transaction /SAPAPO/CDPS_REPT. Make sure that the flag for 'convert time portion in hours' is not set. This functionality is not suited for the reporting of historical capacity consumption, because the capacity is released with the goods receipt of the purchase order. If the goods receipt is too early by more than the planned delivery time, the capacity might be booked twice. Another restriction for the use of the transport resource as supplier capacity is that the PP/DS optimiser does not consider the bucket capacities.

The advantage of using transport resources to production resources at the supplier for the modelling of the supplier capacity is – besides avoiding additional master data – that the quantities are available at any time within the bucket, so that the complete amount can be ordered at the beginning of the period, whereas in a modelling with production resources the total quantity is only available at the end of the period.

21 Subcontracting

21.1 Subcontracting Process Overview

The idea of subcontracting is to outsource production steps. Differing from normal procurement, the subcontractor receives the required components from the customer, processes those and sends the new product back to the customer. Examples for subcontracting are products with irregular demand (e.g. displays resp. kits for promotions in the consumer products industry), production steps which require costly equipment or specialised knowledge (e.g. hardening or electroplating) or production steps with high manual efforts, which can be performed cheaper in countries with a lower wage standard.

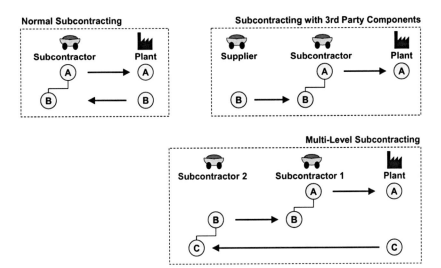

Fig. 21.1. Material flow for subcontracting process variants

J. T. Dickersbach, *Supply Chain Management with APO*,
DOI: 10.1007/978-3-540-92942-0_21, © Springer-Verlag Berlin Heidelberg 2009

Therefore the subcontracting process contains the procurement of an assembled (or otherwise produced) product from the subcontractor and the supply of the required components to the subcontractor. To model this, a purchase requisition resp. purchase order for subcontracting causes a demand for the components.

In some cases the components for the subcontractor are not produced by the customer but externally procured. In this case it might be favourable to send the component immediately from the supplier to the subcontractor. Another variant is a multi-level subcontracting, when the component for the subcontractor is produced by another subcontractor. Figure 21.1 shows the material flow for the normal subcontracting process and the two process variants.

The subcontracting is not supported by all applications in the same way. The limitations are listed at the end of the chapter.

• Modelling Alternatives for Standard Subcontracting
There are two alternative ways to model subcontracting in SAP APO™, depending where the production step is modelled. The first alternative is analogous to SAP ERP™ and uses a PDS or PPM in the receiving plant to calculate the component demands. The advantage of this solution is that only a minimum of master data maintenance is required. If one product is procured from different subcontractors, it is however not possible to separate the component requirements. If fix or minimum lot sizes are used, the production planning might become inaccurate. The second alternative models the production step at the supplier location, which enables separate planning per subcontractor, but requires some additional master data maintenance.

21.2 Modelling of the Production at the Receiving Plant

The modelling of the production process of the subcontracting as a production within the plant corresponds to the modelling of subcontracting in SAP ERP™. This process is however only supported by PP/DS and not by SNP.

In combination with the PDS (transfer only as BOM) the effort for the maintenance of additional master data is very low – only the transport method has to be maintained within the transportation lane and a transport resource has to be assigned for scheduling purposes.

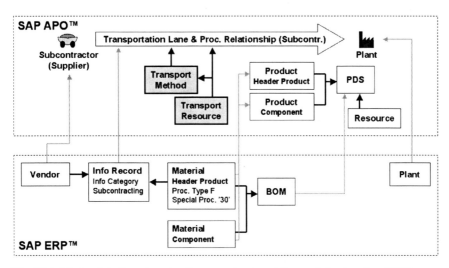

Fig. 21.2. Master data for subcontracting – production at the receiving location (PDS)

If the PPM is used, additionally to the BOM for subcontracting a production version has to be created in SAP ERP™ including a dummy routing and a dummy work center. Figure 21.3 provides an overview of the required master data.

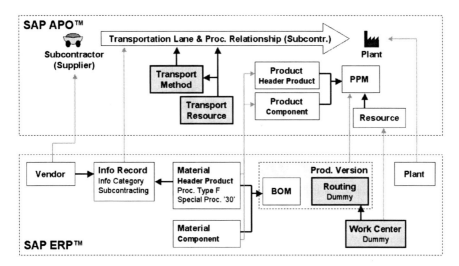

Fig. 21.3. Master data for subcontracting – production at the receiving location (PPM)

• *Order Structure*

The order structure for this alternative is shown in figure 21.4. The planned order in the subcontracting segment is necessary to determine the demand for the component, but does not appear anywhere else. Only the demands for the components are transferred from SAP APO™ to SAP ERP™. The planned order is linked via fixed pegging to the purchase requisition.

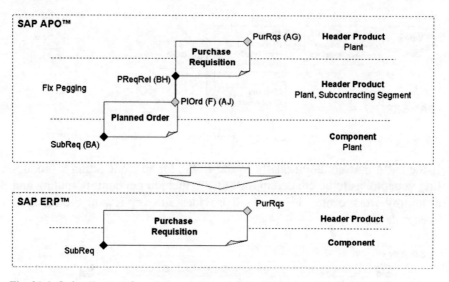

Fig. 21.4. Order structure for subcontracting – production at the receiving location

For the scheduling the maximum of planned delivery time from the procurement relationship and the transport duration from the transportation lane is taken into account plus the duration of the planned order.

The conversion from the purchase requisition to the purchase order is triggered from SAP APO™ or SAP ERP™ like in the standard procurement process. After conversion the categories in the subcontracting segment change from BH to BI for the requirement and from AJ to AI for the receipt, but the pegging remains fixed. The stock for the component is kept as special stock for subcontracting. At the goods receipt of the header product the demand element for the component is deleted and the subcontracting stock for the component is reduced by the according quantity.

21.3 Modelling of the Production at the Supplier

In the second alternative the production is modelled at the supplier, which matches the physical process. If more than one subcontractor is used for the same product, this way of modelling offers the possibility to plan the component requirements and inventories separately per subcontractor. On the other hand the effort for the master data maintenance is higher.

• *Master Data*
Figure 21.5 gives an overview of the master data in SAP ERP™ and in SAP APO™. The master data objects, which have to be created manually in SAP APO™, are displayed as dark elements.

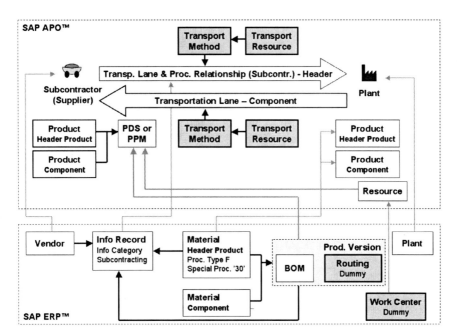

Fig. 21.5. Master data for subcontracting – production at the supplier

For PP/DS both master data (PDS and PPM) can be used, for SNP this kind of master data set-up is only possible for the PDS. This kind of transfer requires
- assignment of the production version to the info record in the 'purchase organisation data'-view on plant level and
- transfer of the PDS resp. the PPM for subcontracting.

The advantage is that the manual creation of the PPM in SAP APO™ is avoided and that only one work center and one routing are required per plant. Even the transportation lane for the component from the plant to the supplier is created automatically at the transfer of the PPM resp. the PDS. Figure 21.6 shows these settings in the info record resp. in the integration model.

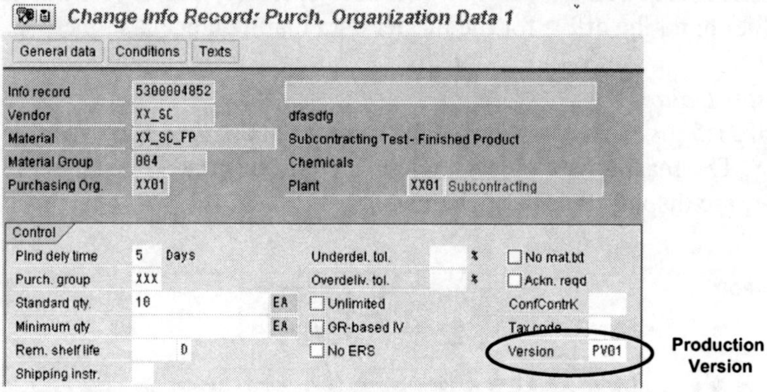

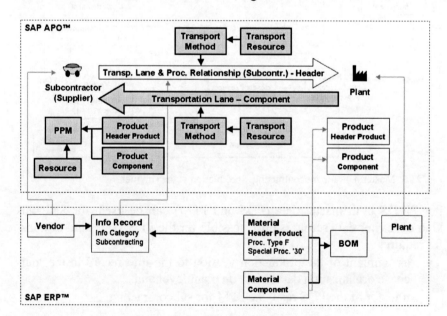

Fig. 21.6. Assignment of the production version to the info record

Alternatively it is possible to avoid either the creation of additional master data in SAP ERP™ or the use of the PDS for SNP for the price of increased master data maintenance in SAP APO™, figure 21.7:

Fig. 21.7. Master data – production at the supplier w/o production version

• *Order Structure*
In all cases the order structure for subcontracting contains three orders for
this alternative, figure 21.8.

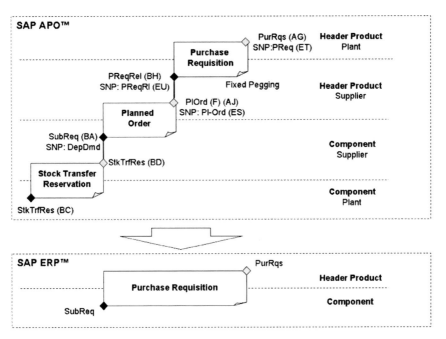

Fig. 21.8. Order structure for subcontracting – production at the supplier

The planned order to determine the component demand is now created at
the supplier and not hidden anymore. Analogous to the first alternative, it
is automatically created and linked by fix pegging to the purchase requisition. The fix pegging is created for both PP/DS and SNP orders. The two
orders are transferred to SAP ERP™ as the subcontracting purchase requisition. The subsequent execution is like in the first alternative (production at
plant), with the exception that the subcontracting stock is transferred to the
supplier location. To transfer the demand for the components to the plant
for further planning steps, a separate production planning step has to be
carried out in the supplier location. The according order for the stock transfer reservation is not transferred to SAP ERP™. The stock transfer reservation is the same with PP/DS and SNP.

For the scheduling the transport duration is used instead of the order
duration of the planned order which was created with the dummy PPM.

Note 379006 describes a BAdI to match the planned delivery time to the transport duration.

21.4 Subcontracting Process Variants

In this chapter the process variants subcontracting with scheduling agreements, third party component supply and multi-level subcontracting are covered.

• *Subcontracting with Scheduling Agreements*
Since SAP APO™ 4.0 subcontracting scheduling agreements are supported as well.

• *Third Party Component Supply*
With third party component supply the component is sent from a supplier directly to the subcontractor, and the invoice is sent to the purchase organisation of the plant. This process is modelled by having the subcontractor's address as the delivery address for the purchase order (and the requisition). Therefore the product flow is only represented in SAP APO™ and not in SAP ERP™. Figure 21.9 shows the required master data in SAP APO™ for this process.

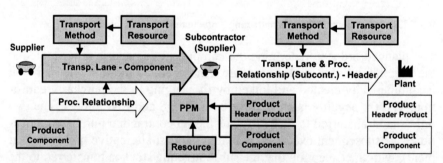

Fig. 21.9. Master data for subcontracting – third party component supply

The peculiarity in this setting is that the procurement relationship has to be assigned to the transportation lane in the product specific view.

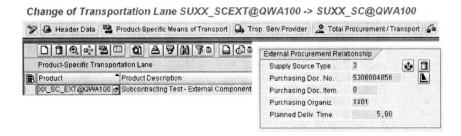

Fig. 21.10. Assignment of the procurement relationship to the transport lane

For the components it is necessary to maintain the storage location for external procurement in the MRP2-view of the material master in SAP ERP™. The publication types for external procurement and reservations have to be maintained for the subcontractor location in transaction /SAPAPO/CP1. The order structure for the third party supply is shown in figure 21.11.

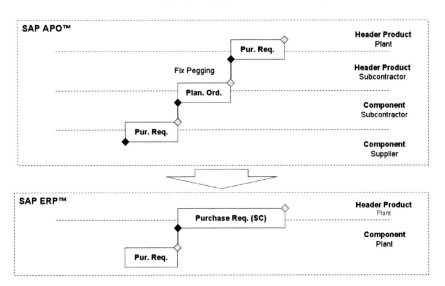

Fig. 21.11. Order structure for subcontracting with third party component supply

The shortage at the supplier location does not need to be resolved. The purchase requisition from the subcontractor to the supplier is transferred as a purchase requisition from the plant to the supplier, but the delivery address of the plant is substituted by the delivery address of the subcontractor, figure 21.12.

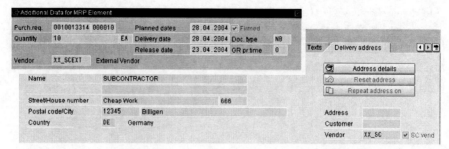

Fig. 21.12. Purchase requisition with the delivery address to the subcontractor

• *Multi-level Subcontracting*

Like in the process with the third party component supply, the delivery address of the subcontracting purchase requisition for the assembly group is the subcontractor for the header product. The main difference is that in both cases subcontracting info records are used. Figure 21.13 shows the order structure for this case.

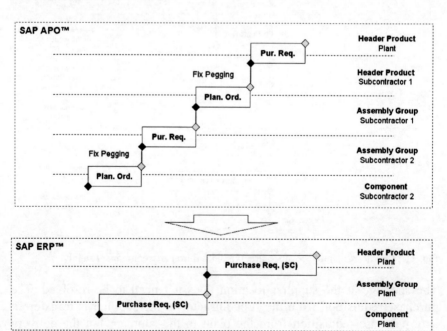

Fig. 21.13. Order structure for multi-level subcontracting

• Subcontracting for Multiple Plants
If the same product is procured from the same subcontractor for more than
one plant, locations of the type subcontractor (1050) have to be created in
SAP APO™. In the location of the type subcontractor the plant and the vendor are assigned.

21.5 Subcontracting in SNP

With SAP APO™ 4.1 the subcontracting process is fully integrated into
SNP and CTM. The prerequisite for SNP is that in the planning book key
figures are defined for the subcontracting distribution receipt. An important point is that the key figures need to have the correct key figure function (the key figure function is assigned to the key figure in the key figure
details of the planning area).

SAP offers the standard planning book 9ASNPSBC for subcontracting
with the following key figures:

Table 21.1. SNP planning area settings for subcontracting

Description	Key Figure	Key Figure Function
Distribution Receipt (Subcontracting Planned)	9APSHIPSBC	4015
Distribution Demand (Subcontracting)	9ADMDDISBC	4016
Production (Subcontracting Planned)	9APPRODSBC	2005

With SAP APO™ 4.1 subcontracting is supported by the SNP optimiser as
well, and interactive changes for SNP subcontracting orders are possible.
With the transfer of the subcontracting purchase requisition to SAP ERP™
the production version is sent if the PDS is used.

If the order conversion from SNP to PP/DS is used, the SNP subcontracting orders are converted into PP/DS subcontracting orders.

Part VII – Cross Process Topics

22 Stock and Safety Stock

22.1 Stock Types in SAP APO™

Different kinds of stock are represented in SAP APO™ by different ATP categories. The standard stock categories in SAP APO™ are listed in table 22.1. Whether a stock category is relevant for planning or not depends on its order type in the live cache.

Table 22.1. Stock categories

SAP APO™ Category	Order Type	Planning Relevance
CA – Stock in Transfer	0F	No
CC – Unrestricted Stock	0B	Yes
CD – Consignment Stock	0B	Yes
CE – Subcontracting Stock	0B	Yes
CF – Stock in Quality Inspection	0E	Yes
CG – Consignment Stock in Quality Inspection	0E	Yes
CH – Subcontracting Stock in Quality Inspection	0E	Yes
CI – Blocked Stock	0C	No
CK – Restricted Use Stock	0D	No

Note 487166 describes how to change the order type of a stock category, if it is required that certain stock categories change their relevance for planning.

The master for all material movements is SAP ERP™ and the inventory information is merely transferred to SAP APO™ (except for returns). This is a one way procedure. A major difference regarding the modelling of special stocks in SAP APO™ and SAP ERP™ is that these stocks are assigned to their physical location in SAP APO™, whereas in SAP ERP™ the ownership is decisive for the location assignment. Figure 22.1 visualises the assignment of the stocks in SAP APO™ and SAP ERP™.

J. T. Dickersbach, *Supply Chain Management with APO*,
DOI: 10.1007/978-3-540-92942-0_22, © Springer-Verlag Berlin Heidelberg 2009

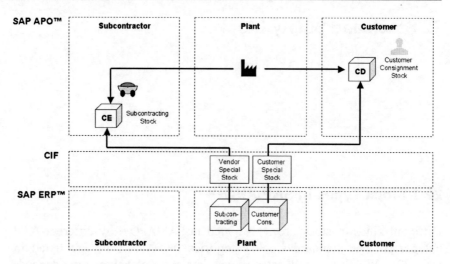

Fig. 22.1. Special stock assignment to the locations

The prerequisite to keep the subcontracting stock at the supplier location and the consignment stock at the customer location is that the according product masters exist in these locations. If this is not the case, they are assigned to the plant like in SAP ERP™.

The storage location and the batch number are transferred to SAP APO™ as additional information and correspond to the entities 'sub location' resp. 'version'. These can be used in ATP (cf. chapter 7), but have no relevance for planning. The most detailed information about the stock situation of a locationproduct is available in the 'stock'-view of the product view (transaction /SAPAPO/RRP3).

In SNP it is defined per location which stock categories are used for net requirements calculation. The relevant categories are grouped into a category group and assigned to the location master. Only these are displayed as initial stock (using the macro function INITIAL_STOCK) and used for planning in SNP. In PP/DS all stock categories which are relevant for planning are used for production planning as well.

22.2 Safety Stock

Safety stock is used to buffer exceptions on the demand side – e.g. an unpredicted increase of sales – and on the supply side. Examples for exceptions on the supply side are lower production output resp. production backlog due to scrap or machine failure and increased transportation times.

The determination of the appropriate safety stock levels in a supply chain is not a trivial task and depends on the deviations of the forecasted demand from the real demand – the forecast error – and the deviations of the supply, the supply chain network and the product structure and the targeted service level. The safety stock levels are either determined by the planner from their experience or calculated using inventory optimisation tools. SAP APO™ offers with the extended safety stock methods a tool for this, but it is also possible to integrate the result of specialised inventory optimisation tools with a justifiable effort. The properties of the extended safety stock methods are described in the online documentation.

Safety stock in SAP APO™ is either modelled as a quantity or as a safety days of supply. The values for the safety stock are either maintained as fix values in the product master or as time dependent values in the SNP planning book. The safety stock method in the product master determines whether quantities, safety days of supply or the maximum of both and whether fix or time dependent values are used, table 22.2.

Table 22.2. Safety stock methods

	Fix	Time Dependent
Quantity	SB	MB
Safety Days of Supply	SZ	MZ
Maximum of the Above	SM	MM

The safety stock method MM is only partly considered by the SNP optimisation – only the independent demands and the dependent demands from fixed orders are taken into account.

Safety stock is not a separate stock category (nor is it a physically separated stock). SAP APO™ considers the safety stock like a demand element that causes either an increase or an earliness of the supply. These supplies are not available in planning to be able to cover unpredicted demands or to compensate unpredicted shortages in the supply.

If the safety stock is maintained as a quantity, in the subsequent planning run the system tries to meet both the total demand and the safety stock. Alternatively the safety days of supply is maintained in the product master. Safety days of supply cause the system to plan the supply for the demand the specified days earlier. If the periods for planning do not match the specified number of days, the supply is split according to the proportions as shown in figure 22.2.

4 Safety Days of Supply, Daily Buckets

Demand									20			
Supply		20										

4 Safety Days of Supply, Weekly Buckets

Demand							20					
Demand int.							4	4	4	4	4	
Supply	16						4					

Fig. 22.2. Calculation of the safety days of supply

SNP is the primary module for planning with safety stock. Nevertheless safety stock is considered in PP/DS and ATP as well

• *Safety Stock in PP/DS*
Basically all safety stock methods – even time dependent safety stocks – are considered in PP/DS, if the planning area is maintained in the customising for the global settings and default parameters with the transaction /SAPAPO/RRPCUST1, see also note 517898. Another prerequisite for using safety stock in PP/DS is the according setting in the version master. The planning mode 'automatic planning' in the product master however ignores the safety stock (note 413822).

If the safety stock is maintained as a quantity, the indicator for 'take safety stock into account' in the version master decides whether a virtual safety stock element or a live cache order is used. For the virtual safety stock an additional demand element of the category CB is created automatically. The downside of this is that this element is not visible in live cache and therefore no pegging takes place with the consequence that the supply for the safety stock is shown as a surplus in the alert monitor. The other option is to create a requirement order for the safety stock in the live cache as described in the next paragraph.

• *Safety Stock as a Live Cache Order*
If the safety stock is modelled as a fix quantity, since SAP APO™ 4.0 it is possible to create a live cache order for the safety stock (ATP category SR) with the heuristic SAP_PP_018 (or using the transaction /SAPAPO/AC08). The requirement date of the safety stock is per default the date when the heuristic is executed. In order to keep the pegging relationship, a certain backward pegging should be allowed. The heuristic should be executed on a regular basis in order to keep the requirement date current. The prerequisite

is that the correct entry for the safety stock consideration is maintained in the version.

● *Safety Stock in ATP*

It is questionable whether ATP needs to consider the safety stock at all. Per default the supply for the safety stock is available for customers. Using live cache orders for the safety stock as described in the previous paragraph it is however possible to consider safety stock by including the safety stock category SR into the scope of check as a requirement.

23 Interchangeability

23.1 Interchangeability Overview

There are two different cases for interchangeability: One is the discontinuation of a product or component and thus related to the product life cycle. Two options exist for the discontinuation, the simple forward interchangeability (i.e. product A is substituted by product B) and the full interchangeability (i.e. for a transition period A is substituted by B, but A may also substitute B). The other case is the form-fit-function class, where planning is performed only for one product, but any product within the form-fit-function class is technically identical. These form-fit-function classes are used to group inventory managed manufacturer parts. Though interchangeability is supported by all modules except TP/VS, there are restriction regarding the extent. Table 23.1 gives a rough classification about the interchangeability functionality per module.

Table 23.1. Interchangeability functions and SAP APO™ modules

Interchangeability Function	DP	SNP	CTM	PP/DS	ATP
Discontinuation (Forward)	Yes	Yes	Yes	Yes	Yes
Discontinuation (Full)	No	Yes	No	Yes	No
Form-Fit-Function Class	No	Yes	No	No	Yes

Additional limitations apply as described in the following paragraphs – e.g. that interchangeability is not compatible with the use of product configuration and characteristic based planning. Another important restriction is that the interchangeability supports only a one-to-one relationship – i.e. no parallel or dependent discontinuation is possible (in contrast to SAP ERP™).

In the following we concentrate on the discontinuation since this is the far more frequent case.

J. T. Dickersbach, *Supply Chain Management with APO*,
DOI: 10.1007/978-3-540-92942-0_23, © Springer-Verlag Berlin Heidelberg 2009

• Interchangeability Master Data

The use of the interchangeability requires a new master data, the inter-changeability group. The interchangeability group is created with the transaction /INCMD/UI. Figure 23.1 shows the maintenance screen for this interchangeability group.

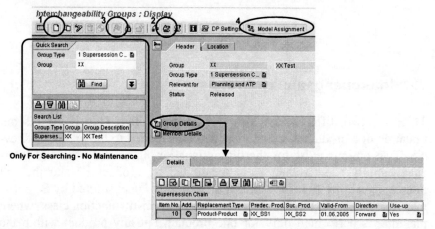

Fig. 23.1. Interchangeability group master

The procedure to maintain the interchangeability group contains four steps:

1. Create the interchangeability group in the right area of the screen and maintain the details – i.e. dates, interchangeability type, members.
2. Check for messages. The interchangeability group is not saved until no messages arise.
3. Release the interchangeability group.
4. Assign the interchangeability group to the model.

Simple discontinuation in SAP ERP™ is transferred via CIF as an inter-changeability group with the according assignment for its use in PP/DS (cf. section 23.4). However, there is no update to the interchangeability group if the discontinuation data changes in SAP ERP™.

23.2 Interchangeability in DP

The life cycle related tasks of interchangeability are already covered in DP by life cycle planning and realignment as described in section 4.4. Therefore DP does not use the interchangeability master directly, but it is possible to generate the like profile and the phase-in and phase-out profiles based on the interchangeability group, figure 23.2.

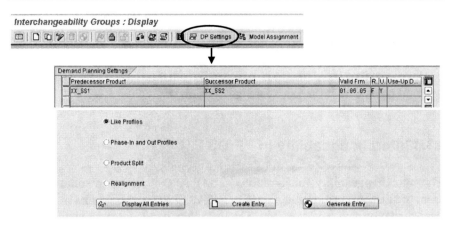

Fig. 23.2. Use of the interchangeability group in DP

The interchangeability group is also used as an input for the realignment.

23.3 Interchangeability in SNP

If there is a demand for a product that has already a valid successor, SNP creates substitution orders for the predecessor product. The nature of these substitution orders is that they transfer the demand from the predecessor product to the successor product and have neither duration nor do they consume capacity. The output of the substitution order is the predecessor product (receipt from substitution, category EO) and the input is the successor product (requirement from substitution, category EN). To display the substitution orders in the SNP planning book, the key figures 9APSUBAB and 9APSUBZU have to be included.

The substitution order exists only in SAP APO™ and is not transferred to SAP ERP™. It is neither possible to convert the SNP substitution orders into PP/DS substitution orders. The difference between forward and full interchangeability is explained in the next chapter.

Note that the use of interchangeability must be activated in the global SNP settings in customising with the path *APO → Supply Chain Planning → SNP → Basic Settings → Maintain Global SNP Settings*. For the SNP heuristic and SNP optimisation in the background the flag to 'add products from supersession chains' has to be set.

• *Distribution Planning and Replenishment with Interchangeability*
To use of interchangeability for distribution planning, the substitution has to take place in the target location and not in the source location. For deployment, first the demand is covered by substitution and the remaining demand takes part at the fair share calculation.

23.4 Interchangeability in PP/DS

The interchangeability functionality in SAP APO™ differs in some aspects from the discontinuation functionality in SAP ERP™ – mainly that no dependent discontinuation is possible. On the other hand, interchangeability in SAP APO™ allows modelling the conditions for the substitution more flexible than in SAP ERP™. Figures 23.3 and 23.4 show the different behaviour, depending on the parameters for forward and full interchangeability and full, restricted or no use-up. The key fields are the 'valid from'-date and the 'use-up'-date.

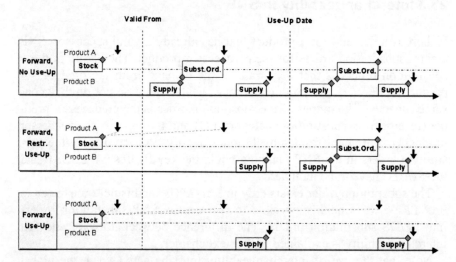

Fig. 23.3. Forward interchangeability

Forward interchangeability will create substitution orders only in one direction, i.e. to substitute the preceding product with the successor product. For a certain time interval there will often be demands for both the preceding product and the successor product. Full interchangeability allows to use the preceding product as well to cover the demand for the successor product and thus to use-up the stock for the preceding product faster, figure 23.4.

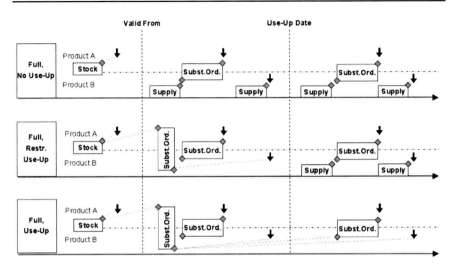

Fig. 23.4. Full interchangeability

The substitution order and the receipt for the substitution requirement are created within the same planning run. The maintenance of the interchangeability group triggers the creation of a planning package that is assigned to all group members in the background. The substitution orders have the category GA for the receipt and GB for the substitution requirement. The substitution orders are not transferred to SAP ERP™ and they cannot be changed interactively.

• *Production Execution: Inclusion of New Component*

The focus for interchangeability in PP/DS lies on the life cycle of components, i.e. the substitution is performed for dependent demand. For the execution of the production order it is not sufficient to have an additional substitution order, but the production order has to be modified to include the substitute component. The trigger to modify the production order is the ATP check, which is usually (but not necessarily) performed with the conversion of the planned order.

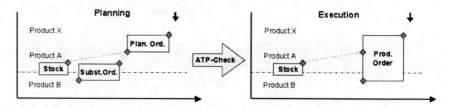

Fig. 23.5. Inclusion of new component

The prerequisite is that the ATP check is performed in SAP APO™ with the settings for rules-based ATP and the use of product interchangeability data as shown in figure 23.6.

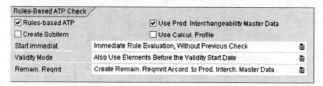

Fig. 23.6. ATP check instructions

The unit of measure has to be maintained in the ATP-view of the product master for the components. The substitution orders (category GA for the receipt and GB for the requirement) are not part of the scope of check. The component is included with the ATP check of the planned order.

Since the substitution orders do not consume any capacity, they are ignored by scheduling. The scheduling strategy has to ensure that the pegging relationship of the substitution orders to the according planned orders is kept.

23.5 Interchangeability in ATP

The interchangeability group can be used in rules-based ATP instead of the product substitution procedure. In the check instructions the flag to use the interchangeability master data has to be set.

24 Exception Reporting

24.1 Basics of Alert Monitoring

Alerts notify the planner about situations which require his attention or interference. A shortage situation caused by higher demand due to increased sales or caused by production shortfall due to increased scrap is a typical example where an alert (in this case of the type 'order has undercoverage') helps the planner to act as early as possible.

Alerts are displayed either in the alert monitor as a stand alone application or are integrated into other planning functionalities. The alert monitor is called with the transaction /SAPAPO/AMON1, figure 24.1:

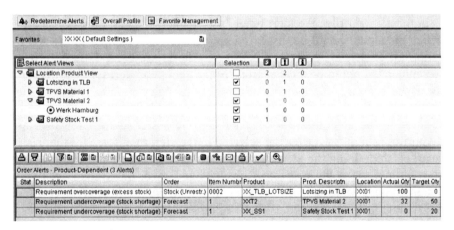

Fig. 24.1. Alert monitor

The alert monitor is structured into two parts: In the area above the objects for the alerts are selected. In the alert view below only those alerts for the objects are shown, which had been selected in the alert profiles and for which current alerts exist. The navigation becomes clearer after regarding the structure of the alert monitor configuration (with the transaction /SAPAPO/AMON_SETTING) in figure 24.2:

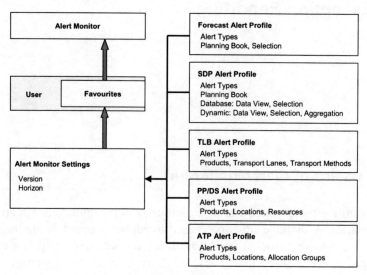

Fig. 24.2. Structure of the alert monitor configuration

There are additional profiles for VS (vehicle scheduling), VMI (vendor managed inventory) and MSP (maintenance and service planning). The two latter areas are not covered in this book.

In the alert profiles the alert types (see chapter 24.2) and the objects for which the alerts should be displayed are selected. Depending on the application different alert types and selection criteria are used. Generally it is recommended not to be too generous in the choice of alert types since performance is an issue in alert monitoring. In the 'alert monitor settings' the version and the horizon for the alert display is defined. To apply these settings in the alert monitor they have to be assigned for each user as favourites. These favourites are selected in the alert monitor. Still there is a small catch defining the alert monitor settings: if you define the name of a new setting, make the respective changes and save, the system asks you whether you want to save the changes. The right answer here is no, else the old settings will be changed.

To each alert there are lots of information – alert priority, alert type, product, location, target and actual value are only a few of them. The selection of the displayed information and their sequence are configured using the 'select layout' button in the header of the alert display box. In the menu bar of 'select layout' a new layout is created, in 'manage layout' it is set as a default. Non-default layouts have to be applied manually. In the alert monitor those alerts are displayed that were valid at the time the alert monitor was called. If the planning situation has changed in the meantime a refresh is necessary.

24.2 Alert Types

SAP APO™ provides some standard alert types and allows for supply and demand planning defining own alert types. The alert types are selected per application area for the according profile.

• *Forecast Alerts*
The alert types in the forecast alert profile are all related to the trespassing of forecast error limits. Therefore the prerequisites for using forecast errors are the selection of the errors in the univariate forecast profile and the definition of the limits for the errors in the respective diagnosis groups, figure 24.3.

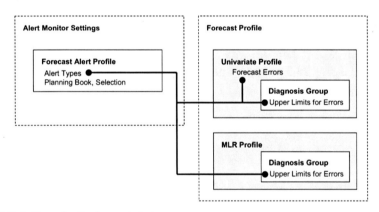

Fig. 24.3. Settings for the forecast alerts

• *SDP Alerts*
SDP alerts are based on the planning book structure and the time series. These alerts are calculated using macros. Technically there are two types of alerts, dynamic alerts and database alerts. In SNP and DP both dynamic and data base alerts can be used, but especially for a huge data amount it is not recommended to use dynamic alerts because of performance disadvantages. Data base alerts are usually calculated by planning jobs (with the structure as described in section 4.8) and relate to the planning situation at the point in time of the planning job. Data base alerts are stored in the data base. Therefore it is recommended first to delete the existing alerts before creating new ones. Figure 24.4 shows a standard macro for backlog

calculation as an example to apply data base alerts syntactically correct. The syntax for the same semantic using a dynamic alert is shown for comparison.

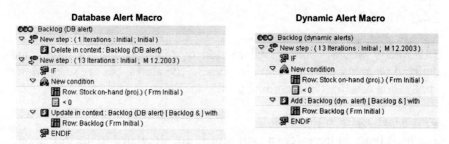

Fig. 24.4. Example for a data base alert macro

Note that the alert semantic is defined in the macro itself and the alert type is only used for organisational purposes. All alert types of the SDP alert profile can be used in the macro builder, but most of the alert types are used in standard macros in the SNP planning book 9ASNP94. It is possible to define alert types with the transaction /SAPAPO/ATYPES_DP including the definition of their priority and their thresholds. With the syntax as shown in figure 24.5 actual values are appended from the assigned rows to the alert text.

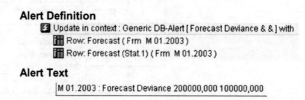

Fig. 24.5. Syntax to append informations to the alert text

It is possible to define alert types which change their priority according to the deviation of the assigned parameters by defining a relational operator (usually 'greater than') in the alert type. The ratio of parameter one divided by parameter two is used. It is possible to call the alert monitor with the according button from the header of the planning book as well. The alerts can be restricted to the version, the planning book or the data selection, but not to the displayed data after drill down. The small icon to the right of the alert button closes the alert monitor within the planning book.

Another macro functionality which is used to attract the attention to deviations in the interactive planning is the colouring of the concerned cells. Figure 24.6 gives an example for such a macro.

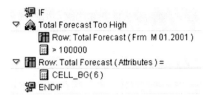

Fig. 24.6. Macro to colour cells

When adding the row to be changed, make sure that the scope of the row is changed from 'values' to 'attributes'.

• *TLB Alerts*
The alerts in the TLB profile relate to both deployment and TLB. Regarding deployment there are alerts for shortages after fair share, for TLB there are alerts for minimum and maximum load violations.

• *PP/DS Alerts*
There is a large variety of PP/DS alerts concerning shortage and surplus, lateness, pegging constraints, order relationship, resource overload, characteristics and other. In contrast to the SDP alerts the logic is predefined.
The example in figure 24.7 explains the interdependency between alerts and pegging for the probably most important PP/DS alert types – shortage, surplus and lateness.

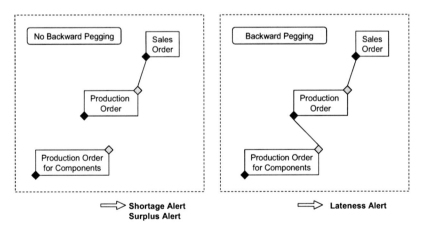

Fig. 24.7. Shortage, surplus and lateness alerts

The alerts for shortage and surplus are calculated depending on the pegging between nodes. If a supply node is too late for the demand node (in figure

24.8 the production order for a component for the secondary demand of the subsequent production order), no pegging exists any more between the nodes. In this case the system creates a shortage alert and a surplus alert, even though no real shortage exists. But if backward pegging is allowed (a setting in the product master), the pegging is kept and a lateness alert is created instead. Another example shows the possible lateness alerts during the order life cycle of in-house production, figure 24.8.

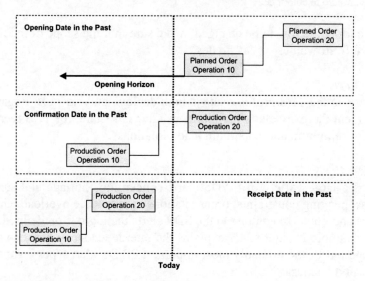

Fig. 24.8. Lateness alerts for the order life cycle

The first lateness refers to the opening date – when a planned order has to be converted into a production order. Consequently this alert only applies to planned orders. After the first operation has slipped into the past, the alert 'confirmation date in the past' is created – accordingly only for production orders. If the last operation lies in the past, the system creates the alert 'receipt date in the past', which applies for both planned order and production order. For all lateness alerts it is possible – and advisable – to define offsets.

The PP/DS applications product view, planning board and product planning table allow to call the alert monitor within the respective transaction. Since the PP/DS applications work with transactional simulations, the alert monitor reflects the planning situation of this transactional simulation which is not necessary identical with the current situation in the active version. Table 24.1 shows the objects and time horizons for the display of the alerts in dependence of the application.

Table 24.1. Alert selection in the PP/DS applications

Application	Objects	Time Horizon
Product View	Selected Product, No Resources	Unlimited
Planning Board	Work Area	Time Horizon of Planning Board
Product Planning Table	Selection	Unlimited

If the alert monitor is called from the planning board, network alerts are automatically displayed although they are rather performance consuming. Note 444832 describes how to exclude these network alerts.

• *ATP Alerts*
A prerequisite to get ATP alerts at all is to flag the checkbox in the check instructions (see chapter 7). Available alert types are shortages in the availability check, in the forecast check and in the allocations check.

The idea of what is considered as a shortage in ATP alert monitoring needs an explanation (see also note 500889). Neither a partial confirmation nor no confirmation at all is regarded as a shortage.

ATP alerts are saved to the data base without the link to the order and are therefore not automatically updated. Due to these restrictions ATP alerts are not used very often.

24.3 Alert Handling

In order to help the planner to keep an overview of the alert situation it is possible to
- sort the alerts according to any column in ascending or descending order,
- confirm the alert, that is to mark it as checked (for example because its resolving is already triggered),
- hide an alert, for example because it is due to some exceptional constraints which are not modelled in the system and
- display hidden alerts again.

Figure 24.9 shows the buttons for these functions in the header bar of the alert monitor.

Check/ Uncheck
Sort Alerts
Hide/ Show
Freeze
Order alerts - Product dependent (6 Alerts)
Stat
Priorität
Descriptn
Order
C...
Product
Produc
1 Shortage Planning FA XX_HEAVY_250 Heavy
1 Shortage Planning FA XX_HEAVY_250 Heavy
1 Shortage Planning FA XX_HEAVY_250 Heavy

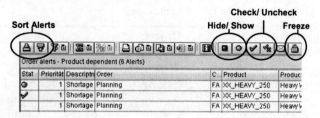

Fig. 24.9. Alert handling functions

Another feature to support the planner in resolving the alert situations is the functionality to freeze the alerts. In this case any change is marked in the status column. The example in figure 24.10 started with a shortage alert where the target quantity was 100 units and the actual quantity zero. After freezing the alert an order for 20 units was created. By changing the situation a new alert with the shortage of 80 units was created instead. After unfreezing, the old alerts vanish after a refresh.

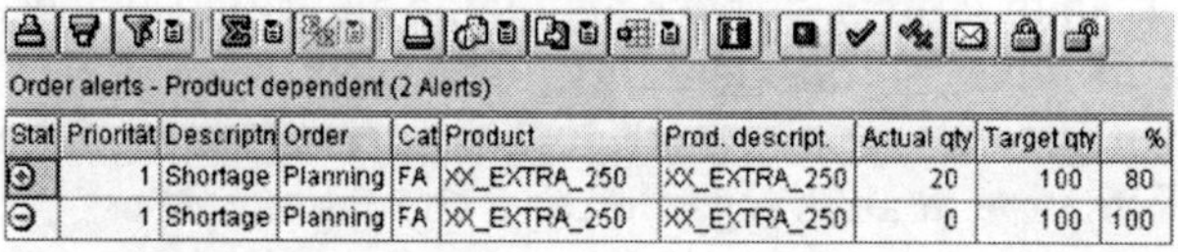

Order alerts - Product dependent (2 Alerts)
Stat
Priorität
Descriptn
Order
Cat
Product
Prod. descript.
Actual qty
Target qty
%
1 Shortage Planning FA XX_EXTRA_250 XX_EXTRA_250 20 100 80
1 Shortage Planning FA XX_EXTRA_250 XX_EXTRA_250 0 100 100

Fig. 24.10. Freezing of alerts

This feature is especially helpful when calling the alert monitor in the transactional simulation mode from a PP/DS application. This way it is possible to experiment with actions and see their impact on the alert situation. Depending on this impact it is possible to decide whether to save or not.

With right mouse click on the alerts it is possible to branch into an application to resolve the alert. This functionality depends on the alert type. It is not recommended to use this functionality from a transactional simulation since in that case the application may represent a different planning situation.

There are three priorities for alerts – 'information' (priority 3), 'warning' (priority 2) and 'error' (priority 1). For some alert types it is possible to set the priority depending on thresholds. Additionally the general priority of some alert types can be changed in customising.

24.4 Alert Calculation in the Background

It is possible to send alerts regularly per batch job either to the SAP office or to a specified e-mail address, figure 24.11. These settings are maintained in the automatic messaging profile with the transaction /SAPAPO/AMONMSG. The actual sending of the alerts is triggered with the transaction /SAPAPO/AMONMSG_SEND.

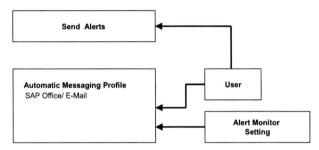

Fig. 24.11. Sending of alerts

24.5 Supply Chain Cockpit

The supply chain cockpit provides an overview about the supply chain structure, both graphical in the right window and listed per elements (locations, products, resources, production and transport lanes) in the object view in the left window. The objects for display are selected with the selection functionality and saved into the work area when leaving the transaction. The supply chain cockpit is called with the transaction /SAPAPO/SCC01, figure 24.12. The benefit of the supply chain cockpit is that it links the supply chain structure with an overview of the alerts. In the bottom line the existence of current alerts is displayed per application and priority. In the object view the number and priority of related alerts is displayed per object.

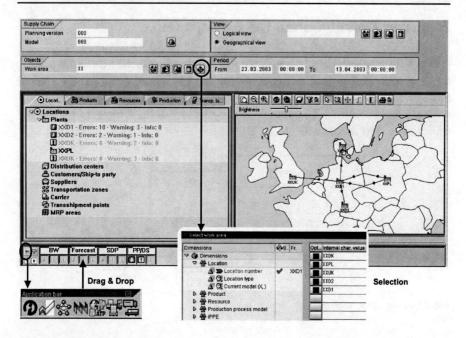

Fig. 24.12. Supply chain cockpit

The supply chain cockpit has to be configured user specifically by each user with the transaction /SAPAPO/SCC_USR_PROF. Figure 24.13 gives an overview of the configuration possibilities.

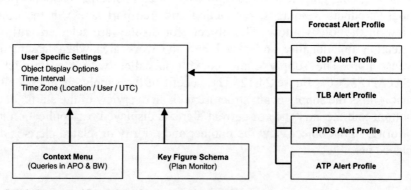

Fig. 24.13. User specific settings for the supply chain cockpit

Another functionality of the supply chain cockpit is to execute queries for planning data in SAP APO™ or for data from SAP BI™. The context menu (transaction /SAPAPO/SCC_CON_MENU) defines the settings for queries in SAP APO™ or SAP BI™.

Part VIII – System Integration

25 Core Interface

25.1 Overview of the Core Interface

The core interface (CIF) is the standard interface between SAP ERP™ and
SAP APO™ for the order related applications PP/DS, SNP and ATP. On
SAP ERP™ side the 'plug-in' has to be implemented to use the CIF, on SAP
APO™ side it is always included. The CIF enables the integration of master
data from SAP ERP™ to SAP APO™ (one way only) and the integration of
transactional data both ways, from SAP ERP™ to SAP APO™ and from SAP
APO™ to SAP ERP™. The basic idea of the integration is to write events
for each planning relevant change – e.g. the creation of an order – and use
these as trigger for the transfer. Technically the transfer is performed via
qRFC. Which objects (i.e. planned orders, stock, …) are transferred is controlled by the integration models, which could be regarded as something
like the master data of the CIF.

25.2 Configuration of the Core Interface

In this chapter the connection of the two systems – SAP ERP™ and SAP
APO™ – is described and the required ALE settings to create change pointers for the data transfer.

• *Application and Change Pointers*
The prerequisite for the integration to SAP APO™ of any kind is setting the
flag for the applications 'ND_APO' and 'NDI' in the transaction BF11 on
SAP ERP™ side (see also note 322800). Further prerequisites on SAP ERP™
side are the activation of the change pointers in general with the transaction BD61 and the definition of the relevant objects with the transaction
BD50. On SAP APO™ side the change pointers must not be activated.

• *System Connection*

The communication between the systems is based on queued remote function calls (qRFC). To enable this communication the ALE settings and the RFC connections have to be created between the two systems and the parameters for the communication have to be configured. The basic settings are shown in figure 25.1.

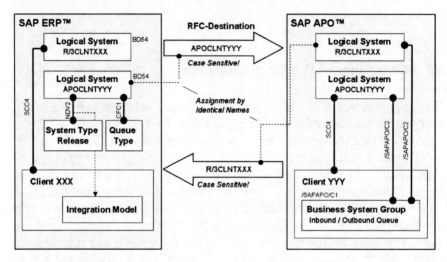

Fig. 25.1. System connection via CIF

The logical systems of both the SAP APO™ and the SAP ERP™ system have to be defined in each system. The convention is the concatenation of the system name plus 'CLNT' plus the client. In the SAP ERP™ system the logical system of the SAP APO™ has to be defined as the target system by assigning the system type and release with the transaction NDV2 and the queue type (see chapter 25.6.4) with the transaction CFC1. If the name of the SAP APO™ system defined in transaction NDV2 on SAP ERP™-side is different from the name of the logical system in SAP APO™ the integration of transactional data from SAP APO™ to SAP ERP™ will have errors. In SAP APO™ the business system group has to be defined and both logical systems assigned to it.

• *RFC Connections*

In the RFC destination the address of the target system and the access data (user for the target system, password and logon language) are defined. The user for the RFC destination has to be a dialog user, since the ATP check requires this type of user. For safety reasons this user should have no authorisation for any transactions. Note also that the RFC destination has to have the same name as the logical system of the target system and that the name is case sensitive. It is necessary to register the inbound queues with the transaction SMQR.

The RFC connection contains options regarding the behaviour in case of connection difficulties (e.g. CPICERR). It is recommended to set the number of retrials to '30' and the time between two tries to '2' in the RFC connection (button 'TRFC').

• *Transfer Parameters per User*
The transfer parameters define per user, whether events for the data transfer are created, the log level and the debugging options. It is recommended to create a default entry for user '*' to avoid transfer restrictions depending on the user.

• *Publication in SAP APO™*
For the transfer of transactional data from SAP APO™ to SAP ERP™ it is necessary to maintain the distribution definition per publication type (e.g. in-house production) and location to the SAP ERP™ system.

Table 25.1. Transactions for the system connection

SAP ERP™		SAP APO™	
Step	Transaction	Step	Transaction
Logical System for SAP ERP™ and SAP APO™	BD54	Logical System for SAP ERP™ and SAP APO™	BD54
Assign SAP ERP™ Client to Logical SAP ERP™ System	SCC4	Assign SAP APO™ Client to Logical SAP APO™ System	SCC4
SAP APO™ Release	NDV2	Business System Group	/SAPAPO/C1
Queue Type	CFC1	Assign Logical System for SAP APO™ and SAP ERP™ to BSG	/SAPAPO/C2
RFC Connection to SAP APO™	SM59	RFC Connection to SAP ERP™	SM59
Register Inbound Queues	SMQR	Register Inbound Queues	SMQR
Transfer Parameters	CFC2	Transfer Parameters	/SAPAPO/C4
Filter & Select Block Size	CFC3	Runtime Information	/SAPAPO/CP3
		Distribution per Publication Type	/SAPAPO/CP1

The transactions to maintain the entities in figure 25.1 are listed in table 25.1. In the runtime information settings on SAP APO™ side and the block size settings on SAP ERP™ side parameters to tune the performance are maintained (among others).

• *Parameters for Filter and Select Block Sizes*
At an initial transfer from SAP ERP™ to SAP APO™ – whether master data or transactional data – there is usually a considerable system load. In transaction CFC3 it is possible to tune the system performance by setting the parameters for the filter block size and the select block size per object type. The filter block size determines how many objects are read from the database in one step, the select block size defines how many objects are transferred to SAP APO™ in one step. Notes 384077 and 436527 give recommendations for the setting of these parameters.

25.3 Integration Models and Data Transfer

Integration models (also called CIF-models) define which objects – master data or orders – are transferred to SAP APO™ and whether the ATP check is carried out in SAP APO™. On the way back – i.e. when SAP APO™ sends orders to SAP ERP™ – it is checked again whether an active integration model exists. The integration models are created on the SAP ERP™ side with the transaction CFM1 and activated with transaction CFM2.

• *Objects*
An integration model contains a set of objects, which are selected per object type and other selection criteria as object name etc. There are two groups of objects, the ones which are related to the material master – like all orders – and the ones which are material independent – as resources and suppliers. Table 25.2 lists some of the objects that can be included into integration models and transferred to SAP APO™. Another material dependent object is the control to perform the ATP check in SAP APO™. For the selection of the material dependent objects the restrictions for the material are taken into account first.

By generating the integration model the objects matching the selection criteria are read and assigned to the model. These objects – the content of the integration model – are displayed with the transaction CFM4.

Table 25.2. Objects for the integration models (extract)

Material Dependent		Material Independent
Master Data	Transactional Data	Master Data
Plant	Stock	Customers
Material	Sales Orders	Work Centers
Planning Material	Planned Indep. Rqmts.	Vendors
Contracts	Planned Orders	Classes/Characteristics
Sched. Agreements	Production Orders	
Info Records	Prod. Campaigns	
PPM	Purch. Orders & Reqs.	
	Manual Reservations	
	Batches	

The key for the integration models consists of the name of the integration model, the target system, the application and its version. The name of the integration model and the application are chosen freely. For the application we propose a naming convention like MASTERDATA, TRANSDATA and ATP. The target system is selected from the SAP APO™ systems that were defined in transaction NDV2. The integration model receives a new version each time it is generated, independent whether its content has changed. The generation date and the user are part of the version name.

• *Data Transfer*

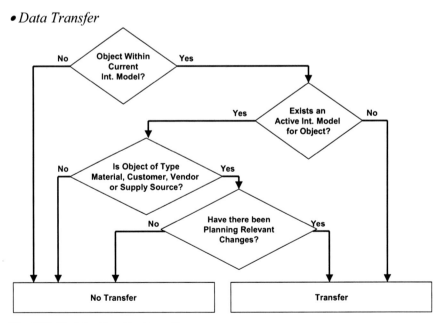

Fig. 25.2. Decision for object transfer

The transfer of the data itself is triggered by activating the model with the transaction CFM2. Now all objects of the integration model are transferred as long as they are not already part of an active integration model. Additionally those materials, customers, vendors and supply sources are updated, which are part of the model and have been changed (planning relevant changes). These rules regarding the transfer of an object are visualised in figure 25.2.

• *Initial and Delta Transfer*

To manage the transfer of new objects an integration model is usually generated periodically with the identical selection criteria. The content of the versions of this integration model however may differ. Since during transfer the CIF checks whether there is already an active integration model for an object, it is a difference whether the (older) active version is deactivated before activating the new version or not. If the older version is deactivated first, all objects within the new version of the integration model are transferred. This procedure is called initial transfer.

If the new version is activated without deactivating the old one first, only those objects are transferred which are not yet included in the old version. This way of activating integration models is called delta transfer. Usually a periodical delta transfer is used for data integration, since the system load is much smaller this way.

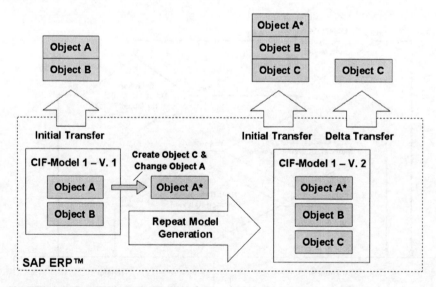

Fig. 25.3. Initial and delta data transfer

The example shown in figure 25.3 visualises the difference between initial and delta transfer (the objects material, customer, vendor and supply source are an exception, cf. section 25.4). Between the generation and the activation of the integration model version 1 and its new generation as version 2 the object C has been created and added to the model, and the object A has been changed. In the initial mode all objects are transferred anew and object A is available in SAP APO™ with its changed properties. Using the delta transfer only the new object C is transferred. Nevertheless we recommend using delta transfer on a regular basis. Changes are transferred with a separate report as described in section 25.4.

Another consequence of the circumstance that the system checks whether active integration models already exist before transferring an object with the activation of the integration model is that overlapping integration model definitions make it very difficult to control the data transfer process, because they might prevent an initial transfer even if it is intended as shown in figure 25.4.

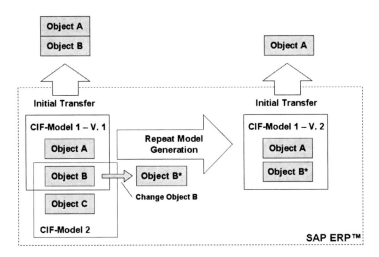

Fig. 25.4. Problem with overlapping CIF-models

Object B is not transferred again with the initial transfer of integration model 1, because it is as well contained in integration model 2, and integration model 2 is still active.

To avoid overlapping CIF model selections it is crucial to set up a central design how to create and tailor the models. With the transaction CFM5 it is possible to search for CIF models – active or inactive – by object. Section 25.6 gives some recommendations how to manage the integration models.

25.4 Master Data Integration

Master data is usually one of the most crucial factors for the success of an implementation. For SAP APO™ this means that two requirements have to be fulfilled:

1. The master data in SAP ERP™ has to have a sufficient quality. If multiple SAP ERP™ systems are connected to SAP APO™, a harmonisation of the master data is required.
2. The master data has to be kept consistent between SAP APO™ and SAP ERP™, else planning and execution drift apart and planning becomes the more inaccurate the higher the level of detail is – therefore mostly PP/DS is affected here.

In this book we deal only with the second requirement, that is keeping the master data consistent between SAP ERP™ and SAP APO™. Though getting the master data in the SAP ERP™ systems into a sufficient shape might often be a major challenge itself, these tasks have to be tackled separately.

• *Master Data Maintenance Strategy for SAP APO™*
Master data is only transferred in one way from SAP ERP™ to SAP APO™. There is no transfer or update of master data from SAP APO™ to SAP ERP™.

In SAP APO™ master data objects exist with and without correspondence to objects in SAP ERP™, and even those with correspondence (as the product master) usually have settings that do not relate to any setting in SAP ERP™. Generally we recommend strongly to use SAP ERP™ as the master for master data maintenance whenever possible and rather transfer some settings via BAdI than to have an additional master data maintenance process in SAP APO™. Table 25.3 lists the most common master data required in SAP APO™ and recommendations regarding the integration.

Table 25.3. Master data transfer strategies

SAP APO™ Master Data	Corresponding Master Data in SAP ERP™	Recommended Strategy
Plant, DC	Plant	initial transfer, maintenance of additional fields in SAP APO™
Product	Material	initial, delta & change transfer, maintenance of additional fields in SAP ERP™-append fields & transfer via user exit[1]
Resource	Work Center/ Resource	initial, delta & change transfer
Shift Model		no transfer, maintenance in SAP APO™
PDS/PPM	Production Version	initial, delta & change transfer, maintenance of additional fields in SAP ERP™ & transfer via user exit/BAdI
Customer	Customer	initial transfer (only for VMI, consignment or TP/VS)
Supplier	Vendor	initial, delta & periodical initial transfer to update changes in the conditions
Procurement Relationship	Info Record, Contract or Scheduling Agreement	initial, delta & periodical initial transfer to update changes in the conditions
Transportation Lane		if Info Record exists, as above else maintenance in SAP APO™

[1] recommendation: additional views for 'APO Sales' & 'APO Logistics' in SAP ERP™

• *Change Transfer for Master Data*

Change transfer is concerned with the transfer of changes in the master data objects on SAP ERP™ side that have already been transferred to SAP APO™ before. For the material master, the customer, the vendor and the supply source changes are transferred additionally in the normal delta transfer or separately with the transaction CFP1. For the material master, the customer and the vendor it is also possible to define an online change transfer with the transaction CFC9. This improves the consistency between the systems and reduces the workload for the periodical transfer as well.

A change of these objects is recognised if the field for the entry which has been changed is configured to create a change pointer. In the transaction BD52 it is possible e.g. for the material master to select the fields which trigger change pointers for the message type CIFMAT, but not all fields offer the option to create change pointers. For the PDS the change transfer is performed with the transaction CURTO_CREATE.

• *Change Transfer for PPMs*

The change transfer for PPMs is a bit more complicated. Planning relevant changes in the BOM, the routing/recipe and the production version create change pointers in the table CIF_PPM_CHANGED. But neither a change in the work center resp. resource nor in the BOM of a phantom causes the creation of a change pointer. This implies that the change of the scheduling formula is not noticed. Changes in the PPM are transferred with the transaction CFP3. Obsolete change pointers are deleted with the transaction CFP4. The PPM is created in SAP APO™ according to the setting for the resolution date of the production version in the integration model. The resolution date might be a defined date, the start or the end date of the validity of the production version or the transfer date. Figure 25.5 shows the history of a production version with two changes, where in each change a new component is added. If the resolution date is defined as the start date of the production version, the changes in the production version are not transferred. Using the option for a specified date as resolution date behaves the same as soon as the specified date is in the past. Changes are only transferred if either the end date or the transfer date is chosen for resolution.

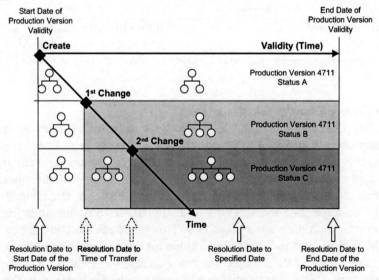

Fig. 25.5. Resolution dates of production versions for the PPM transfer

The notes 508773 and 508779 provide further information regarding the change transfer of PPMs.

• Effectivity and ECM
Engineering change management (ECM) is mostly used for BOMs, but can be used for other master data objects as well. ECM is only supported by the PDS and not by the PPM. Though there are some workarounds to model a way to integrate validity restrictions we do recommend to use the PDS if ECM plays a role.

A workaround for the missing ECM functionality in the PPM – to plan for a change in the production version in the future – is to create a new production version with the same BOM and the same routing. The validity of a new production version starts from the effectivity date of the change, and the validity of the old production version has to be adjusted to end at the effectivity date of the change, cf. figure 25.6.

To ensure that the production versions are resolved the right way at the transfer, the resolution mode 'resolution date according to end date of production version' has to be chosen. The production version has to be valid until the date 31.12.9999 and the same date has to be used as selection date. This triggers the adjustment of the validity of the old PPM to the changed validity of the old production version.

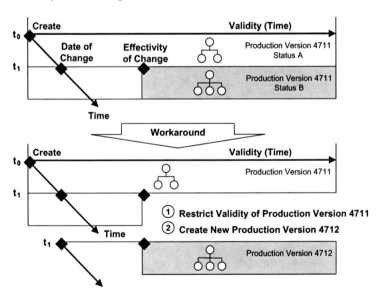

Fig. 25.6. Workaround for ECM with PPMs

Another workaround for ECM only for BOM changes is to transfer the BOM components without key dates as described in note 453198 using the user parameter CF7. One severe restriction for this solution is however that no phantoms are supported any more at all.

• *Deletion of Master Data*
The integration of the master data deletion resp. locking from SAP ERP™ to SAP APO™ is given for the material master and the production version. If a material is marked for deletion, the report RCIFMTDE sets the according product to 'externally planned' in SAP APO™. During the change transfer for PPMs with the report RSPPMCHG a PPM is deactivated if the according production version is locked in SAP ERP™. Of course each master data object can be deleted in SAP APO™ as well. The dependencies to existing orders and to other master data objects have to be considered.

25.5 Transactional Data Integration

In contrast to the master data transfer, the integration of the transactional data is working two ways, from SAP ERP™ to SAP APO™ and from SAP APO™ to SAP ERP™.

• *Transactional Data Transfer from SAP ERP™ to SAP APO™*
From SAP ERP™ to SAP APO™ there is usually one initial upload and from then onward a continuous online transfer of data changes. The strategy profile for the integration should contain an infinite planning mode as described in section 17.2. During the activation of an integration model for materials the flag in the field MARC-APOKZ ('APO-flag') is set. If the material is locked, this flag can however not be set, which causes that transactional data is not transferred from SAP ERP™ to SAP APO™. It is possible to correct the missing flag with the report RAPOKZFIX.

The runtime for a complete transactional data upload should be tested as early as possible, because this is an important constraint for the system recovery scenario.

• *Transactional Data Transfer from SAP APO™ to SAP ERP™*
The prerequisite for the data transfer from SAP APO™ to SAP ERP™ is the maintenance of the publication types per location in the transaction /SAPAPO/CP1.

The default setting for the transfer of transactional data from SAP APO™ to SAP ERP™ is 'immediately' for PP/DS orders and 'periodically' for

SNP orders. These settings can be changed per user with the transaction /SAPAPO/C4 to periodical transfer ('collect changes') or to immediate transfer ('do not collect changes'). In the 'global parameters and default values' setting (transaction /SAPAPO/RRPCUST1) it is possible to inhibit the transfer of planned orders and/or of purchase requisitions to SAP ERP™ without the conversion indicator (option 'transfer only orders with conversion indicator'). If PP/DS is used in SAP APO™, usually planned orders do not have any significance in SAP ERP™. To avoid performance problems, we recommend transferring in-house production orders only with the conversion indicator to SAP ERP™. If the conversion functionality for purchase orders is sufficient in SAP APO™, external procurement should be transferred only with the conversion indicator as well.

The transfer for SNP orders (immediate, periodical or no transfer) is defined in the transaction /SAPAPO/SDP110. There it is also possible to exclude planned orders or stock transfer orders from the transfer. Further restrictions are possible using the method TRANSFER_SET of the BAdI /SAPAPO/SNP_TRANSFER. The periodical transfer itself is triggered with the transaction /SAPAPO/C5 (report /SAPAPO/RDMCPPROCESS). Figure 25.7 gives an overview about the settings that are required for the different order types and the different transfer modes to succeed in transferring of the transactional data to SAP ERP™.

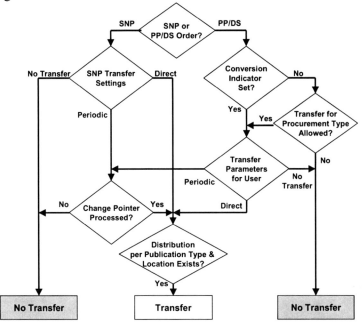

Fig. 25.7. SAP APO™ settings for transactional data transfer to SAP ERP™

These settings describe only the prerequisites on SAP APO™ side to send the data from SAP APO™ to SAP ERP™. On SAP ERP™ side additionally an integration model has to be active for the order type and the according master data. If an order is transferred to SAP ERP™, the order number of the SAP ERP™ order is transferred back to SAP APO™. The adoption of the order number is called 'key completion'.

25.6 Operational Concept

25.6.1 Organisation of the Integration Models

One integration model contains one or more object types, and each object type is restricted according to different selection criteria. The two extremes of handling integration models is to create only one integration model for all object types without any restriction on one hand and to create for each object type e.g. per material separate integration models.

For performance reasons it is recommended to have only a small number (less than 10) of different integration models per object type. If there are disjunctive business areas, having separate integration models for the same object type (e.g. sales order) has the advantage that the impact of corrections in case of a mistake – e.g. a new initial transfer – is limited to a restricted area and does not affect the entire company. The design of the integration models has to keep these two targets – performance and flexibility to apply corrections – in mind.

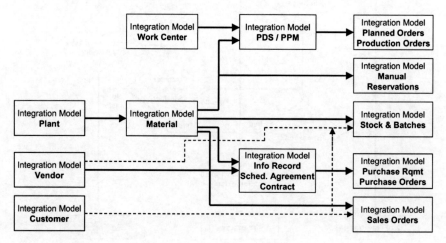

Fig. 25.8. Dependencies in the scheduling of integration models

For the same reason of minimising the impact of corrections and thus increasing the flexibility, we recommend to use separate integration models for different object types whenever sensible (the supply sources are usually combined into one integration model, and production campaigns have to be included with planned orders and production orders).

The dependencies of the master data objects to each other and of the transactional data to the master data have to be regarded for the scheduling of the integration model transfer. Figure 25.8 shows these dependencies for the object types. For the PDS for each PDS-type (PP/DS, SNP, PP/DS-subcontracting, …) it is necessary to create a separate integration model. If classes and characteristics are used in SAP APO™, these have to be transferred in advance.

25.6.2 Organisation of the Data Transfer

The data transfer consists of the initial transfer, the delta transfer and the change transfer. Usually the integration models for the relevant master data are generated and activated in delta mode each night to transfer new objects to SAP APO™. To transfer changes in the purchasing conditions to SAP APO™, for the master data objects vendor, info record, scheduling agreement and contract a periodical initial data transfer is necessary. The changes within the master data objects are transferred either online according to the settings in the transaction CFC9 or periodically with the report RCPTRAN4. Table 25.4 lists the transactions and reports for the data transfer.

Table 25.4. Transactions and reports for data transfer from SAP ERP™ to SAP APO™

Action	Transaction	Report
Generate	CFM1	RIMODGEN
Activate	CFM2	RIMODAC2[1]
Delete Inactive Versions of Integration Models	CFM7	RIMODDEL
Initial Transfer for Work Centers, Classes & Characteristics and Planning Product		RIMODINI
Change Transfer for Material, Vendor, Supply Source and Customer	CFP1	RCPTRAN4[2]
Change Transfer for PPMs	CFP3	RSPPMCHG

[1] choose options 'ignore faulty queue entries' and 'activate newest version'
[2] obsolete if online change transfer is used

The report RSPPMCHG for the PPM change transfer has to be executed before the periodical integration model generation and activation. The integration models for the master data must not be deactivated since for some of the transactional data an active integration model is required for the data transfer (see also note 501718).

To prevent disturbances in the transfer process it is recommended to create the variants for the background jobs with a separate user and flag the option 'protect variant', so that the variant can only be changed by that separate user.

25.6.3 Data Consistency

The consistency regarding the transactional data between SAP ERP™ and SAP APO™ is checked with the CIF delta report from SAP APO™ with the transaction /SAPAPO/CCR. As a result both the objects missing in SAP APO™ and those missing in SAP ERP™ are displayed. A transfer for both of them can be triggered to resolve the inconsistency. If there is no active integration model, the objects are displayed as missing in SAP ERP™.

25.6.4 Queue Monitoring

First some technical aspects of the data transfer are mentioned to understand the possible errors and the tools to monitor and correct them.

• *Excursion: Queued RFC*
It is possible to call a function from a different system using the RFC technique. The prerequisite for this is that the called function supports RFC and that the RFC connection is maintained in the RFC settings with the transaction SM59.

To reduce the sensitivity to network problems, RFCs are usually sent as transactional RFCs (tRFC). For mass processing the concept of writing the tRFC requests into a queue and group them to logical units of work (LUWs) offers the advantage of processing the RFCs in the right sequence and considering the dependencies between LUWs.

For the communication the qRFC (queued RFC) requests are always sent from the outbound queue of the sending system to the inbound queue of the receiving system. Nevertheless there are two concepts of processing the qRFCs, depending whether the scheduling of the LUWs is performed by the sending or by the receiving system. The processing is done by the receiving system anyhow. These two concepts are referred to as 'outbound queue' and 'inbound queue', because this is where the scheduler is in charge and the information about the qRFCs are displayed. The outbound queue is displayed with the transaction SMQ1 and the inbound queue with the transaction SMQ2.

• *Outbound Queue Concept vs. Inbound Queue Concept*
A problem with the use of outbound queues is that it is only possible to control the work processes for the scheduling system – in this case the sending system. This implies that it is possible that the work processes of the receiving system are completely occupied by processing the qRFC requests of the sending system. Since the effort for scheduling is usually far lower than the effort for processing, this situation happens quite often when many qRFC requests are created – for example after a production planning run. The result of this situation is that the system response time for any other application – e.g. dialogue requests – becomes unacceptable. Figure 25.9 visualises this behaviour.

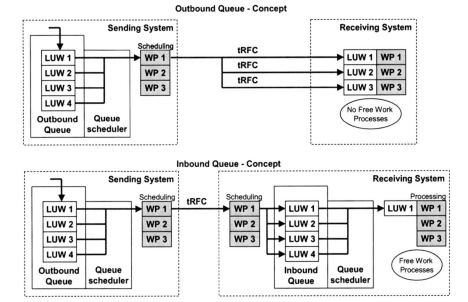

Fig. 25.9. Outbound queue and inbound queue concept

The notes 388528, 388001, 430725 and 416475 describe the configuration of the inbound queues. Inbound queues are the standard setting.

The qRFC monitor is an application with its own release cycle which is mostly independent of the basis release. The current version is displayed in the transaction SMQ1 (for outbound queues) resp. SMQ2 (for inbound queues) with the menu path '*Info → Version*'.

• *Queue Names*

If a queue is stuck with an error (status SYSFAIL) it is helpful for resolving to know what kind of object is causing the problem. The name of the queue gives an indication for this, table 25.5.

Table 25.5. Queue names

Data Content	Queue Name from SAP ERP™ to SAP APO™	Queue Name from SAP APO™ to SAP ERP™
Initial Load	CF_ADC_LOAD	
Material Change Transfer	CFMAT*	
Classes	CFCLA*	
Characteristics	CFCHR*	
Stock	CFSTK*	
Planned Independent Requirements	CFPIR*	
Production	CFPLO*	CFIP*
Manual Reservations	CFRSV*	
Production Confirmation	CFCNF*	
Campaign	CFPCM*	CFPC*
Procurement	CFPO*	CFEP*
Sales Order	CFSLS*	CFCO*
Shipment	CFSHP*	CFSH*

The queue names from SAP APO™ to SAP ERP™ have by default after the first four digits the sum of the digits of the order GUID. With note 440735 it is possible to have the order GUID in the queue name instead. There are several parallel queues for transactional data from SAP ERP™ to SAP APO™ and from SAP APO™ to SAP ERP™, but only one queue for initial load from SAP ERP™ to SAP APO™. It is possible to parallelise the initial load as well. The status of the channels is displayed with the transaction CFP2 in SAP ERP™.

• *Queue Status*

The queue status informs about the processing of the according LUW. The queues may have the statuses listed in table 25.6 as described in note 378903.

Table 25.6. Queue statuses

Queue Status	Description	Required Action
READY	LUW ready for processing	no action required
RUNNING	LUW is being executed	no action required
EXECUTED	LUW has been executed	no action required
WAITING	LUW has dependency to another queue	
WAITUPDA	LUW waits for update of previous queue	
WAITSTOP	LUW has dependency to blocked queue	
SYSLOAD	No dialog work process in the system available	
STOP	LUW is blocked explicitly	remove block
NOSENDS	LUW is not sent	check transfer parameters
SYSFAIL	Error during execution of LUW in the target system	check & correct error
CPICERR	Network or connection problem	check technical connection

- *Queue Logging*
The qRFCs are logged according to the user parameters in the sending system (normal, detailed or not at all). The according log is stored in the receiving system and is displayed with the transaction CFG1 in SAP ERP™ and /SAPAPO/C3 in SAP APO™. The exception to this are the queues for master data transfer, which are logged in the sending system, i.e. in SAP ERP™. The appropriate selection criterion for the logs is the transaction identifier (TID) of the queue. For the investigation of transfer problems it is also possible to search for strings – for example order numbers – within the logs with the transaction /SAPAPO/C7 in SAP APO™ resp. with the transaction CFG3 in SAP ERP™.

- *Queue Monitoring via Queue Manager*
If the processing of a queue is terminated because of an error, this queue remains in the queue monitor of the receiving system (if inbound queues are used) or in the queue monitor of the sending system (if outbound queues are used). After the error has been corrected (e.g. by adjustment of the customising or the master data), the queue can be activated to process the transactions.

A very useful tool to provide an overview about the status of the queue entries is the queue manager in SAP APO™. The queue manager is called with the transaction /SAPAPO/CQ and combines the information of the SAP

APO™ system and one or more connected SAP ERP™ systems in one
screen, figure 25.10.

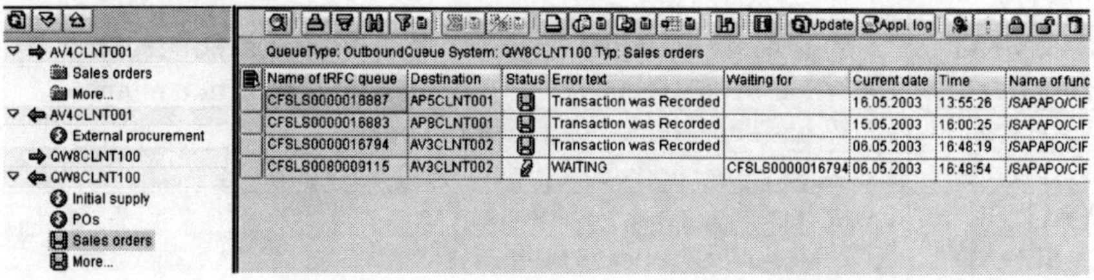

Fig. 25.10. Queue manager

Except from an improved visualisation, the queue manager groups the
queue entries according to their categories, e.g. for sales orders, and enables to branch into the log file via double click on the queue entry. The
queue manager is available for both inbound and outbound queues.
Though the queue manager improves the comfort of the queue monitoring,
there might be performance disadvantages.

• CIF Cockpit
The CIF cockpit is new with SAP APO™ 4.1 and it is in a way the further
development of the Queue Manager and provides an overview about the
queue entries as well as about the CIF configuration. The CIF cockpit has
a monitoring view as shown in figure 25.11 and a settings view.

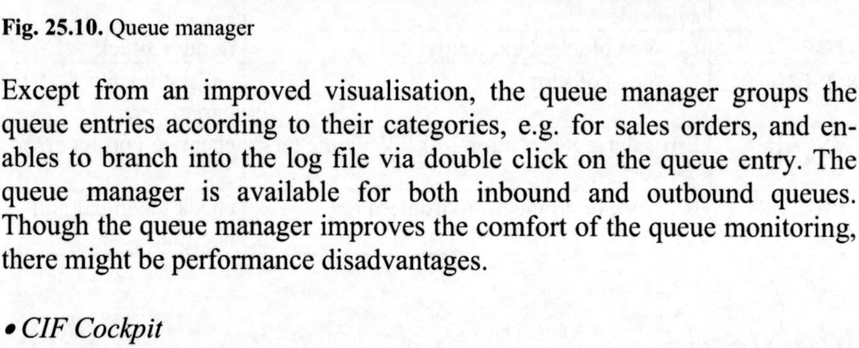

Fig. 25.11. CIF cockpit – monitoring view

The settings view combines the information of technical settings as block
sizes with information like an overview about the existing integration
models. The transaction to call the CIF cockpit is /SAPAPO/CC. Currently
there are no performance restrictions.

• *Queue Alert*

Another tool which supports the monitoring of the queues is the queue alert, which is defined with the transaction /SAPAPO/CW and sends a message either to the system administrator or to the respective user, if a queue has an error status. The queue alert functionality is not event driven, but has to be scheduled as a periodical job.

For monitoring purposes the transaction CFP2 in SAP ERP™ provides an overview of the status of the channels.

26 Integration to DP

26.1 Data Storage in Info Cubes

The usual way to load data into SAP APO™ for Demand Planning is using the info cubes. The info cube consists of the info objects 'key figures', 'characteristics' and 'time characteristics', figure 26.1. The info cube is therefore the analogue to the combination of the planning object structure and the planning area. In SAP APO™ the characteristics 9ALOCNO, 9AMATNR and 9AVERSION are mandatory for the info cube. The info area is merely an entity to group info cubes.

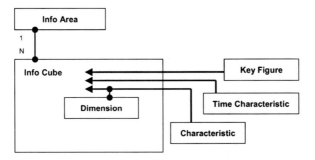

Fig. 26.1. Structure of an info cube

The central transaction for the SAP BI™ related structure set up is the administrator workbench (transaction RSA1), figure 26.2. Here the info objects, the info cubes and further entities which are described later in this chapter are defined.

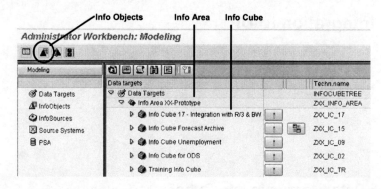

Fig. 26.2. Administrator workbench

The time characteristics, e.g. months, days or weeks (0CALMONTH, 0CALWEEK, 0CALDAY), are provided as standard objects and should not be changed. Own key figures and characteristics can be created if there is no appropriate standard object.

The processing of the objects in the administrator workbench requires in most cases the right mouse click. When creating an info cube, a pop-up asks whether you want to create an info cube for APO or BW. The right answer is BW.

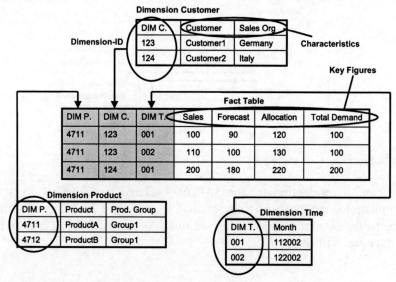

Fig. 26.3. Star scheme using dimensions for characteristics

After selecting the characteristics it is necessary to define dimensions and assign the characteristics to the dimensions. The data is accessed using dimension-IDs, where one dimension-ID represents a characteristic combination of the assigned characteristics, figure 26.3. This so called star scheme has performance advantages.

After assigning the characteristics, the time characteristics and the key figures, as a last step the info cube has to be activated.

26.2 Data Loading Structures

To load data into the info cube the data has to be converted into the req-uired sequence and format. The format conversion for each element is performed within the info objects. Figure 26.4 gives an overview about the involved entities. The right sequence to process these entities is
1. info cube,
2. source system,
3. info source,
4. update rules and finally
5. info package.

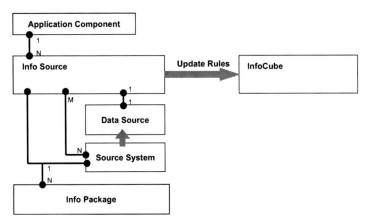

Fig. 26.4. Structure overview for data loading

● *Source System*
The entity 'source system' represents exactly what it is named after. The source system types are 'R/3 system', 'BW system', 'external system' and 'file system'. Technically there is a link to the ALE settings as the logical system and the RFC destination (transaction SM59) and dependencies to the partner functions (transactions WE20, WE30).

The data source is generated from the administrator workbench in the 'source system'-view using the right mouse click on the *Source System* →

Replicate Data Source. If the source system is a file system, the generation of a data source is not necessary. The assignment of the data source to the info source is done in the 'info source'-view by right mouse click → *Assign Data Source*. Here first the source system is selected and then within the source system the data source.

• *Info Sources and Transfer Rules*

The info source translates the input sequence to the required sequence and provides the data for the info cube. Whether and how the data is written into the info cube depends on the update rules.

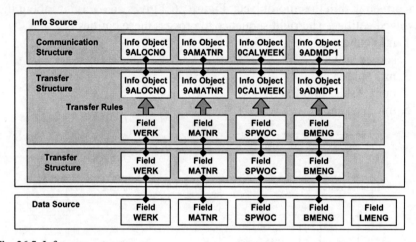

Fig. 26.5. Info source structure

The info source contains three parts, the transfer structure, the communication structure and the transfer rules. In the transfer structure those fields are selected from the data source that are loaded into SAP APO™. The communication structure on the other hand contains the info objects of the target info cube in the same sequence (characteristic, time characteristic, key figure) as in the info cube. In the transfer rules the selected fields of the transfer structure are assigned to the info objects of the communication structure, figure 26.5.

Besides the assignment of a field to an info object there are two other possibilities to fill the info objects. These are to assign a constant value or to use a self defined routine, figure 26.6.

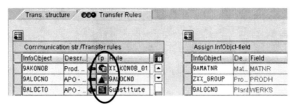

Fig. 26.6. Transfer rules

When defining the routine in the info source any info object from the transfer structure can be used. The parameter definitions for the selected info object are made automatically.

● *Update Rules*
To write the data from the communication structure into the info cube it is necessary to define update rules for the key figures. Analogous to the transfer rules, the data is received from the communication structure of the info source or processed using self defined routines. Update rules are defined in the 'data target'-view using right mouse click → *Create Update Rules* and subsequent assignment of the respective info source.

● *Info Package*
With the scheduling of an info package the data upload is triggered. The selections and parameters (e.g. check whether master data exists) for the data load and the scheduling parameters – immediately or planned as a job – are set in the info package itself. Each data upload adds the data to the info cube. If data is needed from more than one source (which is the usual scenario), the info packages have to be scheduled in a determined order after deleting the info cube content.

Info packages are created in the 'info source'-view by right mouse click on the source system of the target info source. Each data upload is identified by a request number. To check whether the data upload was successful it is possible to branch into a monitor from the info package as shown in figure 26.7 (or with the transaction RSMO). Using the button 'manage' or with right mouse click on the info cube → *Manage*, single requests can be processed (e.g. activated or deleted).

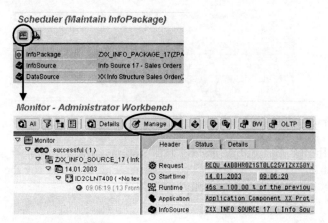

Fig. 26.7. Monitor for data uploads

The content of the info cube is displayed with the transaction LISTCUBE. Among the most common problems during data upload are insufficient table spaces.

26.3 Data Upload

Data upload will be described from SAP ERP™, an external SAP BI™ and from a flat file.

• *SAP ERP™ Info Structures*
The basis for the upload from SAP ERP™ to SAP APO™ is the info structures of the logistics information system (LIS) in SAP ERP™. To connect these info structures the LIS environment must be set up and the according data source has to be generated from the info structure. This customising is accessed from the source system (of the type 'R/3 system') by right mouse click → *Customising for Extraction* (*Settings for Application Specific Data Sources* → *Logistics* → *LIS* → *Connect Info Structure*) or with the transaction LBW0 in SAP ERP™. As the second step the data source is replicated from the 'source system'-view using right mouse click as well. Figure 26.8 shows these steps.

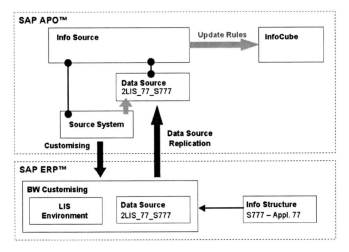

Fig. 26.8. Connecting R/3 info structures to DP

In the same transaction (LBW0) it is possible to define delta updating for the info structure.

● *BW Info Cubes*

To connect an info cube from an external SAP BI™ to DP first an export data source has to be generated in the SAP BI™ system with right mouse click on the info cube → *Generate Export Data Source*. The replication of this data source to SAP APO™ is triggered from the source system in SAP APO™ (system type BW), figure 26.9.

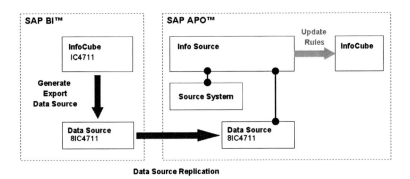

Fig. 26.9. Connecting SAP BI™ info cubes to DP

In case of problems check the extractor (transaction RSA3) and the export data source (transaction RSA6). A possible problem is the missing standard hierarchy 'DM' which is generated with the transaction RSA9.

• *Flat File*

Another common way to load data into SAP APO™ demand planning is via flat files, especially when the data is extracted from non SAP ERP™ systems. In this case the sequence of the columns of the flat file has to match the sequence of the info objects in the transfer structure (which is usually the sequence of the communication structure). The flat file is assigned to the info package as shown in figure 26.10.

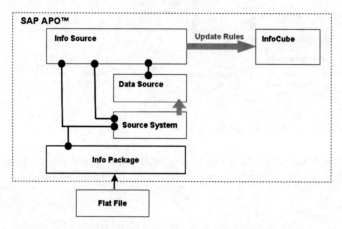

Fig. 26.10. Connecting flat files to DP

If excel is used to create the flat file, it is recommended to save the file with the extension '.csv'. The format for the time characteristics should be WWYYYY resp. MMYYYY (W – week, M – month, Y – year). If header lines are used in the flat file, their number has to be declared in the 'external data parameters'-view of the info package.

Appendix

References

Literature

Dickersbach, J. Th.:
Characteristic Based Planning with mySAP SCM.
Springer-Verlag, Berlin Heidelberg 2005

Dickersbach, J. Th.:
Einsatz von SCM-Software in der Praxis.
Intelligente Systeme zur Entscheidungsunterstützung.
Teilkonferenz zur Multikonferenz Wirtschaftsinformatik, München 2008

Knolmayer, G., Mertens, P., Zeier, A, Dickersbach, J. Th.:
Supply Chain Management Based on SAP Systems.
Springer-Verlag, Berlin Heidelberg 2009

Leitz, A.:
Supplier Collaboration mit dem mySAP SCM Inventory Collaboration Hub.
Galileo Press GmbH, Bonn 2005

OSS Notes

76301   Scheduling stock transport order/PReq

159937   Checklist requirements consumption and reduction in APO

185312   Client in APO

198852   Performance: Creation of the planning board

217210   Mapping of alternative sequences in APO System

321956   CIF-PPM: Display of R/3 move time in PPM

322800   Activating/deactivating events for APO integration

323884   PPM Conversion: Description PPM conv. PP/DS -> SNP

332812   Inconsistencies in selection/notes management

333386   Availability date different in APO and R/3

350065   Consultation: User parameters in forecast environment

350583   CIF-PPM: Modelling Net Indicator in APO

357178   Examples for customer enhancements in runtime obj.

362208   Creating orders in De-Allocated Mode

367031   Consulting for Promotion Update:

374307   Object display in the planning board

374819   Not all graphical objects are displayed

376773   User Exit EXIT_/SAPAPO/SAPLBOP_FILT_010

378903   Queue status in SMQ1, SMQ2 and table ARFCRSTATE

379006   Transport duration and planned delivery time in APO and R/3

379567   Performance: Creating the detailed scheduling planning board

380141   R/3 ->APO: Do not transfer wait times and move times

384077   APO: Optimizing CIF Communication

384550   APO 3.0 promotion: Consulting: Reporting

388001   R/3-APO-CIF: Conversion to inbound queues

388528   APO-CIF: Change to inbound queues

394076   Exits in Forecasting

394113   Date shift between R/3 and APO

400434   Authorisations in DP

403050   Consulting note: Release from DP to SNP

403072   Macrobuilder: Documentation/use of new macro functions

409181   Error: All inbound time series have propagated

410680   Unexpected disaggregation result

413526   Consultation: Navigation attributes versus basic charact.

413822   Safety stock in APO PP/DS (documentation)

416475   APO CIF: Customizing for inbound queues

417461   Scheduling error transferring orders to APO

420648   Scheduling of stock transfers and VMI orders

420650   Optimizer provides unclear results

421940   No reduction of order reservations in APO

426563   CTP: Settings, system behaviour and performance

429131   Consultation: Units of Measure

430725   Inbound queues for PI.2001.1

433166   Macro Builder: Use of DRILL_DOWN, DRILL_UP functions

435160   Finite/infinite pl. with sequence-dep. Setup activities

436395   R/3->APO: Transfer of dummy assemblies

436527   CFC3 - Block Size Recommendations

438766   MacroBuilder: New macro function physical_stock

439596   Notes on Customizing planning processes

440735   Unnecessary queue dependencies in APO outbound processing

441102   Consulting notes in PPDS

441740   Planning time fence in APO: Documentation

444832   Network alerts in alert monitor called via planning board

445899   Problems in planning with sequence-dependent setup times

448085   PPM transfer: Tips and tricks

448960   Net requirements calculation (documentation)

448986   Information About Optimizer Lot Sizes

450674   Processing the date and time in the time series

450761   Problems when planning with the 'Compact planning' strategy

453198   PPM: Transferring BOM components without a key date

453644   No use of navigation attributes in SNP

454433   SNP Optimizer Profile - Long runtime in solution calculation

457723   Planning period with PP heuristics

460107   Fallback strategy to avoid scheduling errors

481906   SNP - PP/DS integration (documentation)

483910   Use rounding value of the location product during transport

487166   Relevance of stocks to MRP

488725   FAQ: Temporary quantity assignments in Global ATP

500889   When are shortage alerts generated for ATP?

501718   Integration models for manufacturing order integration

503294   Info on Optimizer Production Lot Size

506700   CTM: Restrictions concerning component assignment in PPMs

507025   Examples of customer enhancements II

508773   Composite SAP note for CORE - transfer of data changes

508779   Composite SAP note for PPM transfer of data changes

509732   Info. About Optimizer Input and Result Log Files

510313   ATP product substitution and integration with R/3

511782   Information about transport lot sizes in the optimizer

513827   Settings/parallel processing in the PP/DS planning run

513868   Modification: SNP-PPM and change statuses

514842   PPM and output material have different validity dates

514947   Direct processing of SNP stock transfers with the TLB

516260   PPM generation: Functionality description (II)

517264   Documentation: master data functions in APO

517426   Planning time fence in optimization

517898   PP: Standard methods for safety stock planning

519014   Handling Planning Version Management

519070   Planning without final assembly (documentation)

529885   12:00 UTC problem

532979   Pegging in optimization

533824   Umfeldermittlung für dynamische Teilbilder

539797   Collective consulting note on macros

540282   Collective consulting note on promotion planning

544877   Storage cost handling

546079   FAQ: Background jobs in Demand Planning

547328   Validity dates for purchase requisitions

550330   Consulting note about APO resources

557559   ATP integration trees (MLATP/RBA) with PP/DS and SNP

557731   Planning file entry + reuse mode (documentation)

560969   Rescheduling: Bottom-Up (SAP_PP_009)

564186   Restrictions in Collaborative Planning

573127   Creating several char. combinations: /SAPAPO/MC 62

574252   Backorder processing and characteristic evaluation

578352   COPT10 parameter for the "Do not delete orders"

579373   SNP & deployment optimization consulting

590163   Use of products and resources in key figures

601813   FAQ: Characteristic evaluation in ATP check mode

617281   Migration of discontinuation data: SAP_BASIS 610 and above

617283   Migration of discontinuation data :SAP_BASIS 46C and below

643517   Macrobuilder: Fixing key figures with a macro

642593   Life Cycle Planning

660113   Performance: Aufbau der Feinplanungsplantafel zu Rel. 4.0

681451   Macros: Fixed values can be changed

687074   Macros: General notes on fixing with macros

698427   Fixed pegging: Supported document changes and replacements

704583   Fixed pegging in APO: Symptoms and restrictions

705018   Change from PPM to PDS (documentation)

710198   Customer-specific parameters for TLB

Abbreviations

ALE	Application Link Enabling
APO	Advanced Planner and Optimiser
ATD	Available-to-Deploy
ATP	Available-to-Promise
BAdI	Business Add-In
BAPI	Business Application Interface
BOM	Bill of Material
BI	Business uInformation Warehouse
CBF	Characteristic Based Forecasting
CIF	Core Interface
CRM	Customer Relation Management
CTM	Capable-to-Match
CTP	Capable-to-Promise
CVC	Characteristic Value Combination
DC	Distribution Center
DP	Demand Planning
DS	Detailed Scheduling
ECM	Engineering Change Management
EDI	Electronic Data Interface
EM	Event Manager
F&R	Forecast & Replenishment
GR	Goods Receipt
GUID	Global Unique Identifier
ICH	Inventory Collaboration Hub
KF	Key Figure
LIS	Logistic Information System
LP	Linear Programming

LUW	Logical Unit of Work
MAD	Mean Average Deviation
MILP	Mixed-Integer Linear Programming
ML-ATP	Multi-Level ATP
MLR	Multiple Linear Regression
MRP	Material Requirements Planning
PDS	Production Data Structure
PI	Plug-In
PIR	Planned Independent Requirement
PLOB	Planning Object Structure
PP	Production Planning
PP/DS	Production Planning and Detailed Scheduling
PP-PI	Production Planning for Process Industry
PPM	Production Process Model
PRT	Production Resources and Tools
qRFC	Queued Remote Function Call
RBATP	Rules-Based ATP
RFC	Remote Function Call
REM	Repetitive Manufacturing
RTO	Run-Time Object
SCE	Supply Chain Engineer
SCM	Supply Chain Management
SNP	Supply Network Planning
SMI	Supplier Managed Inventory
SOP	Sales and Operations Planning
TA	Transaction
TID	Transaction Identifier
TP/VS	Transportation Planning and Vehicle Scheduling
VMI	Vendor Managed Inventory
XI	Exchange Infrastructure

Transactions and Reports

For the quick access of some functions this chapter provides a list of useful transactions for planning with characteristics. Most of these have been explained in the text.

SAP APO™ Structure & Master Data

System	Description of the Transaction	Transaction
SAP APO™	Location	/SAPAPO/LOC3
SAP APO™	Product	/SAPAPO/MAT1
SAP APO™	Supply Chain Engineer	/SAPAPO/SCC07
SAP APO™	Resource	/SAPAPO/RES01
SAP APO™	Mass Data Maintenance	MASSD
SAP APO™	Version Maintenance	/SAPAPO/MVM
SAP APO™	Version Copy	/SAPAPO/VERCOP
SAP APO™	Time Series Copy	/SAPAPO/TSCOPY
SAP APO™	Planner	/SAPAPO/PLANNER
SAP APO™	ATP Categories	/SAPAPO/ATPC03
SAP APO™	Category Groups	/SAPAPO/SNPCG

Demand Planning

System	Transaction Description	Transaction
SAP APO™	Planning Area and PLOB	/SAPAPO/MSDP_ADMIN
SAP APO™	Consistency Check for PLOB	/SAPAPO/PSTRUCONS
SAP APO™	Consistency Check for PLOB with DP-BOMs	/SAPAPO/PSTRU_PPM_MERGE
SAP APO™	Storage Bucket Profile	/SAPAPO/TR32
SAP APO™	Planning Bucket Profile	/SAPAPO/TR30
SAP APO™	Data Copy from Info Cube	/SAPAPO/TSCUBE
SAP APO™	Calculation of Disaggreg. KF	/SAPAPO/MC8V
SAP APO™	Characteristic Value Comb.	/SAPAPO/MC62
SAP APO™	Info Object	RSD1
SAP APO™	Text for Characteristic	RSDMD

SAP APO™	Realignment	/SAPAPO/RLGCOPY
SAP APO™	Interactive Planning	/SAPAPO/SDP94
SAP APO™	Assign Selection to User	/SAPAPO/MC77
SAP APO™	Planning Book	/SAPAPO/SDP8B
SAP APO™	Assign User to Planning Book	/SAPAPO/SDPPLBK
SAP APO™	Macro Builder	/SAPAPO/ADVM
SAP APO™	Forecast Profile	/SAPAPO/MC96B
SAP APO™	Outlier	/SAPAPO/ FCSTOUTL
SAP APO™	Life Cycle Planning	/SAPAPO/MSDP_FCST1
SAP APO™	Output Length for Product	/SAPAPO/OMSL
SAP APO™	Promotion Key Figure	/SAPAPO/MP33
SAP APO™	Promotion	/SAPAPO/MP34
SAP APO™	Promotion Base	/SAPAPO/MP40
SAP APO™	Promotion Attributes	/SAPAPO/MP31
SAP APO™	Cannibalisation Group	/SAPAPO/MP32
SAP APO™	Promotion Update	/SAPAPO/MP38
SAP APO™	Promotion Management	/SAPAPO/MP42
SAP APO™	Promotion Reporting (Old)	/SAPAPO/MP39
SAP APO™	Promotion Reporting (Def.)	/SAPAPO/MP41A
SAP APO™	Promotion Reporting	/SAPAPO/MP41B
SAP APO™	Generate DP-PDS	/SAPAPO/CURTO_GEN_DP
SAP APO™	DP-PPM	/SAPAPO/SCC05
SAP APO™	Collaboration Partner	/SAPAPO/CLP_SETTINGS
SAP APO™	User-specific settings for Col-laboration	/SAPAPO/CLP_SDP_USER
SAP APO™	Copy Time Series	/SAPAPO/TSCOPY
SAP APO™	Planning Activity	/SAPAPO/MC8T
SAP APO™	Planning Job	/SAPAPO/MC8D
SAP APO™	Planning Job (Change)	/SAPAPO/MC8E
SAP APO™	Planning Job Schedule	/SAPAPO/MC8G
SAP APO™	Planning Log File	/SAPAPO/MC8K
SAP APO™	Forecast Release	/SAPAPO/MC90
SAP APO™	Release Profile	/SAPAPO/MC8S
SAP APO™	Location Split	/SAPAPO/MC7A
SAP APO™	Product Split	/SAPAPO/MC7B
SAP APO™	Transfer Orders to Time Series	/SAPAPO/LCOUT
SAP APO™	Transfer Profile	/SAPAPO/MC8U
SAP ERP™	Requirements Strategy	OPPS
SAP ERP™	Requirements Type	OMP1
SAP ERP™	Requirements Class	OMPO
SAP ERP™	Version	OMP2
SAP ERP™	Planned Independent Requirements	MD63
SAP APO™	Publication Type	/SAPAPO/CP1

Forecast Consumption

System	Description of the Transaction	Transaction
SAP APO™	Forecast Reorganisation	/SAPAPO/MD74
SAP APO™	Forecast Consumption Situation	/SAPAPO/DMP1
SAP APO™	Hierarchy for Planning Product	/SAPAPO/SCCRELSHOW
SAP APO™	ATP Check Mode	/SAPAPO/ATPC06

Sales

System	Descript. of the Transaction	Transaction
SAP ERP™	Sales Order	VA01
SAP ERP™	Delivery	VL01N
SAP ERP™	Customer	VD01
SAP ERP™	Order Type	VOV8
SAP ERP™	Scheduling Agreement	VA31
SAP ERP™	Requirements Type	OVZH
SAP ERP™	Requirements Class	OVZG
SAP ERP™	Assign Strategy to Req. Type	OPPS
SAP ERP™	Assign Req. Type to Req. Class	OMP1, OVZH
SAP ERP™	Assign Item Cat. & MRP Type	OVZI
SAP APO™	Global Settings for the ATP Check	/SAPAPO/ATPC00
SAP APO™	ATP Group	/SAPAPO/ATPC01
SAP APO™	Business Event	/SAPAPO/ATPC02
SAP APO™	ATP Category	/SAPAPO/ATPC03
SAP APO™	Check Control	/SAPAPO/ATPC04_05
SAP APO™	Check Mode	/SAPAPO/ATPC06
SAP APO™	Check Instructions	/SAPAPO/ATPC07
SAP APO™	Requirements Profile (ATP)	/SAPAPO/ATPC08
SAP APO™	Business Transact. for ML ATP	/SAPAPO/ATPC09
SAP APO™	Correlation Profile	/SAPAPO/ATPC10
SAP APO™	Simulative ATP Check	/SAPAPO/AC04
SAP APO™	ATP Time Series	/SAPAPO/AC05
SAP APO™	Availability Overview	/SAPAPO/AC03
SAP APO™	Temporary Qty. Assignments	/SAPAPO/AC06
SAP APO™	Copy CVC from DP	/SAPAPO/ATPQ_PAREA_K
SAP APO™	Display CVC	/SAPAPO/ATPQ_CHKCHAR
SAP APO™	Read Allocation Qty. from DP	/SAPAPO/ATPQ_PAREA_R
SAP APO™	Write Order Qty. to DP	/SAPAPO/ATPQ_PAREA_W
SAP APO™	CVCs for Collective Alloc.	/SAPAPO/ATPQ_COLLECT
SAP APO™	Define Mask Characters	RSKC
SAP APO™	Integrated Rule Maintenance	/SAPAPO/RBA04

SAP APO™	Field Catalogue (RBATP)	/SAPCND/AO01
SAP APO™	Condition Table (RBATP)	/SAPCND/AO03
SAP APO™	Access Sequence (RBATP)	/SAPCND/AO07
SAP APO™	Condition Type (RBATP)	/SAPCND/AO06
SAP APO™	Strategy (RBATP)	/SAPCND/AO08
SAP APO™	Rule Determination (RBATP)	/SAPCND/AO11
SAP APO™	Field Catalogue (Scheduling)	/SAPCND/AO01
SAP APO™	Condition Table (Scheduling)	/SAPCND/AO03
SAP APO™	Access Sequence (Scheduling)	/SAPCND/AO07
SAP APO™	Condition Type (Scheduling)	/SAPCND/AO06
SAP APO™	Strategy (Scheduling)	/SAPCND/AO08
SAP APO™	Rule Determination (Sched.)	/SAPCND/AO11
SAP APO™	Scheduling Test	/SAPAPO/SCHED_TEST
SAP APO™	Backorder Processing	/SAPAPO/BOP
SAP APO™	Display Worklist	/SAPAPO/BOP_WORKLIST
SAP APO™	Filter Type	/SAPAPO/BOPC_FILTER
SAP APO™	Sort Profile	/SAPAPO/BOPC_SORT
SAP APO™	Special Sorting	/SAPAPO/ORDER
SAP APO™	BOP Result Display	/SAPAPO/BOP_RESULT
SAP APO™	Interactive BOP	/SAPAPO/BOPI
SAP APO™	Check Unchecked Deliveries	VL06U, VL10U
SAP APO™	Invoice Without Delivery	VF08

Transportation Planning

System	Description of the Transaction	Transaction
SAP APO™	Transportation Zone Hierarchy	/SAPAPO/SCCRELSHOW
SAP APO™	Transport Zone Coordinates	/SAPAPO/LOCTZCALC
SAP APO™	GIS Distances	/SAPAPO/TR_IGS_BPSEL
SAP APO™	Supply Chain Engineer	/SAPAPO/SCC07
SAP APO™	Itinerary	SAPAPO/TTW1
SAP APO™	Validity Period	/SAPAPO/TTV1
SAP APO™	Schedule	/SAPAPO/TTC1
SAP APO™	Publishing Type	/SAPAPO/CP1
SAP APO™	TP/VS Planning Board	/SAPAPO/VS01
SAP APO™	TP/VS Optimiser	/SAPAPO/VS05
SAP APO™	Optimiser Profile	/SAPAPO/VS021
SAP APO™	Cost Profile	/SAPAPO/CTRP
SAP APO™	Compatibility	/SAPAPO/VS12
SAP APO™	Carrier Selection Profile	/SAPAPO/CSPRF
SAP APO™	Transfer Shipments	/SAPAPO/VS551
SAP APO™	Check Transfer of Shipments	/SAPAPO/VS60

Distribution and Supply Chain Planning Overview

System	Description of the Transaction	Transaction
SAP APO™	Lot Size Profile	/SAPAPO/SNP112
SAP APO™	SNP Transfer Settings	/SAPAPO/SDP110
SAP ERP™	Close Period	MMPV
SAP ERP™	Post Goods Receipt	MB1C
SAP ERP™	Create Stock Transfer Order	ME21N
SAP ERP™	Outbound Delivery	VL10B
SAP ERP™	Picking & Goods Issue	VL02N
SAP ERP™	Inbound Delivery	VL31N
SAP APO™	CTM Customising	/SAPAPO/CTMCUST
SAP ERP™	Stock Transfer Customising	OMGN
SAP APO™	Location Mapping	/SAPAPO/LOCALI
SAP APO™	SNP Interactive Planning	/SAPAPO/SNP94

Integrated Distribution & Production Planning

System	Description of the Transaction	Transaction
SAP APO™	SNP Optimisation	/SAPAPO/SNPOP
SAP APO™	Costs of an Optimisation Result	/SAPAPO/SNP106
SAP APO™	Cost Profile	/SAPAPO/SNP107
SAP APO™	Resource Capacity Variants	/SAPAPO/RESC01
SAP APO™	Cost Function	/SAPAPO/SNPCOSF
SAP APO™	Optimisation Log	/SAPAPO/SNPOPLOG
SAP APO™	Trace File Customising	/SAPAPO/OPT10
SAP APO™	Trace File Display	/SAPAPO/OPT11
SAP APO™	CTM Profile	/SAPAPO/CTM
SAP APO™	Master Data Selection	/SAPAPO/CTMMSEL
SAP APO™	Categorisation Profile	/SAPAPO/CTMSCPR
SAP APO™	Supply Categories	/SAPAPO/SUPCAT
SAP APO™	Inventory Limits	/SAPAPO/CTM02
SAP APO™	Search Strategy	/SAPAPO/CTMSSTRAT
SAP APO™	CTM Explanation profile	/SAPAPO/CTMEXPL
SAP APO™	CTM Customising	/SAPAPO/CTMCUST
SAP APO™	Supply Distribution	/SAPAPO/CTM10

Distribution Planning

System	Description of the Transaction	Transaction
SAP APO™	Transportation Lanes	/SAPAPO/SCC_TL1
SAP APO™	Calendars	/SAPAPO/CALENDAR
SAP ERP™, SAP APO™	Factory Calendar	SCAL
SAP ERP™	Time Zone	STZAC
SAP APO™	Quota Arrangement	/SAPAPO/SCC_TQ1

Replenishment

System	Description of the Transaction	Transaction
SAP APO™	Deployment	/SAPAPO/SNP02
SAP APO™	Quota Arrangement	/SAPAPO/SCC_TQ1
SAP APO™	Demand Profile (Product Master)	/SAPAPO/SNP101
SAP APO™	Supply Profile (Product Master)	/SAPAPO/SNP102
SAP APO™	Deployment Profile (Product Master)	/SAPAPO/SNP111
SAP APO™	Deployment Optimiser	/SAPAPO/SNP03
SAP APO™	Optimisation Log File	/SAPAPO/SNP106
SAP APO™	TLB Parameters	/SAPAPO/TLBPARAM
SAP APO™	TLB Profile	/SAPAPO/TLBPRF
SAP APO™	TLB Run	/SAPAPO/SNPTLB
SAP APO™	TLB Run (Background)	/SAPAPO/SNP04

Production Planning Overview

System	Transaction Description	Transaction
SAP ERP™	Work Centre	CR01
SAP ERP™	Resource	CRC1
SAP ERP™	Capacity	CR11
SAP ERP™	BOM	CS01
SAP ERP™	Routing	CA01
SAP ERP™	Recipe	C201
SAP ERP™	Production Version (Mass Maint.)	C223
SAP ERP™	Where-Used List	WUSL
SAP ERP™	Use of R/3-Capacity in APO	CFC9
SAP ERP™	MRP Based DS	CFDS
SAP ERP™	Change Transfer for PDS	CURTO_CREATE

SAP APO™	Resource	/SAPAPO/RES01
SAP APO™	PDS Display	/SAPAPO/CURTO_SIMU
SAP APO™	PPM	/SAPAPO/SCC05
SAP APO™	Define Mode Combinations	/SAPAPO/OO_PPM_CONV
SAP APO™	PPM Gen. w/o Lot Size	/SAPAPO/PPM_CONV
SAP APO™	PPM Gen. with Lot Size	/SAPAPO/PPM_CONV_310
SAP APO™	PPM Generation Log	/SAPAPO/PPM_CONV_LOG
SAP APO™	Supply Chain Engineer	/SAPAPO/SCC07

Rough Cut Production Planning

System	Description of the Transaction	Transaction
SAP APO™	Order Category Groups	/SAPAPO/SNPCG
SAP APO™	SNP BOM Levels	/SAPAPO/SNPLLC
SAP APO™	Planning File	/SAPAPO/RRP_NETCH
SAP APO™	Capacity Levelling	/SAPAPO/SNP05
SAP APO™	Order Conversion SNP to PP/DS	/SAPAPO/RRP_SNP2PPDS
SAP APO™	SNP Transfer	/SAPAPO/SDP110

Detailed Production Planning

System	Description of the Transaction	Transaction
SAP ERP™	Create Planned Order	MD11
SAP ERP™	Mass Maintenance for Planned Order	MD16
SAP ERP™	Receipt and Requirements List	MD04
SAP APO™	Product View	/SAPAPO/RRP3
SAP APO™	Product View – o.k.-code to display the ATP Categories	gt_io
SAP APO™	Global Settings for PP/DS	/SAPAPO/RRPCUST1
SAP APO™	PP/DS Heuristics	/SAPAPO/CDPSC11
SAP APO™	Planning File	/SAPAPO/RRP_NETCH
SAP APO™	Availability Situation	/SAPAPO/AC03
SAP APO™	Forecast Consumption Situation	/SAPAPO/DMP1
SAP APO™	Requirements View	/SAPAPO/RRP1
SAP APO™	Receipt View	/SAPAPO/RRP4
SAP APO™	Product Overview	/SAPAPO/POV1
SAP APO™	Product Planning Table	/SAPAPO/PPT1
SAP APO™	Order and Resource Reporting	/SAPAPO/CDPS_REPT
SAP APO™	Production List	/SAPAPO/PPL1
SAP APO™	Plan Monitor	/SAPAPO/PMON
SAP APO™	Plan Monitor Definition	/SAPAPO/PMONDEF

Sales in a Make-to-Order Environment

System	Description of the Transaction	Transaction
SAP APO™	Strategy Profile	/SAPAPO/CDPSC1
SAP APO™	Order and Resource Reporting	/SAPAPO/CDPS_REPT
SAP APO™	ATP Tree Display	/SAPAPO/ATREE_DSP
SAP APO™	Global Settings for PP/DS	/SAPAPO/RRPCUST1
SAP APO™	ATP Tree Conversion	/SAPAPO/ATP2PPDS

Detailed Scheduling

System	Description of the Transaction	Transaction
SAP APO™	Detailed Planning Board	/SAPAPO/CDPS0
SAP APO™	Overall Profile	/SAPAPO/CDPSC0
SAP APO™	Strategy Profile	/SAPAPO/CDPSC1
SAP APO™	Planning Board Profile	/SAPAPO/CDPSC2
SAP APO™	Work Area	/SAPAPO/CDPSC3
SAP APO™	Time Profile	/SAPAPO/CDPSC4
SAP APO™	PP/DS Heuristic	/SAPAPO/CDPSC11
SAP APO™	Heuristic Profile	/SAPAPO/CDPSC13
SAP ERP™	Set-Up Group and Key	OP18
SAP ERP™	Set-Up Group and Key (PP-PI)	OP43
SAP APO™	Set-Up Group and Key	/SAPAPO/CDPSC6
SAP APO™	Set-Up Matrix	/SAPAPO/CDPSC7
SAP ERP™	Legacy System Migration Workb.	LSMW
SAP APO™	Background Production Planning	/SAPAPO/CDPSB0
SAP APO™	PP/DS Optimisation	/SAPAPO/OPTB0
SAP APO™	PP/DS Optimisation Profile	/SAPAPO/CDPSC5
SAP APO™	Optimiser Log	/SAPAPO/OPT11
SAP APO™	Optimisation Set-Up	/SAPAPO/COPT01
SAP APO™	Optimisation User Administration	/SAPAPO/OPT03
SAP APO™	Optimiser Version Display	/SAPAPO/OPT09

Production Execution

System	Description of the Transaction	Transaction
SAP APO™	Planned Order Conversion	/SAPAPO/RRP7
SAP APO™	Global Settings for PP/DS	/SAPAPO/RRPCUST1
SAP ERP™	Production Order	CO01
SAP ERP™	Progress Confirmation	CO1F
SAP ERP™	Goods Receipt for Production Order	MB31
SAP ERP™	Order Duration Adjustment	CFO1

Modelling of Special Production Conditions

System	Description of the Transaction	Transaction
SAP APO™	Pegging Overview	/SAPAPO/PEG1
SAP APO™	Push Production	/SAPAPO/PUSH

Purchasing

System	Description of the Transaction	Transaction
SAP ERP™	Vendor	MK01
SAP ERP™	Info Record	ME11
SAP ERP™	Purchase Order	ME21N
SAP APO™	Conversion of Purchase Requisitions	/SAPAPO/RRP7
SAP APO™	Procurement Relationships	/SAPAPO/PWBSRC1
SAP APO™	Receipt View	/SAPAPO/RRP4
SAP APO™	Quota Arrangements	/SAPAPO/SCC_TQ1
SAP ERP™	Scheduling Agreement	ME31L
SAP ERP™	Schedule Line Overview/ASN	ME38
SAP APO™	Release Creation Profile	/SAPAPO/PWB_CM1
SAP APO™	Release	/SAPAPO/PWBSCH1
SAP APO™	Notification Trigger	/SAPAPO/PWBSCH2
SAP APO™	Scheduling Agreement Status	/SAPAPO/PWBSCH3

Cross Process Topics

System	Description of the Transaction	Transaction
SAP APO™	Create Safety Stock Order	/SAPAPO/AC08
SAP APO™	Interchangeability Group	/INCMD/UI
SAP APO™	Alert Monitor	/SAPAPO/AMON1
SAP APO™	Alert Monitor Configuration	/SAPAPO/AMON_SETTING
SAP APO™	Alert Types for DP	/SAPAPO/ATYPES_DP
SAP APO™	Alert Messaging Profile	/SAPAPO/AMONMSG
SAP APO™	Sending of Alerts	/SAPAPO/AMONMSG_SEND
SAP APO™	Supply Chain Cockpit	/SAPAPO/SCC01
SAP APO™	SCC User Profile	/SAPAPO/SCC_USR_PROF
SAP APO™	Context Menu	/SAPAPO/SCC_CON_MENU

Core Interface

System	Description of the Transaction	Transaction
SAP ERP™	BTE Application Indicator	BF11
SAP ERP™	Change Pointer Activation	BD61
SAP ERP™	Flag Objects for Change Pointers	BD50
SAP ERP™	Define Fields for Change Pointer	BD52
SAP ERP™	System Type & Release	NDV2
SAP ERP™	Queue Type	CFC1
SAP ERP™, SAP APO™	Register Inbound Queues	SMQR
SAP ERP™, SAP APO™	Logical System	BD54
SAP ERP™, SAP APO™	Assign Client to Logical System	SCC4
SAP ERP™, SAP APO™	RFC Connection	SM59
SAP ERP™	Transfer Parameters	CFC2
SAP ERP™	Filter & Block Sizes	CFC3
SAP APO™	Business System Group	/SAPAPO/C1
SAP APO™	Assign Logical System to BSG	/SAPAPO/C2
SAP APO™	Transfer Parameters	/SAPAPO/C4
SAP APO™	Runtime Information	/SAPAPO/CP3
SAP APO™	Publication Types	/SAPAPO/CP1
SAP ERP™	Integration Model (Create)	CFM1
SAP ERP™	Integration Model (Activate)	CFM2
SAP ERP™	Search for Objects in CIF Models	CFM5
SAP ERP™	Change Transfer for Master Data	CFP1
SAP ERP™	Define Online Transfer for Master	CFC9
SAP ERP™	PPM Change Transfer	CFP3
SAP ERP™	PDS Change Transfer	CURTO_CREATE
SAP ERP™	Delete Change Pointers	CFP4
SAP ERP™	Changes in the PPM	CFP3
SAP APO™	Global Settings for PP/DS	/SAPAPO/RRPCUST1
SAP APO™	SNP Transfer	/SAPAPO/SDP110
SAP APO™	Periodical Transfer to R/3	/SAPAPO/C5
SAP ERP™	Delete Inactive CIF Models	CFM7
SAP APO™	Delta Report	/SAPAPO/CCR
SAP ERP™, SAP APO™	Inbound Queue	SMQ1
SAP ERP™, SAP APO™	Outbound Queue	SMQ2

SAP ERP™	Channel Status	CFP2
SAP ERP™	qRFC Log	CFG1
SAP APO™	qRFC Log	/SAPAPO/C3
SAP APO™	Search within qRFC Logs	/SAPAPO/C7
SAP ERP™	Search within qRFC Logs	CFG3
SAP APO™	Queue Manager	/SAPAPO/CQ
SAP APO™	CIF Cockpit	/SAPAPO/CC
SAP APO™	Queue Alert	/SAPAPO/CW

Integration to DP

System	Description of the Transaction	Transaction
SAP APO™	Administrator Workbench	RSA1
SAP APO™	RFC Connections	SM59
SAP APO™	ALE Partner Functions	WE20, WE30
SAP APO™	Data Upload Monitor	RSMO
SAP APO™	Info Cube Content Display	LISTCUBE
SAP ERP™	LIS Customising for Extraction	LBW0
SAP APO™	Extractor Check	RSA3
SAP APO™	Export Data Source Check	RSA6
SAP APO™	Create Standard Hierarchy DM	RSA9

Development, Basis and System Administration

System	Description of the Transaction	Transaction
SAP ERP™, SAP APO™	Tables	SE11
SAP ERP™, SAP APO™	Tables (Content)	SE16
SAP ERP™, SAP APO™	Table Entry Maintenance	SM31
SAP ERP™, SAP APO™	Class Builder	SE24
SAP ERP™, SAP APO™	Function	SE37
SAP ERP™, SAP APO™	Report	SE38
SAP ERP™, SAP APO™	Modification of Coding	SE06
SAP ERP™, SAP APO™	Background Job Definition	SM36

SAP ERP™, SAP APO™	Background Job Overview	SM37
SAP ERP™, SAP APO™	User-Exit	SMOD
SAP ERP™, SAP APO™	Project for User-Exit	CMOD
SAP ERP™, SAP APO™	BAdI	SE18
SAP ERP™, SAP APO™	Project for BAdI	SE19
SAP ERP™, SAP APO™	IMG Path	SIMGH
SAP ERP™, SAP APO™	Number Ranges	SNUM
SAP ERP™, SAP APO™	Search Transactions	SE93
SAP ERP™, SAP APO™	Show Authoris. for Last Transaction	SU53
SAP ERP™, SAP APO™	Message Class AUTHORITY	SE91
SAP ERP™, SAP APO™	Last Authorisation Object Checked	SU53
SAP ERP™, SAP APO™	Role Maintenance	PFCG
SAP ERP™, SAP APO™	OSS	OSS1
SAP ERP™, SAP APO™	Service Pack Status	SPAM
SAP ERP™, SAP APO™	Resolve Modifications after SP	SPAU
SAP ERP™, SAP APO™	Release Transports	SE09
SAP ERP™, SAP APO™	Search Objects in Requests	SE03
SAP ERP™, SAP APO™	Changeability of Customising	SCC4
SAP ERP™, SAP APO™	Import Cust. from Other Client	SCC1
SAP ERP™, SAP APO™	Compare Customising	SCU0, OY19
SAP ERP™, SAP APO™	Modification of Namespaces	SE06
SAP ERP™, SAP APO™	Messages	SE91

SAP ERP™, SAP APO™	Locks	SM12
SAP ERP™, SAP APO™	Analyse Dumps	ST22
SAP ERP™, SAP APO™	Update Errors	SM13
SAP ERP™, SAP APO™	System Log	SM21
SAP ERP™, SAP APO™	Show Processes	SM50
SAP ERP™, SAP APO™	Server Selection for Own User	SM51
SAP ERP™	Delete Application Log	CFGD
SAP APO™	Delete Application Log	/SAPAPO/C6
SAP APO™	Errors in Automatic Planning	SM58
SAP ERP™, SAP APO™	Gateway	SMGW
SAP APO™	Live Cache Viewer	/SAPAPO/OM16
SAP APO™	Trace Levels	/SAPAPO/OM02
SAP APO™	Trace Files	/SAPAPO/OM01
SAP APO™	Display of COM Version (326494)	/SAPAPO/OM04
SAP APO™	LC Logging	/SAPAPO/OM06
SAP APO™	Recovery Log	/SAPAPO/OM09
SAP APO™	LC Check	/SAPAPO/OM03
SAP APO™	Parallelisation Profile	/SAPAPO/SDP_PAR
SAP APO™	Application Log	/SAPAPO/SNPAPLOG

Reports

System	Report Description	Report
SAP APO™	Availability of Resource	/SAPAPO/CRES_CREATE_LC_RES
SAP APO™	Planning Area Initialisation	/SAPAPO/TS_PAREA_INITIALIZE
SAP APO™	Planning Area Check	/SAPAPO/TS_LCM_DELTA_SYNC
SAP APO™	DP Planning Jobs	/SAPAPO/TS_BATCH_RUN
SAP APO™	Create Template Info Cube	/SAPAPO/RMDP_BW_INITIALIZE
SAP APO™	Repair Number Ranges (BW)	RSD_CLIENT_COPY_REPAIR_NUMBR
SAP APO™	Activate BW Content	/SAPAPO/TS_D_OBJECTS_COPY

SAP APO™	Delete Entries from InfoCube	RSDRD_DELETE_FACTS
SAP APO™	Generate PLOB anew	/SAPAPO/TS_PSTRU_GEN
SAP APO™	Delete PLOB content	/SAPAPO/TS_PSTRU_CONTENT_DEL
SAP APO™	Deletion of Zero Quantities	/SAPAPO/OM_REORG_ATP
SAP APO™	Resource Status	/SAPAPO/CRES_CREATE_LC_RES
SAP APO™	Resource Status	/SAPAPO/OM_RESOURCE_GET_ALL
SAP APO™	Set-Up Matrix Transport	/SAPAPO/SETUP_MATRIX_COPY
SAP ERP™	Delete Production Orders	PP_SFC_ACT
SAP ERP™	APO-flag Correction	RAPOKZFX
SAP ERP™	Re-Start Queues	RSARFCEX
SAP APO™	Periodical Transfer	/SAPAPO/RDMCPPROCESS
SAP ERP™	Master Data Change	RCPTRAN4
SAP ERP™	PPM Change Transfer	RSPPMCHG
SAP APO™	Delete orders in APO	/SAPAPO/RLCDELETE
SAP APO™	Delete Sales Orders in APO	SAPAPO/SDORDER_DEL
SAP APO™	Delete Production Orders	/SAPAPO/DELETE_PP_ORDER
SAP APO™	DP Consistency Check	/SAPAPO/TS_PSTRU_TOOLS
SAP APO™	DP Planning Structure	/SAPAPO/TS_PSTRU_GEN

Index

Access Strategy 125
Administrator Workbench 462 f.
Aggregate 38
Aggregation 34, 49, 202
Alert 425 ff.
Alert Handling 433 f.
Alert Monitor 427 ff.
Alert Types 429 ff.
Allocation 114 ff., 161, 326
Allocation Group 116
Allocation Procedure 119
Alternative Mode 269, 374
Alternative Resource 269, 373 ff.
Alternative Sequences 275, 375
Assembly Groups 91
ATD 217
ATP Category 173, 297, 415
ATP Check 100, 109 ff., 146,
  369 f., 382, 419, 425 f.
ATP Group 105, 110
ATP Tree 338 f.
Authorisation 52

Backorder Processing 134 ff.
Backward Pegging 28, 361
Base Unit 272
Basis Methods 101
Batch Determination 370, 381 f.
Beer Game 4
Big Bang 7
Block Size 442
Bottom-Up Heuristic 310 f.
Bucket Factor 287

Bucket-Oriented CTP 332 ff.
Buffer 264 f., 275
Bullwhip Effect 5
Business Case 6
Business Event 105 f., 122
Business System Group 440
Business Transaction 130

Calendar 211 f.
Calendar Resource 266, 361
Cannibalisation 70 f.
Capable-to-Match 193 ff., 285 f.,
  322, 377
Capable-to-Promise 138, 326 ff.
Capacity Levelling 282 f.
Capacity Reservation 322
Capacity Profile 398
Capacity View 278
Carrier Selection 160 f.
Category Group 89, 93, 179, 217,
  279
CBF 38
Change Pointer 439 f.
Change Transfer for Master
  Data 447 f.
Change Transfer for PPM 448
Characteristic 36 ff. , 461
Characteristic Evaluation 114,
  382
Characteristic Value Combina-
  tions 43 ff., 120
Check Control 112
Check Instruction 107, 123, 328 ,
  339

Check Mode 106, 328, 339, 382
Check of Confirmation 136 ff.
Check of Request 136 ff.
Checking Horizon 113
Checking Rule 105
CIF 439 ff.
CIF-Cockpit 458
CIF-Model 440 ff., 450 f.
Collaboration Partner 75, 161
Collaborative Forecasting 75 f.
Compatibility 158 f.
Connection to Planning Area
    120
Condition 157 f.
Confirmation 371
Consignment Stock 416
Consistency Check 39
Consistent Planning 34
Consumption Based Planning
    308
Consumption Mode 87
Constraints 188 f.
Continuous Flow 378
Control Key 272
Conversion Indicator 367 f.
Conversion Rule 369
Core Interface 16 , 150, 439 ff.
Corrected History 61
Cost Function 187
Cost Profile 158 f
Costs 187 f.
Cross Plant Planning 307
Cumulation 111, 116

Data Base Alert 430
Data Consistency 456
Data History 58 f.
Data Loading 463 ff.
Data Selection 47
Data Transfer 443 f.
Data Upload 466 ff.
De-Allocation 252, 296, 304, 363

Deletion of Master Data 450
Delivery 103, 145 ff.
Delivery Window 159
Delta Report 452
Delta Transfer 444
Delta Update 467
Demand Planning 33 ff.
Deployment 217 ff.
Deployment Heuristic 218 ff.
Deployment Horizon 219 f.
Deployment Optimisation 225 ff.
Deployment Pull Horizon 219 f.
Deployment Push Horizon 219 f.
Deployment Strategy 219 ff.
Detailed Scheduling 341 ff.
Dimension 462
Disaggregation 34, 42 ff., 49
Discontinuation 275, 421, 424
Discretisation 189 ff., 284 f.
Distribution Key 275
Distribution Planning 165 ff.,
    198, 207, 424
Document Changes 380
DP-BOM 38, 72 f.

Engineering Change Manage-
    ment 449
Error-Tolerant Scheduling 348 f.
Exception Reporting 427
Exponential Smoothing 58 f.
Ex-Post Forecast 61

Fair Share 222 f., 224 f.
Field Catalogue 120
Finiteness Level 349
Fixed Pegging 27, 379 ff.
Fixing 56 f., 343
Fixing Horizon 306, 361
Flat File 468
Forecast After Constraints 81 f.
Forecast Check 122
Forecast Consumption 87 ff, 121

Forecast Profile 63
Forecast Segment 92
Form-Fit-Function Class 421
Forward Interchangeability 424
Full Interchangeability 424 f.

Gantt Chart 341
Genetic Algorithm 359
Geo-Coding 152 f.
Goods Issue 210
Goods Receipt 210, 390

Handling Restrictions 214
Heuristic 155, 300 ff., 309 f.,
    347 ff., 379, 393, 418

Inbound Queue 455 f.
Initial Stock 416
Initial Transfer 444
Info Cube 15, 41, 461 ff.
Info Package 465 f.
Info Record 391
Info Source 464 f.
Info Structure 466 f.
Integrated Planning 183 ff.
Integration Model 442 ff., 452 f.
Interactive Backorder Process-
    ing 142
Interactive CTP 335
Interactive Planning (SNP) 279 f.
Interactive Production Planning
    312 f.
Interchangeability 421 ff.
Interchangeability Group 422

Job Scheduling 78 f.

Key Figure 36 ff., 63, 461
Key Figure Details 179
Key Figure Function 411
Key Figure Semantics 180

Labour 375
Late Demand Fulfilment 201
Life Cycle Planning 64 ff., 422
Like Profile 65, 423
Linear Programming 184
Load Balancing 227 f.
Loading Group 228
Location 20, 207, 416
Location Determination Activity
    124
Location Split 79
Locking 28, 41
Logging 78, 192, 307, 364, 457
Logical System 440
Lot Size 169, 224, 245 ff., 327
Lot Size Profile 165
Low Level Code 281, 304 f.

Macro 52 ff., 430 f.
Make-to-Stock 85
Make-to-Order 86, 131, 325
Master Data 17 f.
Master Data Selection 199
Material Flow 309
Means of Transport 150, 224
Merge 29
Mixed-Integer Linear Program-
    ming 189
Mixed Resources 292
Mode 146, 268
Mode Linkage 270, 375 f.
Mode Priority 297
Model 23 f., 262
MRP Area 319 f.
MRP Heuristic 300 f., 304 f.
MRP-Planner 276
Multi Item Single Delivery
    Location 132
Multi-level ATP 336 ff.
Multi Level Scheduling Frame-
    work 352
Multi-level Subcontracting 410

Multi Resources 265 f.
Multiple Linear Regression 60 f.

Navigation Attributes 38
New Distribution 136 ff.
Net Change 201, 281, 305
Net Segment 92

Offline Planning 51
Opening Horizon 368
Opening Period 389
Operational Concept 452 ff.
Optimisation 156 f., 184 ff., 225,
    283, 329 f., 356 ff.
Optimiser Profile 158, 182, 359
Order Category 25
Order Context 312
Order Due List 143
Order Fulfilment 97
Order Selection 360
Order Status 296, 370
Order Type 415
Organisation of Integration
    Models 452 f.
Outbound Queue 455 f.
Outlier 62 f.
Output Firmed 296
Overlapping Production 378 f.

Parallelisation 307, 479 f.
PDS 22, 253 ff., 257 ff., 268 ff.,
    274
Pegging 25 ff., 357, 358
Pegging Irrelevant 85
Pegging Network 114, 310
Pegging Overview 381
Period Factor 287
Periodical Transfer 450
Phantom 272
Pick-Up Window 159
Plan Monitor 317

Planned Delivery Time 389 f.,
    407
Planned Order 279 f., 296
Planned Order Conversion 293,
    367 f.
Planner 24, 276
Planning Activity 77
Planning Area 39 ff., 120
Planning Board 332, 341 ff.
Planning Board Profile 344 f.
Planning Book 36 f., 46 ff.,
    174 ff., 277 ff., 410, 423
Planning File 305
Planning Job 77
Planning Levels 34
Planning Mode 207 f.
Planning Object Structure 37 ff.
Planning Procedure 299, 330, 339
Planning Segment 86, 89, 92
Planning Strategies 85 ff., 202 ff.
Planning with Final Assembly
    87 ff.
Planning Product 91
Planning Without Final Assem-
    bly 89 ff.
PP/DS Horizon 291, 337
PP/DS Optimisation 356 ff.
PP Firmed 296
PP Heuristic 300 ff., 379, 393, 418
PPM 22, 253 ff., 257 ff., 267 ff.,
    275
PPM Generation 260 f.
PP-PI 273 ff.
Procurement Relationship 391 f.,
    393, 404, 408
Product Overview 314 f.
Product Planning Table 315 f.
Product Split 79
Product View 311 ff.
Production in a Different Loca-
    tion 321
Production Execution 367 ff., 425

Production Order 296, 425 f.
Production Planning 199, 277 ff.,
   290, 300 ff.
Production Version 271
Project Management 7
Project Organisation 8
Promotion 67 ff.
Promotion Attribute 69
Promotion Base 69
Prototyping 8
Publication Type 150, 441 f., 450
Purchase Order 385 f., 389
Purchase Requisition 387 f.
Purchase Requisition Conver-
   sion 390 f.
Push Production 383

qRFC 440, 454
Queue Alert 459
Queue Logging 457
Queue Manager 457 f.
Queue Monitoring 454 ff., 457 f.
Queue Names 456
Queue Status 456 f.
Quota Arrangement 215, 393

Real Quantity 300
Real-Time Deployment 224
Realignment 44 f.
Receipts View 314
Recipe 274
Re-Create Receipts 336
Relationship 269, 379
Release 394 f., 396 f.
Release Creation Profile 396 f.
Reporting 316 ff., 333, 398
Requirements Class 106
Requirements Strategy 85 ff., 92 .
Requirements Type 84, 106
Requirements View 314
Reservation from Alternative
   Plant 321

Resource 21, 252 f., 255 ff.,
   262 ff., 330
Resource Classification 374 f.
Resource Downtime 346
Resource Hierarchy 275
Resource Load 340
Resource Type 266
Re-Use Mode 304
Re-Use Strategy 302 f.
RFC 440 f.
Rough-Cut Production Planning
   277 ff.
Rule Control 123
Rule Determination 129 f.
Rules 122 ff.
Rules-Based ATP 122 ff., 328,
   426
Runtime Lane 159

Safety Days of Supply 417 f.
Safety Stock 176, 416 ff.
Safety Stock Method 417
Sales Order 99, 101 f.
Schedule 151
Schedule Line 101 f., 394
Scheduling 159, 200, 286 ff.,
   346 ff., 389, 405
Scheduling Agreement 102 f.,
   394 ff., 408
Scheduling Heuristic 349 ff.
Scheduling of Stock Transfers
   175
Scheduling Mode 297 f., 333,
   346 f.
Scheduling Status 296, 360
Scheduling Strategy 346 f.
Scope of Check 112, 419, 426
SCOR 3
Scrap 247 ff.
Search Strategy 197
Seasonal Model 59
Secondary Resource 273, 377

Sequence Dependent Set-Up 331, 351 ff.
Service Heuristic 309
Shift Sequence 264 f.
Shipment 145 ff.
Shipment Simplification 152
Shipping Calendar 210, 390
Shipping Notification 395
Simulative ATP Check 100
SNP Checking Horizon 219
SNP Heuristic 210, 280 ff.
SNP Horizon 291
SNP Integration to PP/DS 289 ff.
SNP Integration to SAP ERP 294
SNP Optimiser 181 ff., 283, 288
SNP Planning Book 178 ff., 277 ff., 410, 423
Source System 463 f.
Sourcing 215
Special Procurement Key 275, 321
Special Stock 415 f.
Stack 110
Stacking Factor 231 f.
Start of Production 328
Stock in Transit 212 f.
Stock Transfer Across Company Codes 176
Stock Transfer Cross System 177 f.
Stock Transfer Order 103 f., 124 f., 140, 171 ff., 202, 208 f.
Stock Type 415 f.
Storage Location 416
Storage Restrictions 213 f.
Strategy Profile 297 ff.
Subcontracting 401 ff.
Subcontracting for Multiple Plants 411
Subcontracting Segment 403
Subcontracting Stock 416

Substitution Order 413, 425 f.
Substitution Procedure 124, 426
Supplier 391 f.
Supplier Capacity 397 f.
Supplier Selection 392 f.
Supply Categorisation 195
Supply Chain Cockpit 435
Supply Chain Engineer 20
Supply Distribution 206 f.
System Connection 440
System Management Concept 8
System Integration 439 ff.

Technically Completed 295
Temporary Quantity Assignment 29, 114
Third Party Component Supply 408
Time Bucket 34, 39, 107 f.
Time Profile 346
Time Series 15, 45, 114
Time Stream 199, 211 f.
Time Zone 107 f., 212
TLB 227 ff.
Top-Down Heuristic 310
TP/VS Planning Board 154 f.
TP/VS Optimisation 156 f.
Transactional Data Integration 450 ff.
Transactional Simulation 29, 340
Transfer Parameter 441
Transfer Profile 83
Transfer Reservation 316
Transfer Rule 464 f.
Transport Calendar 210
Transport Duration 172, 208, 389, 404, 407
Transport Matrix 371
Transport Method 208
Transport Resource 209, 389
Transportation and Shipment Scheduling 132 f.

Transportation Group 152
Transportation Lane 149, 153,
    207 f., 390, 398, 405, 408
Transportation Planning 145 ff.
Transportation Zone 148
Trend Model 59

Univariate Models 58
Update Rule 465
Use-up Date 424
User Exit Macro 55
User Function Macro 55
Utilisation 197, 258, 328 f.

Vehicle Modelling 150
Vehicle Resource 150 f.
Version 23 f., 113, 428

Wiggle Factor 159
Work Area 345 f.
Work-List 135 f.

Lightning Source UK Ltd.
Milton Keynes UK
UKOW050613050313

207161UK00001B/11/P